METAL FORMING, FOURTH EDITION

This book is designed to help the engineer understand the principles of metal forming and to analyze forming problems – both the mechanics of forming processes and how the properties of metals interact with the processes. In this book, an entire chapter is devoted to forming limit diagrams and various aspects of stamping and another to other sheet forming operations. Sheet testing is covered in a separate chapter. Coverage of sheet metal properties has been expanded. Interesting end-of-chapter notes have been added throughout, as well as references. More than 200 end-of-chapter problems are also included.

William F. Hosford is a Professor Emeritus of Materials Science and Engineering at the University of Michigan. Professor Hosford is the author of more than 80 technical articles and numerous books, including *Mechanics of Crystals and Textured Polycrystals*; *Physical Metallurgy, Second Edition*; *Mechanical Behavior of Materials, Second Edition*; *Materials Science: An Intermediate Text*; *Materials for Engineers*; *Reporting Results* (with David Van Aken); and *Wilderness Canoe Tripping.*

Robert M. Caddell was a Professor of Mechanical Engineering at the University of Michigan, Ann Arbor.

METAL FORMING

Mechanics and Metallurgy

FOURTH EDITION

WILLIAM F. HOSFORD

University of Michigan, Ann Arbor

ROBERT M. CADDELL

CAMBRIDGE
UNIVERSITY PRESS

32 Avenue of the Americas, New York NY 10013-2473, USA

Cambridge University Press is part of the University of Cambridge.

It furthers the University's mission by disseminating knowledge in the pursuit of education, learning and research at the highest international levels of excellence.

www.cambridge.org
Information on this title: www.cambridge.org/9781107670969

First published 2011
First paperback edition 2014

A catalogue record for this publication is available from the British Library

Library of Congress Cataloguing in Publication data

Hosford, William F., 1928–
Metal Forming : Mechanics and Metallurgy / William F. Hosford, Robert M. Caddell. – 4th edition.
p. cm
Includes bibliographical references and index.
ISBN 978-1-107-00452-8 (hardback)
1. Metal-work. 2. Deformations (Mechanics) 3. Metals – Plastic properties.
I. Caddell, Robert M., 1925– II. Title.
TS213.H66 2011
671.3–dc22 2010044821

ISBN 978-1-107-00452-8 Hardback
ISBN 978-1-107-67096-9 Paperback

Contents

Preface to the Fourth Edition

My coauthor, Robert Caddell, died in 1990, and I have greatly missed working with him.

The most significant changes from the third edition are a new chapter on friction and lubrication and a major rearrangement of the last third of the book dealing with sheet forming. Most of the chapters in the last part of the book have been modified, with one whole chapter devoted to hydroforming. A new section is devoted to incremental forming. No attempt has been made to introduce numerical methods. Other books treat numerical methods. We feel that a thorough understanding of a process and the constitutive relations that are embedded in a computer program to analyze it are necessary. For example, the use of Hill's 1948 anisotropic yield criterion leads to significant errors.

I wish to acknowledge my membership in the North American Deep Drawing Research Group from whom I have learned so much about sheet forming. Particular thanks are due to Alejandro Graf of ALCAN, Robert Wagoner of the Ohio State University, John Duncan formerly with the University of Auckland, and Thomas Stoughton of General Motors.

William F. Hosford

1 Stress and Strain

An understanding of stress and strain is essential for the analysis of metal forming operations. Often the words *stress* and *strain* are used synonymously by the nonscientific public. In engineering usage, however, stress is the intensity of force and strain is a measure of the amount of deformation.

1.1 STRESS

Stress σ is defined as the intensity of force at a point.

$$\sigma = \partial F/\partial A \quad \text{as} \quad \partial A \rightarrow 0, \tag{1.1}$$

where F is the force acting on a plane of area, A.

If the stress is the same everywhere in a body,

$$\sigma = F/A. \tag{1.2}$$

There are nine components of stress as shown in Figure 1.1. A normal stress component is one in which the force is acting normal to the plane. It may be tensile or compressive. A shear stress component is one in which the force acts parallel to the plane.

Stress components are defined with two subscripts. The first denotes the normal to the plane on which the force acts and the second is the direction of the force.* For example, σ_{xx} is a tensile stress in the x-direction. A shear stress acting on the x-plane in the y-direction is denoted by σ_{xy}.

Repeated subscripts (e.g., σ_{xx}, σ_{yy}, σ_{zz}) indicate normal stresses. They are tensile if both the plane and direction are positive or both are negative. If one is positive and the other is negative they are compressive. Mixed subscripts (e.g., σ_{zx}, σ_{xy}, σ_{yz}) denote shear stresses. A state of stress in tensor notation is expressed as

$$\sigma_{ij} = \begin{vmatrix} \sigma_{xx} & \sigma_{yx} & \sigma_{zx} \\ \sigma_{xy} & \sigma_{yy} & \sigma_{zx} \\ \sigma_{xz} & \sigma_{yz} & \sigma_{zz} \end{vmatrix}, \tag{1.3}$$

* The use of the opposite convention should cause no problem because $\sigma_{ij} = \sigma_{ji}$.

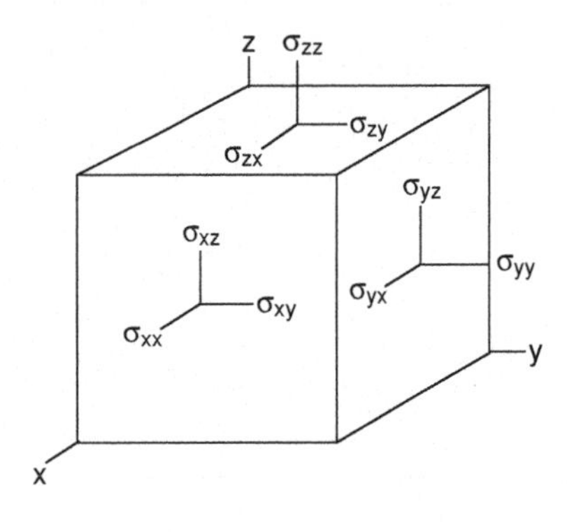

Figure 1.1. Nine components of stress acting on an infinitesimal element.

where i and j are iterated over x, y, and z. Except where tensor notation is required, it is simpler to use a single subscript for a normal stress and denote a shear stress by τ. For example, $\sigma_x \equiv \sigma_{xx}$ and $\tau_{xy} \equiv \sigma_{xy}$.

1.2 STRESS TRANSFORMATION

Stress components expressed along one set of orthogonal axes may be expressed along any other set of axes. Consider resolving the stress component, $\sigma_y = F_y/A_y$, onto the x' and y' axes as shown in Figure 1.2.

The force, $F_{y'}$, acts in the y' direction is $F_{y'} = F_y \cos\theta$ and the area normal to y' is $A_{y'} = A_y/\cos\theta$, so

$$\sigma_{y'} = F_{y'}/A_{y'} = F_y \cos\theta/(A_y/\cos\theta) = \sigma_y \cos^2\theta. \tag{1.4a}$$

Similarly

$$\tau_{y'x'} = F_{x'}/A_{y'} = F_y \sin\theta/(A_y/\cos\theta) = \sigma_y \cos\theta \sin\theta. \tag{1.4b}$$

Note that transformation of stresses requires two sine and/or cosine terms.

Pairs of shear stresses with the same subscripts that are in reverse order are always equal (e.g., $\tau_{ij} = \tau_{ji}$). This is illustrated in Figure 1.3 by a simple moment balance

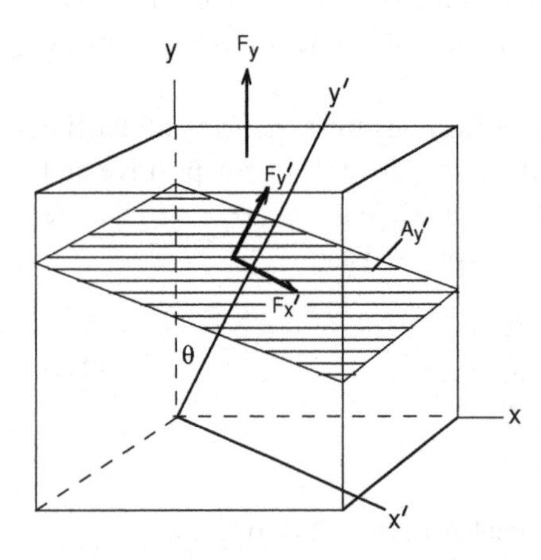

Figure 1.2. The stresses acting on a plane, A', under a normal stress, σ_y.

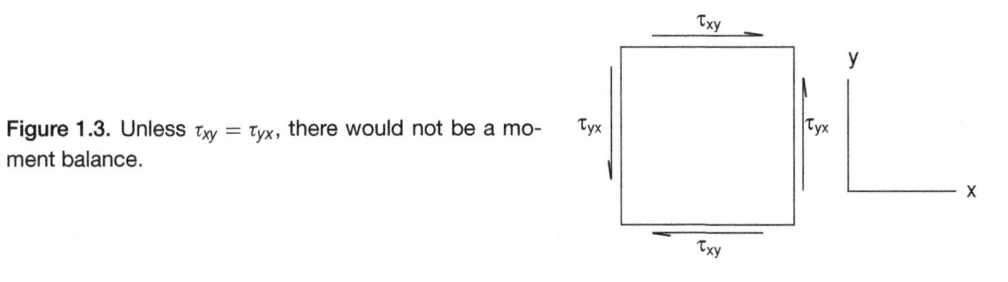

Figure 1.3. Unless $\tau_{xy} = \tau_{yx}$, there would not be a moment balance.

on an infinitesimal element. Unless $\tau_{ij} = \tau_{ji}$, there would be an infinite rotational acceleration. Therefore

$$\tau_{ij} = \tau_{ji}. \tag{1.5}$$

The general equation for transforming the stresses from one set of orthogonal axes (e.g., *n*, *m*, *p*) to another set of axes (e.g., *i*, *j*, *k*), is

$$\sigma_{ij} = \sum_{n=1}^{3}\sum_{m=1}^{3} \ell_{im}\ell_{jn}\sigma_{mn}. \tag{1.6}$$

Here, the term ℓ_{im} is the cosine of the angle between the i and the m axes and the term ℓ_{jn} is the cosine of the angle between the j and n axes. This is often written more simply as

$$\sigma_{ij} = \ell_{in}\ell_{jn}\sigma_{mn}, \tag{1.7}$$

with the summation implied. Consider transforming stresses from the x, y, z axis system to the x', y', z' system shown in Figure 1.4.

Using equation 1.6,

$$\begin{aligned}\sigma_{x'x'} &= \ell_{x'x}\ell_{x'x}\sigma_{xx} + \ell_{x'x}\ell_{x'y}\sigma_{xy} + \ell_{x'x}\ell_{x'z}\sigma_{xz} + \ell_{x'x}\ell_{x'z}\sigma_{xz}\\ &\quad + \ell_{x'y}\ell_{x'z}\sigma_{yz} + \ell_{x'z}\ell_{x'z}\sigma_{zz}\end{aligned} \tag{1.8a}$$

and

$$\begin{aligned}\sigma_{x'y'} &= \ell_{x'x}\ell_{y'x}\sigma_{xx} + \ell_{x'y}\ell_{y'x}\sigma_{xy} + \ell_{x'z}\ell_{y'z}\sigma_{xz}\\ &\quad + \ell_{x'x}\ell_{y'y}\sigma_{xx} + \ell_{x'y}\ell_{y'y}\sigma_{yy} + \ell_{x'z}\ell_{y'y}\sigma_{yz}\\ &\quad + \ell_{x'x}\ell_{y'z}\sigma_{xz} + \ell_{x'y}\ell_{y'z}\sigma_{yz} + \ell_{x'z}\ell_{y'z}\sigma_{zz}.\end{aligned} \tag{1.8b}$$

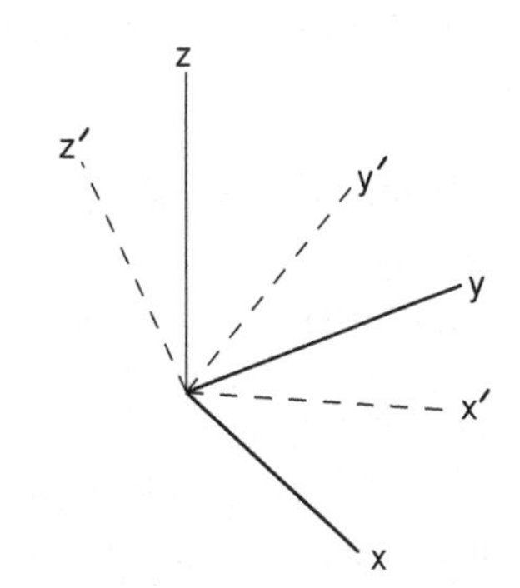

Figure 1.4. Two orthogonal coordinate systems.

These can be simplified to

$$\sigma_{x'} = \ell_{x'x}^2\sigma_x + \ell_{x'y}^2\sigma_y + \ell_{x'z}^2\sigma_z + 2\ell_{x'y}\ell_{x'z}\tau_{yz} + 2\ell_{x'z}\ell_{x'x}\tau_{zx} + 2\ell_{x'x}\ell_{x'y}\tau_{xy} \quad (1.9a)$$

and

$$\tau_{x'y'} = \ell_{x'x}\ell_{y'x}\sigma_x + \ell_{x'y}\ell_{y'y}\sigma_y + \ell_{x'z}\ell_{y'z}\sigma_z + (\ell_{x'y}\ell_{y'z} + \ell_{x'z}\ell_{y'y})\tau_{yz} + (\ell_{x'z}\ell_{y'x} + \ell_{x'x}\ell_{y'z})\tau_{zx} + (\ell_{x'x}\ell_{y'y} + \ell_{x'y}\ell_{y'x})\tau_{xy} \quad (1.9b)$$

1.3 PRINCIPAL STRESSES

It is always possible to find a set of axes along which the shear stress terms vanish. In this case σ_1, σ_2 and σ_3 are called the principal stresses. The magnitudes of the principal stresses, σ_p, are the roots of

$$\sigma_p^3 - I_1\sigma_p^2 - I_2\sigma_p - I_3 = 0, \quad (1.10)$$

where I_1, I_2 and I_3 are called the *invariants* of the stress tensor. They are

$$I_1 = \sigma_{xx} + \sigma_{yy} + \sigma_{zz},$$
$$I_2 = \sigma_{yz}^2 + \sigma_{zx}^2 + \sigma_{xy}^2 - \sigma_{yy}\sigma_{zz} - \sigma_{zz}\sigma_{xx} - \sigma_{xx}\sigma_{yy} \quad \text{and} \quad (1.11)$$
$$I_3 = \sigma_{xx}\sigma_{yy}\sigma_{zz} + 2\sigma_{yz}\sigma_{zx}\sigma_{xy} - \sigma_{xx}\sigma_{yz}^2 - \sigma_{yy}\sigma_{zx}^2 - \sigma_{zz}\sigma_{xy}^2.$$

The first invariant, $I_1 = -p/3$ where p is the pressure. I_1, I_2 and I_3 are independent of the orientation of the axes. Expressed in terms of the principal stresses they are

$$I_1 = \sigma_1 + \sigma_2 + \sigma_3,$$
$$I_2 = -\sigma_2\sigma_3 - \sigma_3\sigma_1 - \sigma_1\sigma_2 \quad \text{and} \quad (1.12)$$
$$I_3 = \sigma_1\sigma_2\sigma_3.$$

EXAMPLE 1.1: Consider a stress state with $\sigma_x = 70$ MPa, $\sigma_y = 35$ MPa, $\tau_{xy} = 20$, $\sigma_z = \tau_{zx} = \tau_{yz} = 0$. Find the principal stresses using equations 1.10 and 1.11.

SOLUTION: Using equations 1.11, $I_1 = 105$ MPa, $I_2 = -2{,}050$ MPa, $I_3 = 0$. From equation 1.10, $\sigma_p^3 - 105\sigma_p^2 + 2{,}050\sigma_p + 0 = 0$, $\sigma_p^2 - 105\sigma_p + 2{,}050 = 0$.

The principal stresses are the roots, $\sigma_1 = 79.1$ MPa, $\sigma_2 = 25.9$ MPa and $\sigma_3 = \sigma_z = 0$.

EXAMPLE 1.2: Repeat Example 1.1, with $I_3 = 170{,}700$.

SOLUTION: The principal stresses are the roots of $\sigma_p^3 - 65\sigma_p^2 + 1750\sigma_p + 170{,}700 = 0$. Since one of the roots is $\sigma_z = \sigma_3 = -40$, $\sigma_p + 40 = 0$ can be factored out. This gives $\sigma_p^2 - 105\sigma_p + 2{,}050 = 0$, so the other two principal stresses are $\sigma_1 = 79.1$ MPa, $\sigma_2 = 25.9$ MPa. This shows that when σ_z is one of the principal stresses, the other two principal stresses are independent of σ_z.

1.4 MOHR'S CIRCLE EQUATIONS

In the special cases where two of the three shear stress terms vanish (e.g., $\tau_{yx} = \tau_{zx} = 0$), the stress, σ_z, normal to the xy plane is a principal stress and the other two principal stresses lie in the xy plane. This is illustrated in Figure 1.5.

For these conditions $\ell_{x'z} = \ell_{y'z} = 0$, $\tau_{yz} = \tau_{zx} = 0$, $\ell_{x'x} = \ell_{y'y} = \cos\phi$ and $\ell_{x'y} = -\ell_{y'x} = \sin\phi$. Substituting these relations into equations 1.9 results in

$$\begin{aligned} \tau_{x'y'} &= \cos\phi\sin\phi(-\sigma_x + \sigma_y) + (\cos^2\phi - \sin^2\phi)\tau_{xy}, \\ \sigma_{x'} &= \cos^2\phi\sigma_x + \sin^2\phi\sigma_y + 2\cos\phi\sin\phi\tau_{xy}, \quad \text{and} \\ \sigma_{y'} &= \sin^2\phi\sigma_x + \cos^2\phi\sigma_y + 2\cos\phi\sin\phi\tau_{xy}. \end{aligned} \tag{1.13}$$

These can be simplified with the trigonometric relations,

$$\sin 2\phi = 2\sin\phi\cos\phi \quad \text{and} \quad \cos^2\phi = \cos^2\phi - \sin^2\phi \quad \text{to obtain}$$

$$\tau_{x'y'} = -\sin 2\phi(\sigma_x - \sigma_y)/2 + \cos 2\phi\tau_{xy}, \tag{1.14a}$$

$$\sigma_{x'} = (\sigma_x + \sigma_y)/2 + \cos 2\phi(\sigma_x - \sigma_y) + \tau_{xy}\sin 2\phi, \quad \text{and} \tag{1.14b}$$

$$\sigma_{y'} = (\sigma_x + \sigma_y)/2 - \cos 2\phi(\sigma_x - \sigma_y) + \tau_{xy}\sin 2\phi. \tag{1.14c}$$

If $\tau_{x'y'}$ is set to zero in equation 1.14a, ϕ becomes the angle θ between the principal axes and the x and y axes. Then

$$\tan 2\theta = \tau_{xy}/[(\sigma_x - \sigma_y)/2]. \tag{1.15}$$

The principal stresses, σ_1 and σ_2, are then the values of $\sigma_{x'}$ and σ_y,

$$\sigma_{1,2} = (\sigma_x + \sigma_y)/2 \pm [(\sigma_x - \sigma_y)/\cos 2\theta] + \tau_{xy}\sin 2\theta \quad \text{or}$$

$$\sigma_{1,2} = (\sigma_x + \sigma_y)/2 \pm (1/2)\left[(\sigma_x - \sigma_y)^2 + 4\tau_{xy}^2\right]^{1/2}. \tag{1.16}$$

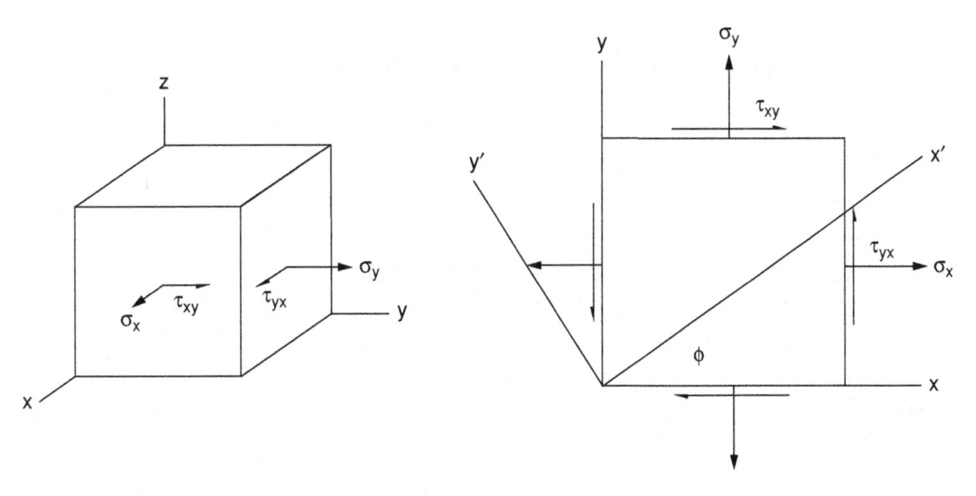

Figure 1.5. Stress state for which the Mohr's circle equations apply.

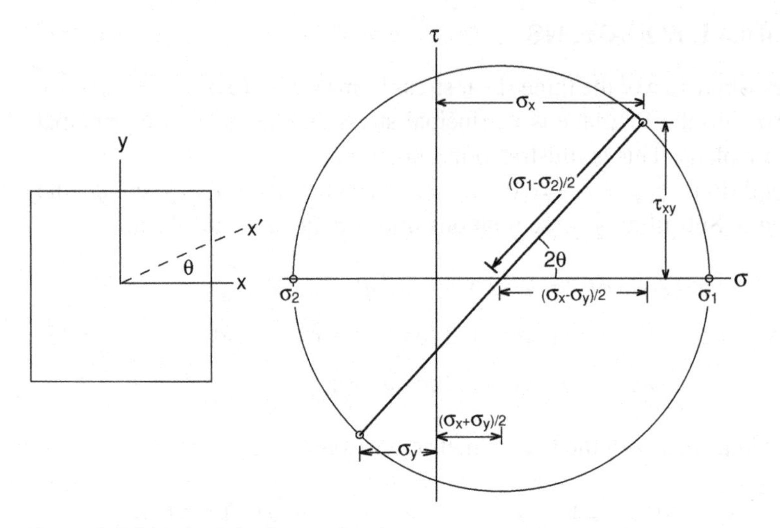

Figure 1.6. Mohr's circle diagram for stress.

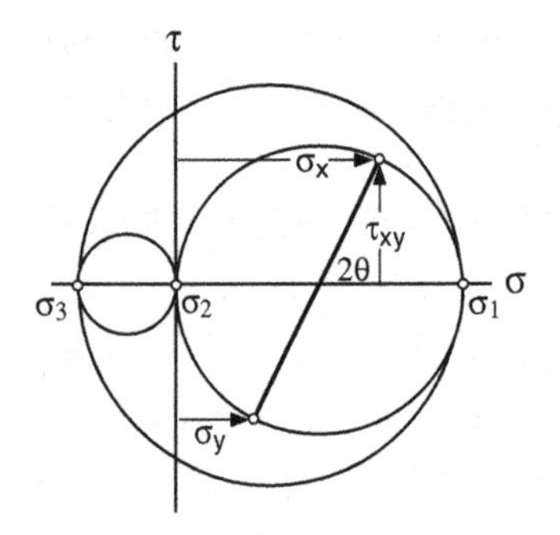

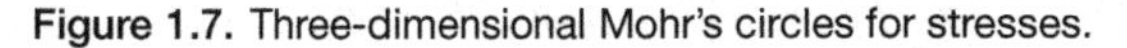

Figure 1.7. Three-dimensional Mohr's circles for stresses.

A Mohr's* circle diagram is a graphical representation of equations 1.15 and 1.16. They form a circle of radius $(\sigma_1 - \sigma_2)/2$ and with the center at $(\sigma_1 + \sigma_2)/2$ as shown in Figure 1.6. The normal stress components are plotted on the ordinate and the shear stress components are plotted on the abscissa.

Using the Pythagorean theorem on the triangle in Figure 1.6,

$$(\sigma_1 - \sigma_2)/2 = \left\{[(\sigma_x + \sigma_y)/2]^2 + \tau_{xy}^2\right\}^{1/2} \tag{1.17}$$

and

$$\tan(2\theta) = \tau_{xy}/[(\sigma_x + \sigma_y)/2]. \tag{1.18}$$

A three-dimensional stress state can be represented by three Mohr's circles as shown in Figure 1.7. The three principal stresses σ_1, σ_2 and σ_3 are plotted on the ordinate. The circles represent the stress state in the 1–2, 2–3 and 3–1 planes.

EXAMPLE 1.3: Construct the Mohr's circle for the stress state in Example 1.2 and determine the largest shear stress.

* O. Mohr, *Zivilingeneur* (1882), p. 113.

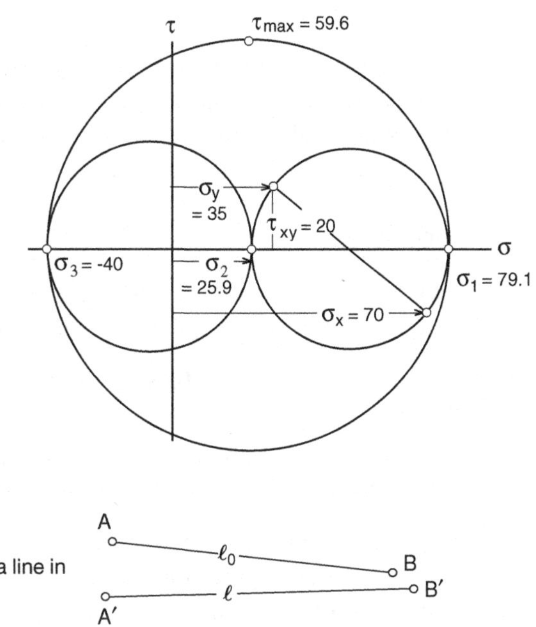

Figure 1.8. Mohr's circle for stress state in Example 1.2.

Figure 1.9. Deformation, translation, and rotation of a line in a material.

SOLUTION: The Mohr's circle is plotted in Figure 1.8. The largest shear stress is $\tau_{\max} = (\sigma_1 - \sigma_3)/2 = [79.1 - (-40)]/2 = 59.6$ MPa.

1.5 STRAIN

Strain describes the amount of deformation in a body. When a body is deformed, points in that body are displaced. Strain must be defined in such a way that it excludes effects of rotation and translation. Figure 1.9 shows a line in a material that has been deformed. The line has been translated, rotated, and deformed. The deformation is characterized by the *engineering* or *nominal strain*, e

$$e = (\ell - \ell_0)/\ell_0 = \Delta\ell/\ell_0. \tag{1.19}$$

An alternative definition* is that of *true* or *logarithmic strain*, ε, defined by

$$d\varepsilon = d\ell/\ell, \tag{1.20}$$

which on integrating gives $\varepsilon = \ln(\ell/\ell_0) = \ln(1 + e)$

$$\varepsilon = \ln(\ell/\ell_0) = \ln(1 + e). \tag{1.21}$$

The true and engineering strains are almost equal when they are small. Expressing ε as $\varepsilon = \ln(\ell/\ell_0) = \ln(1 + e)$ and expanding, so as $e \to 0$, $\varepsilon \to e$.

There are several reasons why true strains are more convenient than engineering strains. The following examples indicate why.

* True strain was first defined by P. Ludwig, *Elemente der Technishe Mechanik*, Springer, 1909.

EXAMPLE 1.4:

(a) A bar of length, ℓ_0, is uniformly extended until its length, $\ell = 2\,\ell_0$. Compute the values of the engineering and true strains.
(b) What final length must a bar of length ℓ_0, be compressed if the strains are the same (except sign) as in part (a)?

SOLUTION:

(a) $e = \Delta\ell/\ell_0 = 1.0$, $\varepsilon = \ln(\ell/\ell_0) = \ln 2 = 0.693$
(b) $e = -1 = (\ell - \ell_0)/\ell_0$, so $\ell = 0$. This is clearly impossible to achieve.

$$\varepsilon = -0.693 = \ln(\ell/\ell_0),\ \text{ so } \ell = \ell_0 \exp(0.693) = \ell_0/2.$$

EXAMPLE 1.5: A bar 10 cm long is elongated to 20 cm by rolling in three steps: 10 cm to 12 cm, 12 cm to 15 cm, and 15 cm to 20 cm.

(a) Calculate the engineering strain for each step and compare the sum of these with the overall engineering strain.
(b) Repeat for true strains.

SOLUTION:

(a) $e_1 = 2/10 = 0.20$, $e_2 = 3/12 = 0.25$, $e_3 = 5/15 = 0.333$, $e_{\text{tot}} = 0.20 + .25 + .333 = 0.833$, $e_{\text{overall}} = 10/10 = 1$.
(b) $\varepsilon_1 = \ln(12/10) = 0.182$, $\varepsilon_2 = \ln(15/12) = 0.223$, $\varepsilon_3 = \ln(20/15) = 0.288$, $\varepsilon_{\text{tot}} = 0.693$, $\varepsilon_{\text{overall}} = \ln(20/10) = 0.693$.

With true strains, the sum of the increments equals the overall strain, but this is not so with engineering strains.

EXAMPLE 1.6: A block of initial dimensions, ℓ_0, w_0, t_0, is deformed to dimensions of ℓ, w, t.

(a) Calculate the volume strain, $\varepsilon_v = \ln(v/v_0)$ in terms of the three normal strains, ε_ℓ, ε_w and ε_t.
(b) Plastic deformation causes no volume change. With no volume change, what is the sum of the three normal strains?

SOLUTION:

(a) $\varepsilon_{\text{v}} = \ln[(\ell\, wt)/(\ell_0 w_0 t_0)] = \ln(\ell/\ell_0) + \ln(w/w_0) + \ln(t/t_0) = \varepsilon_\ell + \varepsilon_w + \varepsilon_t$.
(b) If $\varepsilon_{\text{v}} = 0$, $\varepsilon_\ell + \varepsilon_w + \varepsilon_t = 0$.

Examples 1.4, 1.5 and 1.6 illustrate why true strains are more convenient than engineering strains.

1. True strains for an equivalent amount of tensile and compressive deformation are equal except for sign.
2. True strains are additive.
3. The volume strain is the sum of the three normal strains.

If strains are small, true and engineering strains are nearly equal. Expressing true strain as $\varepsilon = \ln(\frac{\ell_0 + \Delta\ell}{\ell_0}) = \ln(1 + \Delta\ell/\ell_0) = \ln(1 + e)$ and taking the series expansion, $\varepsilon = e - e^2/2 + e^3/3! \ldots.$, it can be seen that as $e \to 0$, $\varepsilon \to e$.

EXAMPLE 1.7: Calculate the ratio of ε/e for $e = 0.001$, 0.01, 0.02, 0.05, 0.1 and 0.2.

SOLUTION:

For $e = 0.001$, $\varepsilon = \ln(1.001) = 0.0009995$; $\varepsilon/e = 0.9995$.
For $e = 0.01$, $\varepsilon = \ln(1.01) = 0.00995$, $\varepsilon/e = 0.995$.
For $e = 0.02$, $\varepsilon = \ln(1.02) = 0.0198$, $\varepsilon/e = 0.99$.
For $e = 0.05$, $\varepsilon = \ln(1.05) = 0.0488$, $\varepsilon/e = 0.975$.
For $e = 0.1$, $\varepsilon = \ln(1.1) = 0.095$, $\varepsilon/e = 0.95$.
For $e = 0.2$, $\varepsilon = \ln(1.2) = 0.182$, $\varepsilon/e = 0.912$.

As e gets larger the difference between ε and e become greater.

1.6 SMALL STRAINS

Figure 1.10 shows a small two-dimensional element, $ABCD$, deformed into $A'B'C'D'$ where the displacements are u and v. The normal strain, e_{xx}, is defined as

$$e_{xx} = (A'D' - AD)/AD = A'D'/AD - 1. \tag{1.22}$$

Neglecting the rotation

$$e_{xx} = A'D'/AD - 1 = \frac{\mathrm{d}x - u + u + (\partial u/\partial x)\,\mathrm{d}x}{\mathrm{d}x} - 1 \quad \text{or}$$

$$e_{xx} = \partial u/\partial x. \tag{1.23}$$

Similarly, $e_{yy} = \partial v/\partial y$ and $e_{zz} = \partial w/\partial z$ for a three-dimensional case.

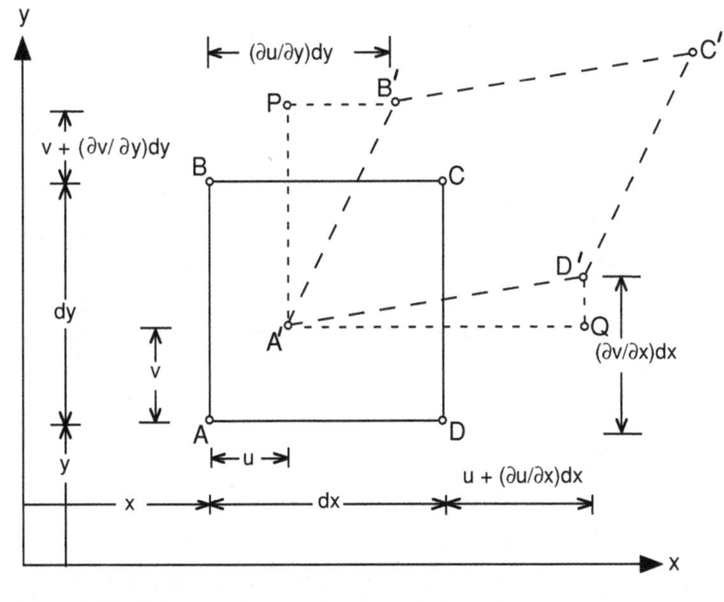

Figure 1.10. Distortion of a two-dimensional element.

The shear strain are associated with the angles between AD and $A'D'$ and between AB and $A'B'$. For small deformations

$$\angle_{A'D'}^{AD} \approx \partial v/\partial x \quad \text{and} \quad \angle_{A'B'}^{AB} \approx \partial u/\partial y \tag{1.24}$$

The total shear strain is the sum of these two angles,

$$\gamma_{xy} = \gamma_{yx} = \frac{\partial u}{\partial y} + \frac{\partial v}{\partial x}. \tag{1.25a}$$

Similarly,

$$\gamma_{yz} = \gamma_{zy} = \frac{\partial v}{\partial z} + \frac{\partial w}{\partial y} \quad \text{and} \tag{1.25b}$$

$$\gamma_{zx} = \gamma_{xz} = \frac{\partial w}{\partial x} + \frac{\partial u}{\partial z}. \tag{1.25c}$$

This definition of shear strain, γ, is equivalent to the simple shear measured in a torsion of shear test.

1.7 THE STRAIN TENSOR

If tensor shear strains, ε_{ij}, are defined as

$$\varepsilon_{ij} = (1/2)\gamma_{ij}, \tag{1.26}$$

small shear strains form a tensor,

$$\varepsilon_{ij} = \begin{vmatrix} \varepsilon_{xx} & \varepsilon_{yx} & \varepsilon_{zx} \\ \varepsilon_{xy} & \varepsilon_{yy} & \varepsilon_{zy} \\ \varepsilon_{xz} & \varepsilon_{yz} & \varepsilon_{zz} \end{vmatrix}. \tag{1.27}$$

Because small strains form a tensor, they can be transformed from one set of axes to another in a way identical to the transformation of stresses. Mohr's circle relations can be used. It must be remembered, however, that $\varepsilon_{ij} = \gamma_{ij}/2$ and that the transformations hold only for small strains. If $\gamma_{yz} = \gamma_{zx} = 0$,

$$\varepsilon_{x'} = \varepsilon_x \ell_{xx}^2 + \varepsilon_y \ell_{xy}^2 + \gamma_{xy} \ell_{x'x} \ell_{x'y} \tag{1.28}$$

and

$$\gamma_{x'y'} = 2\varepsilon_x \ell_{x'x} \ell_{y'x} + 2\varepsilon_y \ell_{x'y} \ell_{y'y} + \gamma_{xy}(\ell_{x'x} \ell_{y'y} + \ell_{y'x} \ell_{x'y}). \tag{1.29}$$

The principal strains can be found from the Mohr's circle equations for strains,

$$\varepsilon_{1,2} = \frac{\varepsilon_x + \varepsilon_y}{2} \pm (1/2)\left[(\varepsilon_x - \varepsilon_y)^2 + \gamma_{xy}^2\right]^{1/2}. \tag{1.30}$$

Strains on other planes are given by

$$\varepsilon_{x,y} = (1/2)(\varepsilon_1 + \varepsilon_2) \pm (1/2)(\varepsilon_1 - \varepsilon_2)\cos 2\theta \tag{1.31}$$

and

$$\gamma_{xy} = (\varepsilon_1 - \varepsilon_2)\sin 2\theta. \tag{1.32}$$

1.8 ISOTROPIC ELASTICITY

Although the thrust of this book is on plastic deformation, a short treatment of elasticity is necessary to understand springback and residual stresses in forming processes.

Hooke's laws can be expressed as

$$\begin{aligned} e_x &= (1/E)[\sigma_x - \upsilon(\sigma_y + \sigma_z)], \\ e_y &= (1/E)[\sigma_y - \upsilon(\sigma_z + \sigma_x)], \\ e_z &= (1/E)[\sigma_z - \upsilon(\sigma_x + \sigma_y)], \end{aligned} \tag{1.33}$$

and

$$\begin{aligned} \gamma_{yz} &= (1/G)\tau_{yz}, \\ \gamma_{zx} &= (1/G)\tau_{zx}, \\ \gamma_{xy} &= (1/G)\tau_{xy}, \end{aligned} \tag{1.34}$$

where E is Young's modulus, υ is Poisson's ratio and G is the shear modulus. For an isotropic material, E, υ and G are inter-related by

$$E = 2G(1 + \upsilon) \quad \text{or} \tag{1.35}$$

$$G = E/[2(1 + \upsilon)]. \tag{1.36}$$

EXAMPLE 1.8: In Example 1.2 with $\sigma_x = 70$ MPa, $\sigma_y = 35$ MPa, $\tau_{xy} = 20$, $\sigma_z = \tau_{zx} = \tau_{yz} = 0$, it was found that $\sigma_1 = 79.1$ MPa and $\sigma_2 = 25.9$ MPa. Using $E = 61$ GPa and $\upsilon = 0.3$ for aluminum, calculate ε_1 and ε_1 by

(a) First calculating ε_x, ε_y and γ_{xy} using equations 1.33 and then transforming these strains to the 1, 2 axes with the Mohr's circle equations.

(b) By using equations 1.33 with σ_1 and σ_2.

SOLUTION:

(a)

$$\begin{aligned} e_x &= (1/61 \times 10^9)[70 \times 10^6 - 0.30(35 \times 10^6 + 0)] = 0.9754 \times 10^{-3} \\ e_y &= (1/61 \times 10^9)[35 \times 10^6 - 0.30(70 \times 10^6 + 0)] = 0.2295 \times 10^{-3} \\ \gamma_{xy} &= [2(1.3)/61 \times 10^9](20 \times 10^6) = 0.853 \times 10^{-3} \end{aligned}$$

Now using the Mohr's strain circle equations.

$$\begin{aligned} e_{1,2} &= (e_x + e_y)/2 \pm (1/2)[(e_x - e_y)^2 + \gamma_{xy}^2]^{1/2} \\ &= 0.603 \times 10^{-3} \pm (1/2)[(0.1391 \times 10^{-3/2} + (0.856 \times 10^{-3})^2]^{1/2} \\ &= 1.169 \times 10^{-3}, 0.0361 \times 10^{-3} \end{aligned}$$

(b)

$$\begin{aligned} e_1 &= (1/61 \times 10^9)[79.9 \times 10^6 - 0.30(25.9 \times 10^6)] = 1.169, \\ e_2 &= (1/61 \times 10^9)[25.9 \times 10^6 - 0.30(79.9 \times 10^6)] = 0.0361. \end{aligned}$$

1.9 STRAIN ENERGY

If a bar of length, x and cross-sectional area, A, is subjected to a tensile force F_x, which caused an increase in length, dx, the incremental work, dW, is

$$\mathrm{d}W = F_x \mathrm{d}x. \tag{1.37}$$

The work per volume, dw, is

$$\mathrm{d}w = \mathrm{d}W/A = F_x \mathrm{d}x/(\mathrm{A}x) = \sigma_x \mathrm{d}e_x. \tag{1.38}$$

For elastic loading, substituting $\sigma_x = Ee_x$ into equation 1.38 and integrating

$$w = \sigma_x e_x/2 = Ee_x^2/2. \tag{1.39}$$

For multiaxial loading

$$dw = \sigma_x de_x + \sigma_y de_y + \sigma_z de_z + \tau_{yz} d\gamma_{yz} + \tau_{zx} d\gamma_{zx} + \tau_{xy} d\gamma_{xy}. \tag{1.40}$$

and if the deformation is elastic,

$$dw = (1/2)(\sigma_1 de_1 + \sigma_2 de_2 + \sigma_3 de_3). \tag{1.41}$$

1.10 FORCE AND MOMENT BALANCES

Many analyses of metal forming operations involve force or moment balances. The net force acting on any portion of a body must be zero. External forces on a portion of a body are balanced by internal forces acting on the arbitrary cut through the body. As an example, find the stresses in the walls of thin wall tube under internal pressure (Figure 1.11). Let the tube length be L, its diameter D and its wall thickness t and let the pressure be P (Figure 1.11a). The axial stress, σ_y, can be found from a force balance on a cross section of the tube. Since in $P\pi D^2/4 = \pi Dt\sigma_y$,

$$\sigma_y = PD/(4t). \tag{1.42}$$

The hoop stress, σ_x, can be found from a force balance on a longitudinal section of the tube (Figure 1.11b). $PDL = 2\sigma_x tL$ or

$$\sigma_x = \frac{1}{2}PD/t \ \sigma_y = PD/(2t). \tag{1.43}$$

A moment balance can be made about any axis through a material. The internal moment must balance the external moment. Consider a cylindrical rod under torsion. A moment balance relates the torque, T, to the distribution of shear stress, τ_{xy} (Figure 1.12). Consider an annular element of thickness dr at a distance r from the axis. The shear force on this element is the shear stress times the area of the element, $(2\pi r)\tau_{xy}dr$. The moment caused by this element is the shear force times the distance, r, from the axis so $dT = (2\pi r)\tau_{xy}(r)dr$ so

$$T = 2\pi \int_0^R \tau_{xy} r^2 \, dr. \tag{1.44}$$

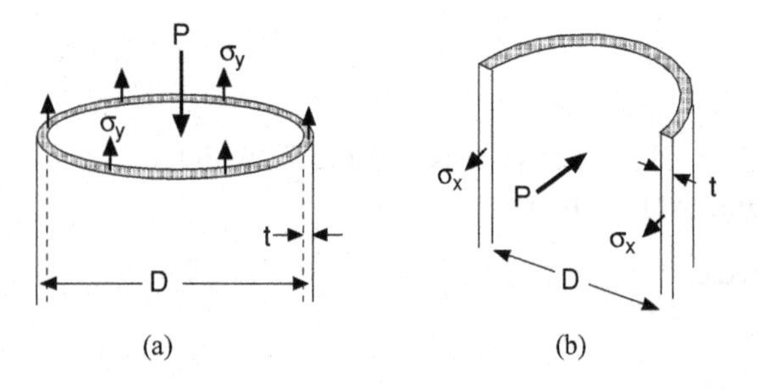

Figure 1.11. Forces acting on cuts through a tube under pressure.

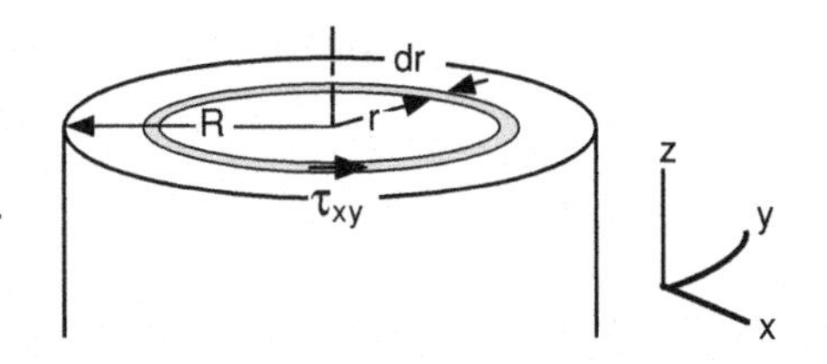

Figure 1.12. Moment balance on an annular element.

An explicit solution requires knowledge of how τ_{xy} varies with r for integration.

1.11 BOUNDARY CONDITIONS

In analyzing metal forming problems, it is important to be able to recognize boundary conditions. Often these are not stated explicitly. Some of these are listed below:

1. A stress, σ_z, normal to a free surface and the two shear stresses in the surface are zero.
2. Likewise there are no shear stresses in surfaces that are assumed to be frictionless.
3. Constraints from neighboring regions: The strains in a region are often controlled by the deformation in a neighboring region. Consider a long narrow groove in a plate (Figure 1.13.) The strain ε_x, in the groove must be the same as the strain in the region outside the groove. However, the strains ε_y and ε_z need not be the same.
4. Saint-Venant's principle states that the constraint from discontinuity will disappear within one characteristic distance of the discontinuity. For example, the shoulder on a tensile bar tends to suppress the contraction of the adjacent region of the gauge section. However this effect is very small at a distance equal to the diameter away from the shoulder. Figure 1.14 illustrates this on sheet specimen.

Bending of a sheet (Figure 1.15) illustrates another example of Saint-Venant's principle. The plane-strain condition $\varepsilon_y = 0$ prevails over most of the material because the bottom and top surfaces are so close. However, the edges are not in plane strain because $\sigma_y = 0$. However, there is appreciable deviation from plane strain only in a region within a distance equal to the sheet thickness from the edge.

EXAMPLE 1.9: A metal sheet, 1 m wide, 3 m long and 1 mm thick is bent as shown in Figure 1.15. Find the state of stress in the surface in terms of the elastic constants and the bend radius, ρ.

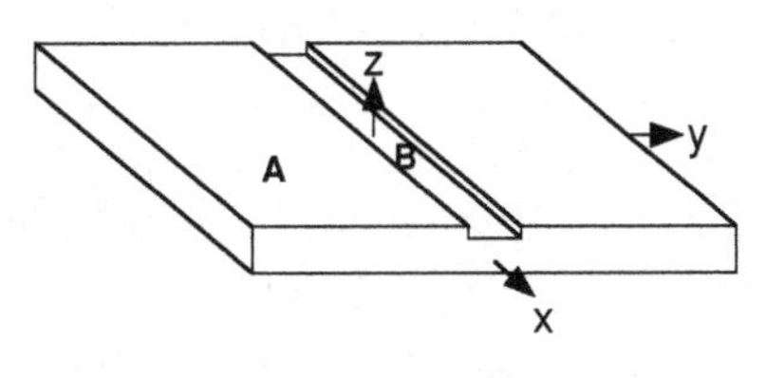

Figure 1.13. Grooved plate. The material outside the groove affects the material inside the groove, $\varepsilon_{xA} = \varepsilon_{xB}$.

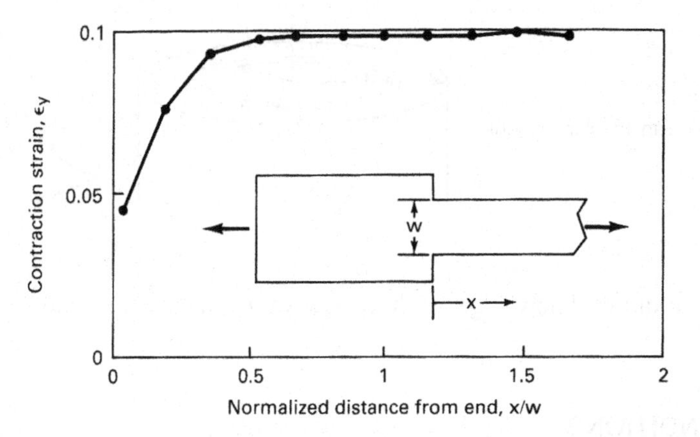

Figure 1.14. The lateral-contraction strain of a sheet tensile specimen of copper as a function to the distance from the shoulder. The strain was measured when the elongation was 27.6%.

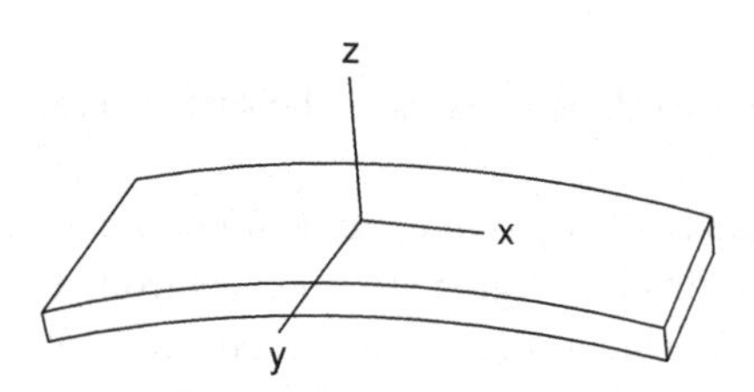

Figure 1.15. In bending of a sheet, plane-strain ($\varepsilon_y = 0$) prevails except within a distance equal to the thickness from the edges where $\sigma_y = 0$.

SOLUTION: $e_y = (1/\mathrm{E})[\sigma_y - \upsilon(\sigma_z + \sigma_x)] = 0$ and $\sigma_z = 0$, so $\sigma_y = \upsilon\sigma_x$. Neglecting any shift of the neutral plane, $e_x = t/(2\rho)$. Substituting into Hooke's law,

$$e_x = t/(2\rho) = (1/E)[\sigma_x - \upsilon(\sigma_y + \sigma_z)] \quad \text{or} \quad t/(2\rho) = (\sigma_x/E)(1 - \upsilon^2).$$

Solving for σ

$$\sigma_x = \frac{Et}{2\rho(1 - \upsilon^2)} \quad \text{and} \quad \sigma_y = \frac{\upsilon Et}{2\rho(1 - \upsilon^2)}.$$

NOTES OF INTEREST

Otto Mohr (1835–1918) made popular the graphical representation of stress at a point (*Civiling*, 1882, p. 113) even though it had previously been suggested by Culman (*Graphische Statik,* 1866, p. 226).

Barré de Saint-Venant was born 1797. In 1813 at the age of 16 he entered L'École Polytechnique. He was a sergeant on a student detachment as the allies were attacking Paris in 1814. Because he stepped out of ranks and declared that he could not in good conscience fight for a usurper (Napoleon), he was prevented from taking further classes at L'École Polytechnique. He later graduated from L'École des Ponts et Chaussées where he taught through his career.

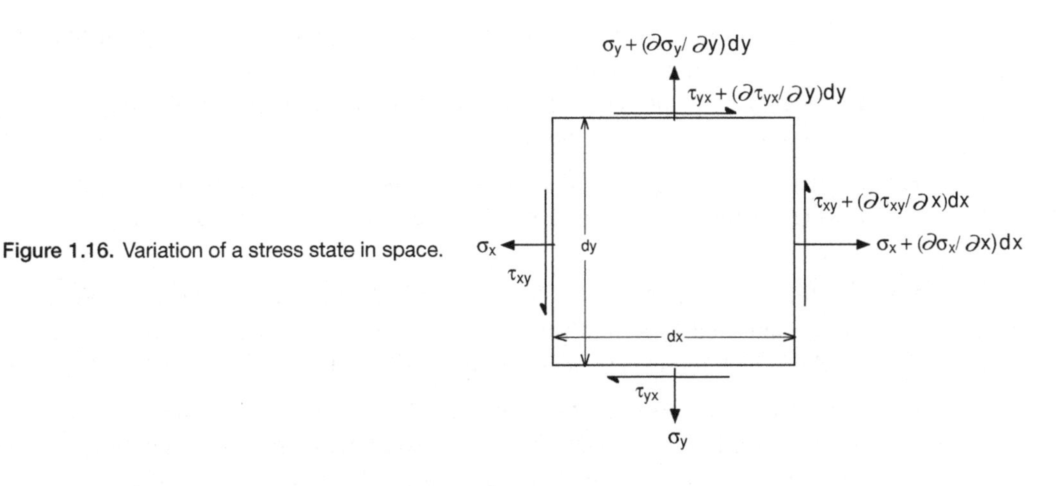

Figure 1.16. Variation of a stress state in space.

REFERENCES

R. M. Caddell, *Deformation and Fracture of Solids*, Prentice-Hall, 1980.
H. Ford, *Advanced Mechanics of Materials*, Wiley, 1963.
W. F. Hosford, *Mechanical Behavior of Materials*, Cambridge University Press, 2004.
W. Johnson and P. B. Mellor, *Engineering Plasticity*, Van Nostrand-Reinhold, 1973.
N. H. Polokowski and E. J. Ripling, *Strength and Stucture of Engineering Materials*, Prentice-Hall, 1966.

APPENDIX – EQUILIBRIUM EQUATIONS

As the stress state varies from one place to another, there are equilibrium conditions, which must be met. Consider Figure 1.16.

An x-direction force balance gives

$$\sigma_x + \tau_{xy} = \sigma_x + \partial\sigma_x/\partial x + \tau_{xy} + \partial\tau_{xy}/\partial y$$

or simply

$$\partial\sigma_x/\partial x + \partial\tau_{xy}/\partial y = 0. \tag{1.45}$$

In three dimensions

$$\begin{aligned} \partial\sigma_x/\partial x + \partial\tau_{xy}/\partial y + \partial\tau_{yz}/\partial z &= 0 \\ \partial\tau_{xy}/\partial x + \partial\sigma_y/\partial y + \partial\tau_{yz}/\partial z &= 0 \\ \partial\tau_{xz}/\partial x + \partial\tau_{yz}/\partial y + \partial\sigma_z/\partial z &= 0. \end{aligned} \tag{1.46}$$

These equations are used in Chapter 9.

PROBLEMS

1.1. Determine the principal stresses for the stress state

$$\sigma_{ij} = \begin{vmatrix} 10 & -3 & 4 \\ -3 & 5 & 2 \\ 4 & 2 & 7 \end{vmatrix}.$$

1.2. A 5-cm-diameter solid shaft is simultaneously subjected to an axial load of 80 kN and a torque of 400 Nm.

a) Determine the principal stresses at the surface assuming elastic behavior.

b) Find the largest shear stress.

1.3. A long thin-wall tube, capped on both ends, is subjected to internal pressure. During elastic loading, does the tube length increase, decrease, or remain constant?

1.4. A solid 2-cm-diameter rod is subjected to a tensile force of 40 kN. An identical rod is subjected to a fluid pressure of 35 MPa and then to a tensile force of 40 kN. Which rod experiences the largest shear stress?

1.5. Consider a long, thin-wall, 5-cm-diameter tube, with a wall thickness of 0.25 mm that is capped on both ends. Find the three principal stresses when it is loaded under a tensile force of 400 kN and an internal pressure of 200 kPa.

1.6. Three strain gauges are mounted on the surface of a part. Gauge A is parallel to the x-axis, and gauge C is parallel to the y-axis. The third gauge, B, is at 30° to gauge A. When the part is loaded, the gauges read

Gauge A	$3{,}000 \times 10^{-6}$
Gauge B	$3{,}500 \times 10^{-6}$
Gauge C	$1{,}000 \times 10^{-6}$

a) Find the value of γ_{xy}.

b) Find the principal strains in the plane of the surface.

c) Sketch the Mohr's circle diagram.

1.7. Find the principal stresses in the part of problem 1.6 if the elastic modulus of the part is 205 GPa and Poisson's ratio is 0.29.

1.8. Show that the true strain after elongation may be expressed as $\varepsilon = \ln(\frac{1}{1-r})$, where r is the reduction of area.

1.9. A thin sheet of steel, 1-mm thick, is bent as described in Example 1.9. Assuming that $E =$ is 205 GPa and $\nu = 0.29$, and that the neutral axis doesn't shift,

a) Find the state of stress on most of the outer surface.

b) Find the state of stress at the edge of the outer surface.

1.10. For an aluminum sheet, under plane stress loading $\varepsilon_x = 0.003$ and $\varepsilon_y = 0.001$. Assuming that $E =$ is 68 GPa and $\nu = 0.30$, find ε_z.

1.11. A piece of steel is elastically loaded under principal stresses, $\sigma_1 = 300$ MPa, $\sigma_2 = 250$ MPa, and $\sigma_3 = -200$ MPa. Assuming that $E =$ is 205 GPa and $\nu = 0.29$, find the stored elastic energy per volume.

1.12. A slab of metal is subjected to plane-strain deformation ($e_2 = 0$) such that $\sigma_1 = 40$ ksi and $\sigma_3 = 0$. Assume that the loading is elastic, and that $E =$ is 205 GPa, and $\nu = 0.29$ (note the mixed units). Find

a) the three normal strains.

b) the strain energy per volume.

2 Plasticity

If a body is deformed elastically, it returns to its original shape when the stress is removed. The stress and strain under elastic loading are related through Hooke's laws. Any stress will cause some strain. In contrast, no plastic deformation occurs until the stress reaches the *yield strength* of the material. When the stress is removed, the plastic strain remains. For ductile metals large amounts of plastic deformation can occur under continually increasing stress.

In this text experimental observations are linked with mathematical expressions. *Yield criteria* are mathematical descriptions of the combination of stresses necessary to cause yielding.

2.1 YIELD CRITERIA

A yield criterion is a postulated mathematical expression of the states of stress that will cause yielding. The most general form is:

$$f(\sigma_x, \sigma_y, \sigma_z, \tau_{yz}, \tau_{zx}, \tau_{xy}) = C. \tag{2.1}$$

For isotropic materials, this can be expressed in terms of principal stresses as

$$f(\sigma_1, \sigma_2, \sigma_3) = C. \tag{2.2}$$

For most isotropic ductile metals the following assumptions are commonly made:

1. The yield strengths in tension and compression are the same. That is any Bauschinger* effect is small enough so it can be ignored.
2. The volume remains constant during plastic deformation.
3. The magnitude of the mean normal stress, does not affect yielding.

$$\sigma_m = \frac{\sigma_1 + \sigma_2 + \sigma_3}{3}, \tag{2.3}$$

* J. Bauschinger, *Civilingenieur*, 27 (1881), p. 289.

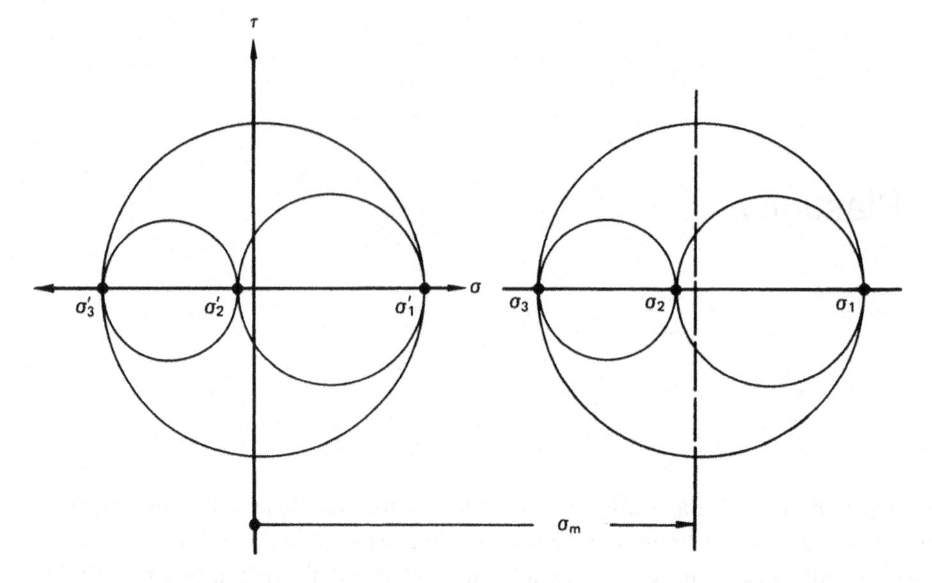

Figure 2.1. Mohr's circles for two stress states that differ only by a hydrostatic stress, σ_m, and are therefore equivalent in terms of yielding.

The yield criteria to be discussed involve these assumptions. Effects of temperature, prior straining, and strain rate will be discussed in later chapters. The assumption that yielding is independent of σ_m is reasonable because deformation usually occurs by slip or twining which are shear mechanisms. Therefore all yield criteria for isotropic materials have the form

$$f[(\sigma_2 - \sigma_3), (\sigma_3 - \sigma_1), (\sigma_1 - \sigma_2)] = C. \quad (2.4)$$

This is equivalent to stating that yielding depends only on the size of the Mohr's circles and not on their positions. Figure 2.1 shows this. If a stress state, $\sigma_1, \sigma_2, \sigma_3$, will cause yielding another stress state, $\sigma_1' = \sigma_1 - \sigma_m$, $\sigma_2' = \sigma_2 - \sigma_m$, $\sigma_3' = \sigma_3 - \sigma_m$ that differs only by σ_m will also cause yielding. The stresses, σ_1', σ_2', σ_3', are called the *deviatoric stresses*.

2.2 TRESCA CRITERION

The Tresca criterion postulates that yielding depends only on the largest shear stress in the body. With the convention, $\sigma_1 \geq \sigma_2 \geq \sigma_3$, this can be expressed as $\sigma_1 - \sigma_3 = C$. The constant C can be found by considering a tension test. In this case, $\sigma_3 = 0$ and $\sigma_1 = Y$, the yield strength at yielding, so $C = Y$. Therefore this criterion can be expressed as

$$\sigma_1 - \sigma_3 = Y. \quad (2.5)$$

Yielding in pure shear occurs when the largest shear stress, $\sigma_1 = k$ and $\sigma_3 = -\sigma_1 = k$, where k is the yield strength in shear.

$$\sigma_1 - \sigma_3 = 2k. \quad (2.6)$$

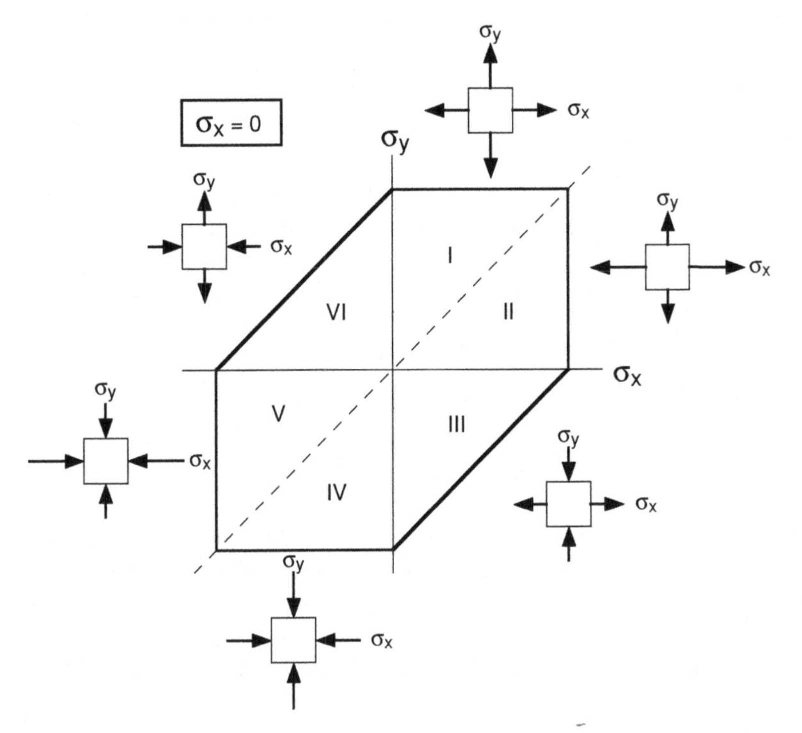

Figure 2.2. The Tresca criterion. In the six sectors the following conditions apply:

I	$\sigma_y > \sigma_x > 0$:	so $\sigma_y = Y$
II	$\sigma_x > \sigma_y > 0$:	so $\sigma_x = Y$
III	$\sigma_x > 0 > \sigma_y$:	so $\sigma_x - \sigma_y = Y$
IV	$0 > \sigma_x > \sigma_y$:	so $\sigma_y = -Y$
V	$0 > \sigma_y > \sigma_x$:	so $\sigma_x = -Y$
VI	$\sigma_y > 0 > \sigma_x$:	so $\sigma_y - \sigma_x = Y$

A yield locus is a plot of a yield criterion. Figure 2.2 is a plot of the Tresca yield locus, σ_x vs. σ_y for $\sigma_z = 0$, where σ_x, σ_y, and σ_z are principal stresses.

EXAMPLE 2.1: A thin-wall tube with closed ends is subjected to a maximum internal pressure of 35 MPa in service. The mean radius of the tube is 30 cm.

(a) If the tensile yield strength is 700 MPa, what minimum thickness must be specified to prevent yielding?
(b) If the material has a yield strength in shear of $k = 280$ MPa, what minimum thickness must be specified to prevent yielding?

SOLUTION:

(a) Hoop stress, $\sigma_1 = Pr/t = 35(30\text{ cm})/t = \sigma_{\max}$, longitudinal stress $= \sigma_2 = Pr/(2t) = (35\text{ MPa})(30\text{ cm})/(2t)$, $\sigma_{\max}$, $=$ thickness stress, $\sigma_3 \approx 0$. Yielding occurs when $\sigma_1 = 700$, or $t = (35\text{ MPa})(30\text{ cm})/700\text{ MPa} = 1.5\text{ cm}$.

(b) $\sigma_1 - \sigma_3 = 2k = 560$ MPa at yielding, so yielding occurs when $t = (35\,\text{MPa})(30\,\text{cm})/(560\,\text{MPa}) = 1.875$ cm.

2.3 VON MISES CRITERION

The von Mises criterion postulates that yielding will occur when the value of the root-mean-square shear stress reaches a critical value. Expressed mathematically,

$$\left[\frac{(\sigma_2 - \sigma_3)^2 + (\sigma_3 - \sigma_1)^2 + (\sigma_1 - \sigma_2)^2}{3}\right] = C_1$$

or equivalently

$$(\sigma_2 - \sigma_3)^2 + (\sigma_3 - \sigma_1)^2 + (\sigma_1 - \sigma_2)^2 = C_2.$$

Again, C_2 may be found by considering a uniaxial tension test in the 1-direction. Substituting $\sigma_1 = Y, \sigma_2 = \sigma_3 = 0$ at yielding, the von Mises criterion may be expressed as

$$(\sigma_2 - \sigma_3)^2 + (\sigma_3 - \sigma_1)^2 + (\sigma_1 - \sigma_2)^2 = 2Y^2 = 6k^2. \tag{2.7}$$

Figure 2.3 is the yield locus with $\sigma_2 = 0$.

In a more general form equation 2.7 may be written as

$$(\sigma_y - \sigma_z)^2 + (\sigma_z - \sigma_x)^2 + (\sigma_x - \sigma_y)^2 + 6\left(\tau_{yz}^2 + \tau_{zx}^2 + \tau_{xy}^2\right) = 2Y^2 = 6k^2. \tag{2.8}$$

The Tresca and von Mises yield loci are plotted together in Figure 2.4 for the same values of Y. Note that the greatest differences occur for $\alpha = -1$, ½ and 2.

Three-dimensional plots of the Tresca and von Mises yield criteria are shown in Figure 2.5. The Tresca criterion is a regular hexagonal prism and the von Mises criterion

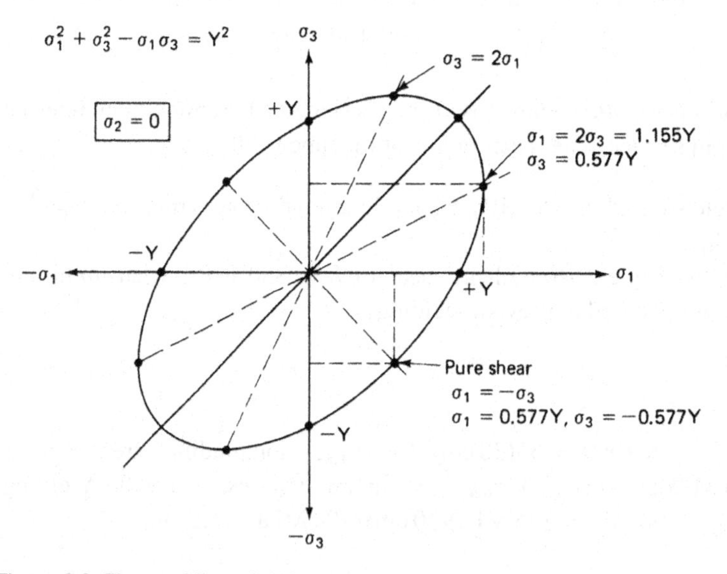

Figure 2.3. The von Mises yield locus.

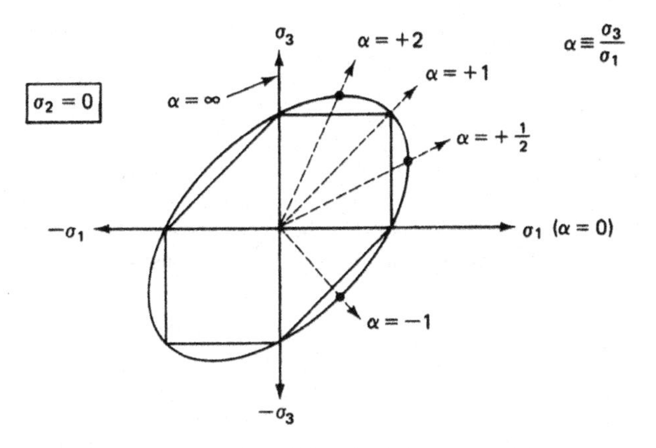

Figure 2.4. Tresca and von Mises loci showing certain loading paths.

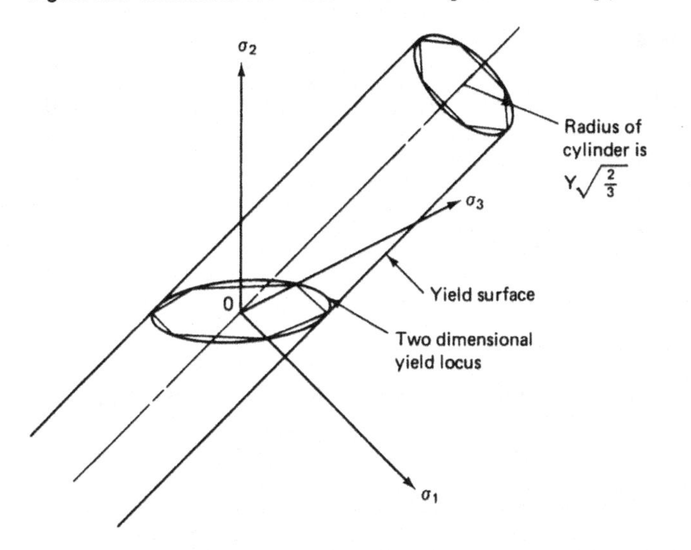

Figure 2.5. Three-dimensional plots of the Tresca and von Mises yield criteria.

is a cylinder. Both are centered on a line, $\sigma_1 = \sigma_2 = \sigma_3$. The projection of these on a plane $\sigma_1 + \sigma_2 + \sigma_3 =$ a constant is shown in Figure 2.6.

EXAMPLE 2.2: Reconsider the capped tube in Example 2.1 except let $t = 1.5$ cm. Use both the Tresca and von Mises criteria to determine the necessary yield strength to prevent yielding.

SOLUTION:
Tresca: $\sigma_1 = Y = (700\text{ MPa})(30\text{ cm})/1.5\text{ cm} = 1400\text{ MPa}$.
Von Mises: $\sigma_1 = (\sqrt{2}/3)Y$. $Y = (3/\sqrt{2})(700\text{ MPa})(30\text{ cm})/1.5\text{cm} = 1212\text{ MPa}$.

2.4 EFFECTIVE STRESS

It is useful to define an *effective stress*, $\bar{\sigma}$, for a yield criterion such that yielding occurs when the magnitude of $\bar{\sigma}$ reaches a critical value. For the von Mises criterion,

$$\bar{\sigma} = \sqrt{(1/2)[(\sigma_2 - \sigma_3)^2 + (\sigma_3 - \sigma_1)^2 + (\sigma_1 - \sigma_2)^2]}. \tag{2.9}$$

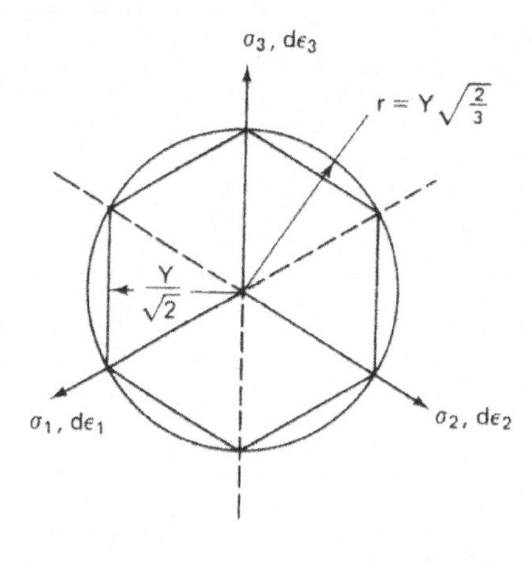

Figure 2.6. Projection of the Tresca and von Mises yield criteria onto a plane $\sigma_1 + \sigma_2 + \sigma_3 =$ a constant.

This can also be expressed as

$$\bar{\sigma} = \sqrt{(1/2)[\alpha^2 + 1 + (1-\alpha)^2]}\sigma_1 = \sqrt{1 - \alpha + \alpha^2}\sigma_1 \tag{2.10}$$

where $\alpha = \sigma_2/\sigma_1$.

For the Tresca criterion,

$$\bar{\sigma} = \sigma_1 - \sigma_3 \text{ where } \sigma_1 \geq \sigma_2 \geq \sigma_3. \tag{2.11}$$

2.5 EFFECTIVE STRAIN

The *effective strain*, $\bar{\varepsilon}$, is defined such that the incremental plastic work per volume is

$$dw = \sigma_1 d\varepsilon_1 + \sigma_2 d\varepsilon_2 + \sigma_3 d\varepsilon_3 = \bar{\sigma}\, d\bar{\varepsilon}. \tag{2.12}$$

The von Mises effective strain may be expressed as

$$d\bar{\varepsilon} = \sqrt{2[(d\varepsilon_2 - d\varepsilon_3)^2 + (d\varepsilon_3 - d\varepsilon_1)^2 + (d\varepsilon_1 - d\varepsilon_2)^2]/3}, \tag{2.13}$$

or more simply as

$$d\bar{\varepsilon} = \sqrt{(2/3)\left(d\varepsilon_1^2 + d\varepsilon_2^2 + d\varepsilon_3^2\right)}. \tag{2.14}$$

For proportional straining with a constant ratio of $d\varepsilon_1 : d\varepsilon_2 : d\varepsilon_3$, the total effective strain is

$$\bar{\varepsilon} = \sqrt{(2/3)\left(\varepsilon_1^2 + \varepsilon_2^2 + \varepsilon_3^2\right)}. \tag{2.15}$$

For the Tresca criterion the effective strain is

$$d\bar{\varepsilon} = |d\varepsilon_i|_{max}, \tag{2.16}$$

where the subscript, i, refers to the principal strains. Thus for Tresca, the effective strain is the absolutely largest principal strain. The relation provides a simple check when evaluating $\bar{\varepsilon}$ for the von Mises criterion.

$$|\varepsilon_i|_{max} \leq \bar{\varepsilon}_{Mises} \leq 1.15|\varepsilon_i|_{max} \tag{2.17}$$

When the von Mises criterion and effective stress is used, the von Mises effective strain must be used. Conversely if the Tresca criterion and effective stress is used, the Tresca effective strain must be used.

It should be realized that in both cases the $\sigma - \varepsilon$ curve in a tension test is the $\bar{\sigma} - \bar{\varepsilon}$ curve, since $\bar{\sigma}$ reduces to σ and $\bar{\varepsilon}$ reduces to ε in a tension test. It is often assumed that strain hardening is described by the $\bar{\sigma} - \bar{\varepsilon}$ curve found in a tension test. However, at large strains there may be deviations from this because of the different changes in crystallographic texture that occur during straining along different paths.

2.6 FLOW RULES

The strains that result from elastic deformation are described by Hooke's law. There are similar relations for plastic deformation, called the *flow rules*. In the most general form the flow rule may be written,

$$d\varepsilon_{ij} = d\lambda(\partial f/\partial\sigma_{ij}), \tag{2.18}$$

where f is the function of σ_{ij} that describes yielding (e.g., the yield criterion.) It is related to what has been called the *plastic potential*. For the von Mises criterion, differentiation results in

$$\begin{aligned} d\varepsilon_1 &= d\lambda[\sigma_1 - (1/2))(\sigma_2 + \sigma_3)] \\ d\varepsilon_2 &= d\lambda[\sigma_2 - (1/2))(\sigma_3 + \sigma_1)] \\ d\varepsilon_3 &= d\lambda[\sigma_3 - (1/2))(\sigma_1 + \sigma_2)]. \end{aligned} \tag{2.19}$$

In these expressions $d\lambda = d\bar{\varepsilon}/\bar{\sigma}$ which varies with position on the $\bar{\sigma} - \bar{\varepsilon}$ curve. However the ratio of the plastic strains remains constant.

$$d\varepsilon_1 : d\varepsilon_2 : d\varepsilon_3 = [\sigma_1 - (1/2))(\sigma_2 + \sigma_3)] : [\sigma_2 - (1/2))(\sigma_3 + \sigma_1)] : [\sigma_3 - (1/2))(\sigma_1 + \sigma_2)]. \tag{2.20}$$

For Tresca, $f = \sigma_1 - \sigma_3$, so the flow rules are simply

$$d\varepsilon_1 = d\lambda, d\varepsilon_2 = 0 \quad \text{and} \quad d\varepsilon_3 = -d\lambda. \tag{2.21}$$

EXAMPLE 2.3: Show that for plastic deformation in (a) uniaxial tension and (b) plane-strain compression ($\varepsilon_2 = 0$, $\sigma_3 = 0$), the incremental work per volume, dw, found from $dw = \bar{\sigma}\, d\bar{\varepsilon}$ is the same as $dw = \sigma_1 d\varepsilon_1 + \sigma_2 d\varepsilon_2 + \sigma_3 d\varepsilon_3 \ldots$

SOLUTION:

(a) Substituting $\sigma_2 = \sigma_3 = 0$ into equation 2.9, $\bar{\sigma} = \sigma_1$ and substituting $d\varepsilon_2 = d\varepsilon_3 = (-1/2)d\varepsilon_1$ into equation 2.13, so $\bar{\sigma}\, d\bar{\varepsilon} = \sigma_1 de_1$
(b) Substituting $\varepsilon_2 = 0$, $\sigma_3 = 0$ into $dw = \sigma_1 d\varepsilon_1 + \sigma_2 d\varepsilon_2 + \sigma_3 d\varepsilon_3 = \sigma_1 d\varepsilon_1$.

Substituting $d\varepsilon_2 = 0$, and $d\varepsilon_3 = -d\varepsilon_1$ into equation 2.13, $d\bar{\varepsilon} = \sqrt{(2/3)(\varepsilon_1^2 + 0 + (-\varepsilon_1)^2)} = (2/\sqrt{3})\varepsilon_1$. From the flow rules with $\varepsilon_2 = 0$ and $\sigma_3 = 0$, into

$\sigma_2 = \sigma_1$. Substituting into equation 2.9, $\bar{\sigma} = (1/\sqrt{2})[(\sigma_1 - \sigma_1/2)^2 + (\sigma_1/2 - 0)^2 + (0 - \sigma_1)^2]^{1/2} = (\sqrt{3}/2)\sigma_1$. Therefore $dw = \bar{\sigma}\,d\bar{\varepsilon} = [(2/\sqrt{3})d\varepsilon_1](\sqrt{3}/2)\sigma_1 = \sigma_1 d\varepsilon_1$.

EXAMPLE 2.4: A circle 1 cm diameter was printed on a sheet of metal prior to a complex stamping operation. After the stamping, it was found that the circle had become an ellipse with major and minor diameters of 1.300 and 1.100 cm.

(a) Determine the effective strain.
(b) If a condition of plane stress ($\sigma_z = 0$) existed during the stamping, and the ratio $\alpha = \sigma_2/\sigma_1$ remained constant what ratio $\sigma_1/\bar{\sigma}$ must have existed?

SOLUTION:

(a) $\varepsilon_1 = \ln(1.3/1) = 0.2624$, $\varepsilon_2 = \ln(1.11) = 0.0953$, $\varepsilon_3 = -\varepsilon_1 - \varepsilon_2 = -0.358$. $\bar{\varepsilon} = \sqrt{(2/3)(\varepsilon_1^2 + \varepsilon_2^2 + \varepsilon_3^2)} = [(2/3)(0.262^2 + 0.0953^2 + 0.358^2)]^{1/2} = 0.3705$
Note that this is larger than 0.358 but not 15% larger.
(b) From the flow rules (equation 2.19) with $\sigma_3 = 0$, $\varepsilon_2/\varepsilon_1 = (2\sigma_2 - \sigma_1)/(2\sigma_1 - \sigma_2)$.

Solving for α, $\alpha = \sigma_2/\sigma_1 = (2\varepsilon_2/\varepsilon_1 + 1)/(\varepsilon_2/\varepsilon_1 + 2) = [2(1.1/1.3) + 1]/[(1.1/1.3) + 2] = 0.946$. Now substituting into equation 2.10, $\bar{\sigma} = \sigma_1\sqrt{1 - 0.946 + 0.946^2}$, $\sigma_1/\bar{\sigma} = 1.027$.

2.7 NORMALITY PRINCIPLE

One interpretation of the flow rules is that the vector sum of the plastic strains is normal to the yield surface*. This is illustrated in three-dimensions in Figure 2.7 and in two-dimensions in Figure 2.8. With isotropic solids, the directions of principal strain and principal stress coincide. As a result the relation

$$\frac{d\varepsilon_2}{d\varepsilon_1} = -\frac{\partial\sigma_1}{\partial\sigma_2}. \tag{2.22}$$

EXAMPLE 2.5: A thin sheet is subjected to biaxial tension $\sigma_1 = \sigma_2 \neq 0$, $\sigma_3 = 0$. The principal strains in the sheet were $\varepsilon_2 = -(1/4)\varepsilon_1$.

(a) Using the principle of normality, determine the stress ratio, $\alpha = \sigma_2/\sigma_1$, using the von Mises and the Tresca criteria.
(b) Show that the normal to the yield locus in both cases corresponds to the answers to (a).

SOLUTION:

(a) $\varepsilon_2/\varepsilon_1 = -0.25 = (\sigma_2 - 0.5\sigma_1)/(\sigma_1 - 0.5\sigma_2)$. Solving for σ_2/σ_1, $\alpha = 2/7$. With the Tresca criterion, $\varepsilon_2 = -(1/4)\varepsilon_1$ can occur only at the uniaxial tension corner, so $\alpha = 0$.
(b) See Figure 2.9.

* See D. C. Drucker, *Proc. 1st U.S. Nat. Congr. Appl. Mech.* (1951), p. 487.

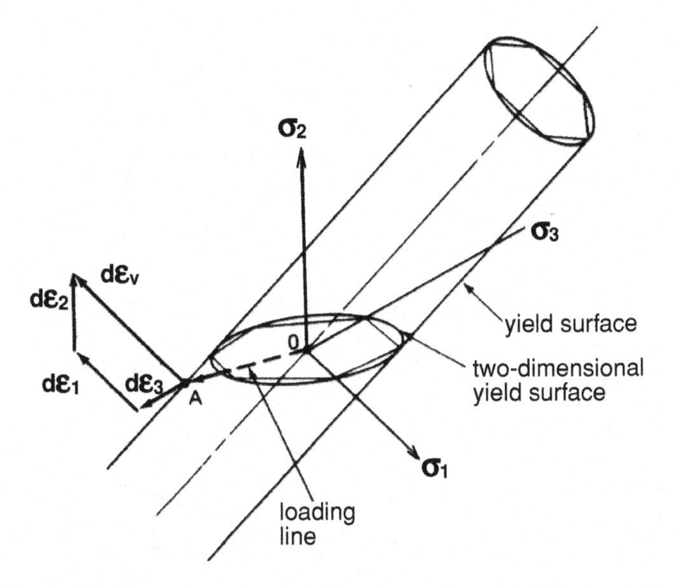

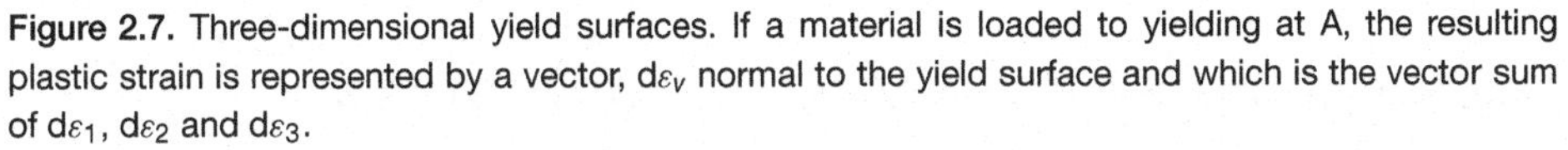
Figure 2.7. Three-dimensional yield surfaces. If a material is loaded to yielding at A, the resulting plastic strain is represented by a vector, $d\varepsilon_v$ normal to the yield surface and which is the vector sum of $d\varepsilon_1$, $d\varepsilon_2$ and $d\varepsilon_3$.

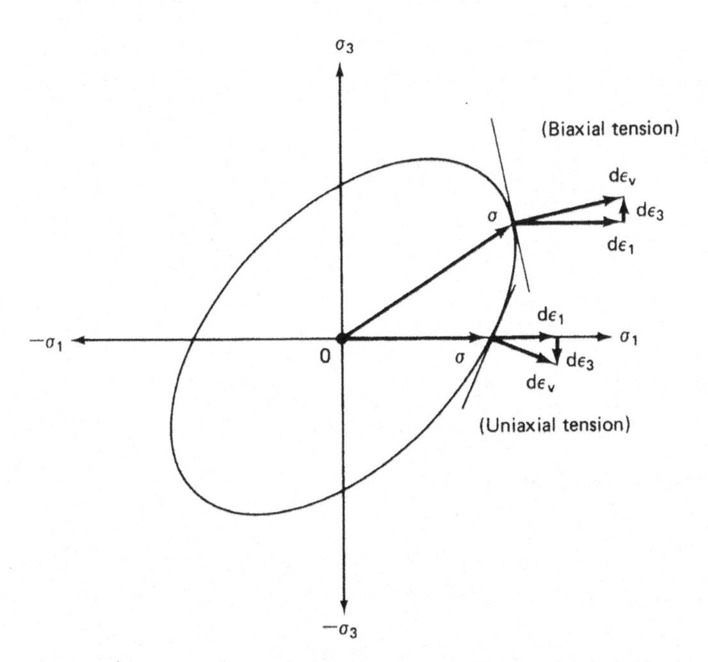

Figure 2.8. Illustration of normality. Note that the ratio *of* $d\varepsilon_3/d\varepsilon_1$ is given by the normal to the yield surface at the point of yielding.

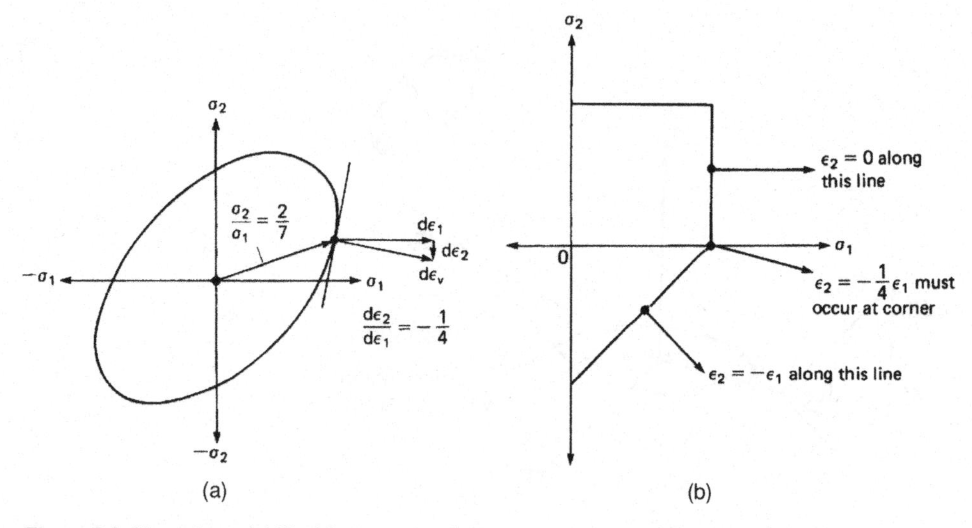

Figure 2.9. Normals to yield loci for $\varepsilon_2/\varepsilon_1 = -0.25$.

2.8 DERIVATION OF THE VON MISES EFFECTIVE STRAIN

The effective stress-strain function is defined such that the incremental work per volume is $dw = \sigma_1 d\varepsilon_1 + \sigma_2 d\varepsilon_2 + \sigma_3 d\varepsilon_3 = \bar{\sigma}\, d\bar{\varepsilon}$. For simplicity consider a stress state with $\sigma_3 = 0$. Then

$$\bar{\sigma}\, d\bar{\varepsilon} = \sigma_1 d\varepsilon_1 + \sigma_2 d\varepsilon_2 = \sigma_1 d\varepsilon_1 (1 + \alpha\rho), \tag{2.23}$$

where $\alpha = \sigma_{2/}\sigma_1$ and $\rho = d\varepsilon_2/d\varepsilon_1$. Then

$$d\bar{\varepsilon} = d\varepsilon_1(\sigma_1/\bar{\sigma})(1 + \alpha\rho). \tag{2.24}$$

From the flow rules, $\rho = d\varepsilon_2/d\varepsilon_1 = [\sigma_2 + (1/2)\sigma_1]/[\sigma_1 + (1/2)\sigma_2] = (2\alpha - 1)/(2 - \alpha)$ or

$$\alpha = (2\rho + 1)/(2 + \rho). \tag{2.25}$$

Combining equations 2.24 and 2.25,

$$d\bar{\varepsilon} = d\varepsilon_1(\sigma_1/\bar{\sigma})[2(1 + \rho + \rho)/(2 + \rho]. \tag{2.26}$$

With $\sigma_3 = 0$, the von Mises expression for $\bar{\sigma}$ is

$$\bar{\sigma} = \left[(\sigma_1^2 + \sigma_2^2 - \sigma_1\sigma_2)\right]^{1/2} = (1 - \alpha + \alpha^2)^{1/2}. \tag{2.27}$$

Combining equations 2.25 and 2.26,

$$\frac{\sigma_1}{\bar{\sigma}} = \left(\frac{2 + \rho}{\sqrt{3}}\right) \Big/ (1 + \rho + \rho^2)^{1/2}. \tag{2.28}$$

Since $\rho = d\varepsilon_2/d\varepsilon_1$,

$$d\bar{\varepsilon} = \left(\frac{2}{\sqrt{3}}\right)\left(d\varepsilon_1^2 + d\varepsilon_1 d\varepsilon_2 + d\varepsilon_2^2\right)^{1/2} \tag{2.29}$$

Now since

$$d\varepsilon_1^2 + d\varepsilon_2^2 + d\varepsilon_3^2 = d\varepsilon_1^2 + d\varepsilon_2^2 + (-d\varepsilon_1 - d\varepsilon_2)^2 = 2\left(d\varepsilon_1^2 + d\varepsilon_2 d\varepsilon_1 + d\varepsilon_3^2\right), \quad (2.30)$$

equation 2.29 becomes

$$d\bar{\varepsilon} = \left[(2/3)\left(d\varepsilon_1^2 + d\varepsilon_2^2 + d\varepsilon_e^2\right)\right]^{1/2}. \quad (2.31)$$

This derivation also holds where $\sigma_3 \neq 0$, since this is equivalent to a stress state $\sigma_1' = \sigma_1 - \sigma_3, \sigma_2' = \sigma_2 - \sigma_3, \sigma_3' = \sigma_3 - \sigma_3 = 0$.

NOTES OF INTEREST

Otto Z. Mohr (1835–1918) worked as a civil engineer designing bridges. At 32, he was appointed a professor of engineering mechanics at Stuttgart Polytecknium. Among other contributions, he also devised the graphical method of analyzing the stress at a point.

He then extended Coulomb's idea that failure is caused by shear stresses into a failure criterion based on maximum shear stress, or diameter of the largest circle. He proposed the different failure stresses in tension, shear, and compression could be combined into a single diagram, in which the tangents form an envelope of safe stress combinations.

This is essentially the Tresca yield criterion. It may be noted that early workers used the term "failure criteria," which failed to distinguish between fracture and yielding.

In 1868, Tresca presented two notes to the French Academy (*Comptes Rendus Acad. Sci. Paris*, 1864, p. 754). From these, Saint-Venant established the first theory of plasticity based on the assumptions that

1) plastic deformation does not change the volume of a material,
2) directions of principal stresses and principal strains coincide,
3) the maximum shear stress at a point is a constant.

The Tresca criterion is also called the *Guest* or the "*maximum shear stress*" criterion.

In letters to William Thompson, John Clerk Maxwell (1831–1879) proposed that "strain energy of distortion" was critical, but he never published this idea and it was forgotten. M. T. Huber, in 1904, first formulated the expression for "distortional strain energy,"

$$\mathrm{U} = [1/(12\mathrm{G})][(\sigma_2 - \sigma_3)^2 + (\sigma_3 - \sigma_1)^2 + (\sigma_1 - \sigma_2)^2] \text{ where } \mathrm{U} = \sigma_{yp}^2/(6\mathrm{G}).$$

The same idea was independently developed by von Mises (*Göttinger. Nachr. Math. Phys.*, 1913, p. 582) for whom the criterion is generally called. It is also referred to by the names of several people who independently proposed it: Huber, Hencky, as well as Maxwell. It is also known as the "maximum distortional energy" theory and the "octahedral shear stress" theory. The first name reflects that the elastic energy in an isotropic material, associated with shear (in contrast to dilatation) is proportional to $(\sigma_2 - \sigma_3)^2 + (\sigma_3 - \sigma_1)^2 + (\sigma_1 - \sigma_2)^2$. The second name reflects that the shear terms, $(\sigma_2 - \sigma_3)$, $(\sigma_3 - \sigma_1)$, and $(\sigma_1 - \sigma_2)$, can be represented as the edges of an octahedron in principal stress space.

REFERENCES

F. A. McClintock and A. S. Argon, *Mechanical Behavior of Materials*, Addison Wesley, 1966.

W. F. Hosford, *Mechanical Behavior of Materials*, Cambridge University Press, 2005.

PROBLEMS

2.1. **a)** If the principal stresses on a material with a yield stress in shear are $\sigma_1 =$ 175 MPa and $\sigma_2 = 350$ MPa, what tensile stress σ_3 must be applied to cause yielding according to the Tresca criterion?

b) If the stresses in **a)** were compressive, what tensile stress σ_3 must be applied to cause yielding according to the Tresca criterion?

2.2. Consider a 6-cm-diameter tube with 1-mm-thick wall with closed ends made from a metal with a tensile yield strength of 25 MPa. After applying a compressive load of 2,000 N to the ends, what internal pressure is required to cause yielding according to **a)** the Tresca criterion and **b)** the von Mises criterion?

2.3. Consider a 0.5-m-diameter cylindrical pressure vessel with hemispherical ends made from a metal for which $k = 500$ MPa. If no section of the pressure vessel is to yield under an internal pressure of 35 MPa, what is the minimum wall thickness according to **a)** the Tresca criterion? **b)** the von Mises criterion?

2.4. A thin-wall tube is subjected to combined tensile and torsional loading. Find the relationship between the axial stress, σ, the shear stress, τ, and the tensile yield strength, Y, to cause yielding according to **a)** the Tresca criterion, and **b)** the von Mises criterion.

2.5. Consider a plane-strain compression test with a compressive load, F_y, a strip width, w, an indenter width, b, and a strip thickness, t. Using the von Mises criterion, find:

a) $\bar{\varepsilon}$ as a function of ε_y.

b) $\bar{\sigma}$ as a function of σ_y.

c) an expression for the work per volume in terms of ε_y and σ_y.

d) an expression in the form of $\sigma_y = f(K, \varepsilon_y, n)$ assuming $\bar{\sigma} = K\bar{\varepsilon}^n$.

2.6. The following yield criterion has been proposed: "Yielding will occur when the sum of the two largest shear stresses reaches a critical value." Stated mathematically,

$$(\sigma_1 - \sigma_3) + (\sigma_1 - \sigma_2) = C \quad \text{if} \quad (\sigma_1 - \sigma_2) > (\sigma_2 - \sigma_3) \quad \text{or}$$
$$(\sigma_2 - \sigma_3) + (\sigma_1 - \sigma_2) = C \quad \text{if} \quad (\sigma_1 - \sigma_2) \leq (\sigma_2 - \sigma_3)$$

where $\sigma_1 > \sigma_2 > \sigma_3$, $C = 2Y$, and $Y =$ tensile yield strength. Plot the yield locus with $\sigma_3 = 0$ in $\sigma_1 - \sigma_2$ space.

2.7. Consider the stress states

$$\begin{vmatrix} 15 & 3 & 0 \\ 3 & 10 & 0 \\ 0 & 0 & 5 \end{vmatrix} \quad \text{and} \quad \begin{vmatrix} 10 & 3 & 0 \\ 3 & 5 & 0 \\ 0 & 0 & 0 \end{vmatrix}.$$

a) Find σ_m for each.
b) Find the deviatoric stress in the normal directions for each.

2.8. Calculate the ratio of $\bar{\sigma}/\tau_{\text{max}}$ for **a)** pure shear, **b)** uniaxial tension, and **c)** plane-strain tension. Assume the von Mises criterion.

2.9. A material yields under a biaxial stress state, $\sigma_3 = -(1/2)\sigma_1$.
a) Assuming the von Mises criterion, find $\mathrm{d}\varepsilon_1/\mathrm{d}\varepsilon_2$.
b) What is the ratio of τ_{max}/Y at yielding?

2.10. A material is subjected to stresses in the ratio, σ_1, $\sigma_2 = 0.3\sigma_1$ and $\sigma_3 = -0.5\sigma_1$. Find the ratio of σ_1/Y at yielding using the **a)** Tresca and **b)** von Mises criteria.

2.11. Plot ε_1 versus ε_2 for a constant level of $\bar{\varepsilon} = 0.10$, according to
a) von Mises.
b) Tresca

2.12. A proposed yield criterion is that yielding will occur if the diameter of Mohr's largest circle plus half of the diameter of the second largest Mohr's circle reaches a critical value. Plot the yield locus with $\sigma_3 = 0$ in $\sigma_1 - \sigma_2$ space.

3 Strain Hardening

When metals are deformed plastically at temperatures lower than would cause recrystallization, they are said to be *cold worked.* Cold working increases the strength and hardness. The terms *work hardening* and *strain hardening* are used to describe this. Cold working usually decreases the ductility.

Tension tests are used to measure the effect of strain on strength. Sometimes other tests, such as torsion, compression, and bulge testing are used, but the tension test is simpler and most commonly used. The major emphasis in this chapter is the dependence of yield (or flow) stress on strain.

3.1 THE TENSION TEST

The temperature and strain rate influence test results. Generally, in a tension test, the strain rate is in the order of 10^{-2} to 10^{-3}/s and the temperature is between 18 and 25°C. These effects are discussed in Chapter 5. Measurements are made in a gauge section that is under uniaxial tension during the test.

Initially the deformation is elastic and the tensile force is proportional to the elongation. Elastic deformation is recoverable. It disappears when the tensile force is removed. At higher forces the deformation is plastic, or nonrecoverable. In a ductile material, the force reaches a maximum and then decreases until fracture. Figure 3.1 is a schematic tensile load-extension curve.

Stress and strain are computed from measurements in a tension test of the tensile force, F, and the elongation, $\Delta\ell$. The *nominal* or *engineering stress,* s, and *strain,* e, are defined as

$$s = F/A_0 \tag{3.1}$$

and

$$e = \Delta\ell/\ell_0. \tag{3.2}$$

where A_0 is the initial cross-sectional area, and ℓ_0 is the initial gauge length. Since A_0, and ℓ_0 are constants, the shapes of the $s{-}e$ and $F{-}\Delta\ell$ curves are identical. That

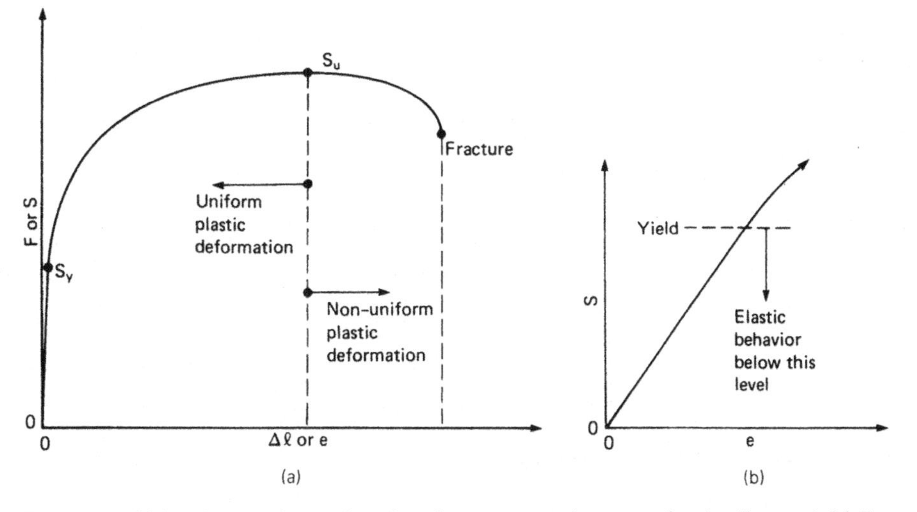

Figure 3.1. (a) Load–extension and engineering stress–strain curve of a ductile metal. (b) Expansion of initial part of the curve.

is why the axes of Figure 3.1 have double notation. The stress at which plastic flow begins, s_y, is called the yield strength, Y, and is defined as

$$Y = F_y/A_0 = s_y. \tag{3.3}$$

The tensile strength or ultimate tensile strength, s_u, is defined as

$$s_u = F_{\max}/A_0. \tag{3.4}$$

Ductility is a measure of the amount that a material can be deformed before failure. Two common parameters are used to describe ductility: % elongation and % reduction of area.

$$\% \text{ elongation} = 100\frac{(\ell_f - \ell_0)}{\ell_0} \tag{3.5}$$

$$\% \text{ reduction of area} = 100\frac{(A_0 - A_f)}{A_0}, \tag{3.6}$$

where A_f and ℓ_f are the cross-sectional area and gauge length at fracture. Although standard values of A_f and ℓ_f are usually used, the % elongation depends on the ratio of the gauge length-to-diameter because the elongation after necking depends on the diameter and the uniform elongation depends on the gauge length. The % reduction of area is much less dependent on specimen dimensions. The % reduction of area in a tension test should not be confused with the reduction of area, r, in a metal working process,

$$r = \frac{(A_0 - A)}{A_0}, \tag{3.7}$$

where A is the cross-sectional area after forming.

Figure 3.2 shows the yielding behavior of an annealed brass and a low-carbon steel. Brass is typical of most ductile metals. Yielding is so gradual that it is difficult

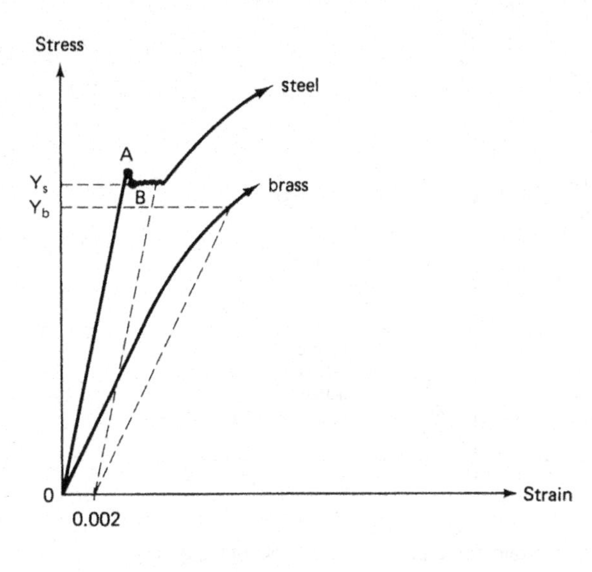

Figure 3.2. Offset method for determining the yield strength.

to define the point at which plastic deformation begins. Therefore the yield strength is usually defined as the stress at which the plastic strain is 0.002 or 0.2%. This can be found as the intersection of a line parallel to the elastic portion of the curve with the stress-strain curve. This 0.2% offset yield strength is shown in Figure 3.2 as Y_b. Low-carbon steels typically have an upper yield stress (point A) and a lower yield stress (B). Because the upper yield strength is very sensitive to the alignment of the tensile specimen, the lower yield stress (Y_b in Figure 3.2) is defined as the yield strength.

3.2 ELASTIC-PLASTIC TRANSITION

The transition from elastic to plastic flow is gradual as illustrated in Figure 3.3 for plane-strain deformation with $\varepsilon_y = 0$ and $\sigma_z = 0$. For elastic deformation, $\alpha = \nu$ and for fully plastic deformation $\alpha = 0.5$. In this figure the ε_x is normalized by the ratio of the yield strength to the modulus. Note that 95% of the change from elastic to plastic deformation occurs when the plastic strain is 3 times the elastic strain.

For a material that strain hardens, there is additional elastic deformation after yielding. The total strain is the sum of the elastic and plastic parts, $\varepsilon = \varepsilon_e + \varepsilon_p$. Even though the elastic strain may be very small relative to the plastic strain, elastic recovery on unloading controls residual stresses and springback.

3.3 ENGINEERING VS. TRUE STRESS AND STRAIN

Figure 3.1 shows that after initial yielding, further deformation requires an increased stress. Although the material strain hardens as it is extended, its cross-sectional area decreases. These two factors control the shape of the load-extension curve. Eventually there is a maximum when the rate of reduction of load-carrying capacity caused by reduction of area equals the rate of strain hardening. Up to this point the deformation along the gauge section is uniform. However after the maximum load is reached the

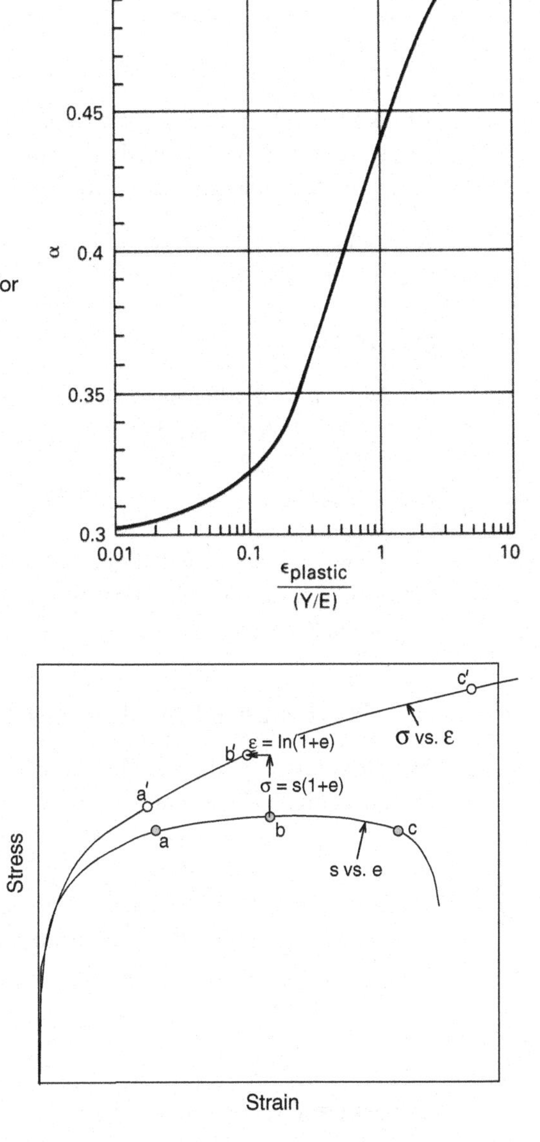

Figure 3.3. Change in the stress ratio, $\alpha = \sigma_y/\sigma_x$, for plane strain, $\varepsilon_y = 0$, as a function of strain.

Figure 3.4. Comparison of engineering and true stress-strain curves.

deformation will localize and a *neck* will form. With continued extension almost all of the deformation is restricted to the necked region.

Figure 3.4 shows the engineering stress-strain curve and the corresponding true stress-strain curve.

The true strain is defined by $d\varepsilon = d\ell/\ell$. Before necking occurs

$$\varepsilon = \ln(1 + e). \tag{3.8}$$

After necking the true strain must be bases on the area, $d\varepsilon = -dA/A$ or

$$\varepsilon = \ln(A_0/A). \tag{3.9}$$

The true stress $\sigma = F/A$ can be found before necking as

$$\sigma = s(1 + e). \tag{3.10}$$

EXAMPLE 3.1: When a tensile specimen with a diameter of 0.505 in. and a gauge length of 2.0 in. was loaded to 10,000 lbs., it was found that the gauge length was 2.519 in. Assuming that the deformation was uniform,

(a) compute the true stress and true strain.
(b) find the diameter.

SOLUTION:

(a) With $d_0 = 0.505$ in., $A_0 = 0.200$ in.^{2} $s = 10{,}000/0.2 = 50{,}000$ psi. $e = (2.519 - 2.000)/2.000 = 0.259$. $\sigma = s(1 + e) = 50{,}000(1.259) = 62{,}950$, $\varepsilon = \ln(1 + 0.259) = 0.23$.
(b) $d = 0.505/\sqrt{[\exp(0.23) = 0.491]}$

After necking the actual measured areas must be used to find the average stress in the necked region. Furthermore the stress state is actually triaxial as discussed further in Section 3.7 so $\bar{\sigma} \neq \sigma$.

3.4 POWER-LAW EXPRESSION

The curve a′b′c′ in Figure 3.4 is the true stress – true strain curve. Often this curve can be approximated by a power-law expression*

$$\sigma = K\varepsilon^n. \tag{3.11}$$

Here σ can be interpreted as the new yield strength after a cold reduction corresponding to $\varepsilon = \ln[1/(1 - r)]$. Because the tensile true stress, σ, and true tensile strain, ε, in a tension test are the effective stress and strain,

$$\bar{\sigma} = K\bar{\varepsilon}^n. \tag{3.12}$$

EXAMPLE 3.2: The plastic behavior of a metal can be expressed as $\bar{\sigma} = 500\bar{\varepsilon}^{0.50}$ MPa. Estimate the yield strength if a bar of this material is uniformly cold worked to a reduction of $r = 0.3$.

SOLUTION: Substituting $\varepsilon = \ln[1/(1 - .3)] = 0.357$, $\bar{\sigma} = 500(0.357)^{0.50} = 299$ MPa.

The values of K and n can be found plotting the true stress-strain curve on log-log coordinates. Noting that $\log \sigma = \log K + n\log \varepsilon$ so the slope equals n and K is the intercept at $\varepsilon = 1$. Figure 3.5 shows such a plot for an aluminum alloy. Note that there are three zones. Zone 1 is the elastic region where $\sigma = E\varepsilon$. Zone II is the region of

* The power-law hardening expression is sometimes called the Hollomon equation because it was first made popular by J. H. Hollomon in *Ferrous Metallurgical Design* by John H. Hollomon and Leonard D. Jaffe, New York, J. Wiley & Sons, Inc.

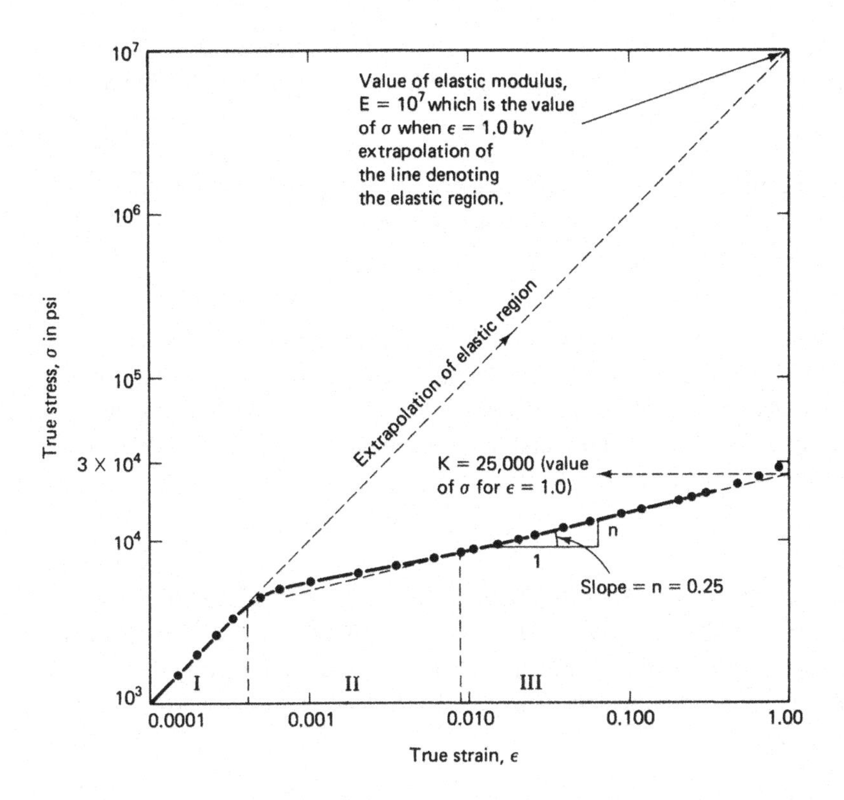

Figure 3.5. True stress-strain curve of aluminum 1100-O plotted on logarithmic coordinates.

transition between elastic and fully plastic behavior and the material in Zone III is fully plastic.

Strictly equations 3.11 and 3.12 apply only to the plastic part of the strain, but since the elastic strain is small relative to the plastic strain after a few percent, that distinction can be ignored.

EXAMPLE 3.3: Show that the maximum load in a tension test starts when $\varepsilon = n$.

SOLUTION: Since $F = \sigma A$, when $\mathrm{d}F = 0, \sigma \mathrm{d}A + A\mathrm{d}\sigma = 0$. Rearranging, $\mathrm{d}\sigma/\sigma = -\mathrm{d}A/A$. Substituting $\mathrm{d}\varepsilon = -\mathrm{d}A/A$, the maximum load corresponds to $\mathrm{d}\sigma/\mathrm{d}\varepsilon = \sigma$. With the power law, $\sigma = K\varepsilon^n$ and $\mathrm{d}\sigma/\mathrm{d}\varepsilon = nK\varepsilon^{n-1}$. Equating and simplifying, $K\varepsilon^n = nK\varepsilon^{n-1}$, $\varepsilon = n$. Thus maximum load and necking start when the strain equals strain-hardening exponent.

The true stress at maximum load can be expressed as

$$\sigma_u = K\varepsilon_u^n = Kn^n. \tag{3.13}$$

Since

$$s_u A_0 = \sigma_u A_u = (Kn^n)A_u, s_u = (Kn^n)A_u/A_0.$$

$$\text{Substituting } A_u/A_0 = \exp(-\varepsilon_u),\ s_u = K(n/\mathbf{e})^n, \tag{3.14}$$

where e is the base of natural logarithms.

EXAMPLE 3.4: In the tension test for Figure 3.5, the tensile strength was experimentally measured as 28,000 psi. Is this consistent with the values of $n = 0.25$ and $K = 50{,}000$?

SOLUTION: Using equation 3.14, $s_u = (0.25/\mathrm{e})^{0.25} = 27{,}500$ psi. This is within 2% so it is reasonable in view of errors in establishing n and K.

EXAMPLE 3.5:

(a) The strain-hardening behavior of an annealed low-carbon steel is approximated by $\bar{\sigma} = 700\bar{\varepsilon}^{0.20}$ MPa. Estimate the yield strength after the bar is cold worked 50%.
(b) Suppose another bar of this same steel was cold worked an unknown amount and then cold worked 15% more and found to have a yield strength of 525 MPa. What was the unknown amount of cold work?

SOLUTION:

(a) The strain was $\ln[1/(1-0.5)] = 0.693$. $\sigma = 700(0.693)^{0.20} = 650$ MPa.
(b) Solving equation 3.12 for $\bar{\varepsilon}$, $\bar{\varepsilon} = (\bar{\sigma}/K)^{1/n} = (525/700)^5 = 0.237$. The 15% cold work corresponds to a strain of $\ln[(1/1 - .15)] = 0.1625$. Subtracting this known strain, the unknown strain must be $0.237 - .1625 = .0745$, which corresponds to a cold work of 0.72%

3.5 OTHER STRAIN-HARDENING APPROXIMATIONS

While the power law is often a good approximation to the stress-strain behavior, sometimes other expressions are better approximations or more convenient mathematically. The stress-strain curves of some aluminum alloys seem to approach an asymptote indicating a saturation of strain hardening. In this case a better approximation is

$$\sigma = \sigma_0[1 - \exp(-B\varepsilon)], \tag{3.15}$$

where σ_0 is the level of σ at an infinite strain. If there has been prior cold work an appropriate expression may be

$$\sigma = K(\varepsilon + \varepsilon_0).^n \tag{3.16}$$

where the amount of prior cold work was equivalent to a strain of ε_0. It should be noted that the more constants in an approximation the better the fit that can be achieved, but the less useful the approximation is. Sometimes it is reasonable to assume linear strain hardening. This is very simple mathematically:

$$\sigma = \sigma_0 + C\varepsilon. \tag{3.17}$$

3.6 BEHAVIOR DURING NECKING

It was noted in Section 3.4 that once necking starts $\bar{\sigma} \neq \sigma$ because the stress state is no longer uniaxial. As the smallest section elongates, its lateral contraction is resisted by adjacent regions that are slightly less stressed. This constraint causes lateral stresses

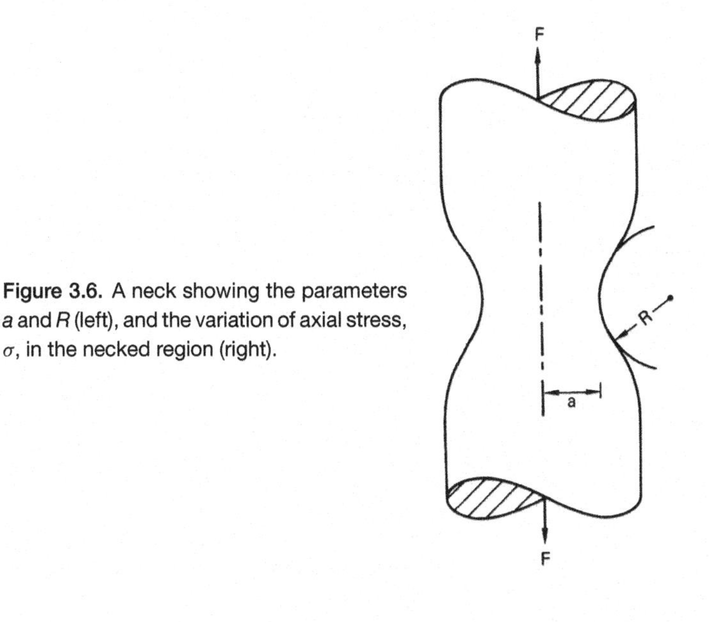

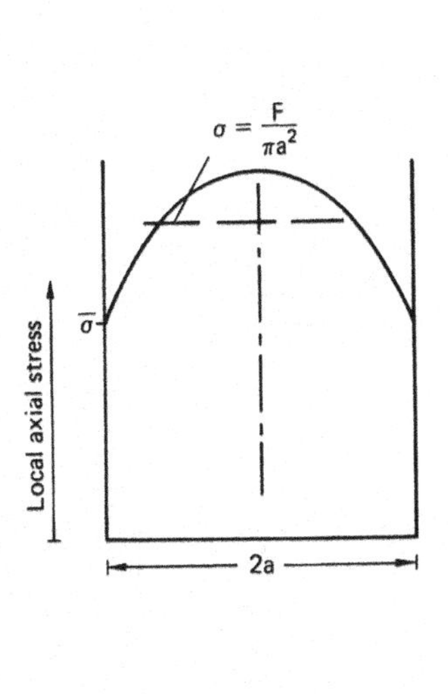

Figure 3.6. A neck showing the parameters a and R (left), and the variation of axial stress, σ, in the necked region (right).

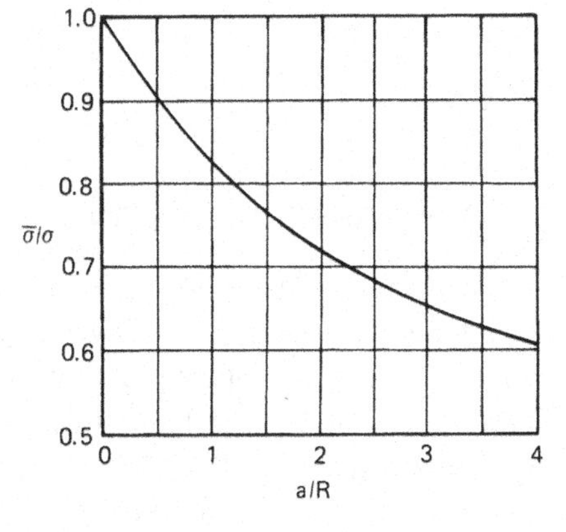

Figure 3.7. Ratio of effective stress to average axial stress as a function of neck geometry. After Bridgman, *ibid.*

that increase from zero on the outside to a maximum in the center. To maintain plastic flow the axial true stress must become larger than the effective stress.

Bridgman,* analyzed this problem and developed an expression for $\bar{\sigma}/\sigma$ as a function of a/R, where $2a$ is the diameter of the minimum section and R is the radius of curvature of the neck as shown in Figure 3.6. He found that

$$\sigma/\bar{\sigma} = (1 + 2R/a)\ln(1 + 2a/R), \tag{3.18}$$

where σ is the average value of the axial stress, $\sigma = F/A$. As the neck becomes sharper the correction becomes greater. Figure 3.7 is a plot of this correction factor.

The use of the Bridgman correction allows effective stress–effective strain data to be collected at large strains. However if a neck becomes too sharp, voids may form at the center of the neck region where the hydrostatic tension is greatest and the voids decrease the load-carrying cross section.

* P. W. Bridgman, *Trans. ASM*, 32 (1944) pp. 553–74.

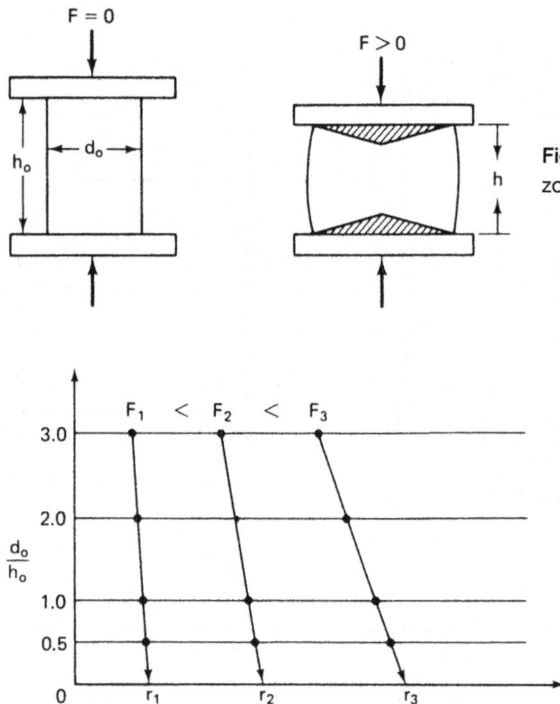

Figure 3.8. Barreling and formation of a dead-metal zone in compression.

Figure 3.9. The Watts and Ford method of finding the true stress-strain behavior in compression.

3.7 COMPRESSION TESTING

The compression test is an apparent solution to obtaining data at high strains without the interference of necking. However friction at the interface between the specimen and the platens complicates the stress state and results in *barreling* as shown in Figure 3.8. The friction tends to prevent radial expansion of the material in contact with the platens. This creates a dead-metal zone at the top and bottom. Thus end effects cause non-uniform stresses and strains.

The problem is lessened as the height-to-diameter ratio, h_0/d_0, of the specimen is increased. However when h_0/d_0 exceeds about 3, the specimens will buckle. An extrapolation method proposed by Cook and Larke* and improved by Watts and Ford† provides a way to obtain true stress–strain data. They tested well-lubricated specimens of the same initial diameter, but varying heights and measured the radius, r, as a function of load. By plotting d_0/h_0 against r at constant load they could find the value of r corresponding to an infinite h_0/d_0 ratio by extrapolation as shown schematically in Figure 3.9. This value of r was used to find the true strain.

3.8 BULGE TESTING

The hydraulic bulge test is a method of testing a sheet in balanced biaxial tension. A thin disc is clamped around the edges and subjected to increasing fluid pressure on

* M. Cook and E. C. Larke, *J. Inst. Metals* 71 (1945), pp. 371–90.

† A. B. Watts and H. Ford, *Proc. Inst. Mech. Eng.* 169 (1955), pp. 114–49.

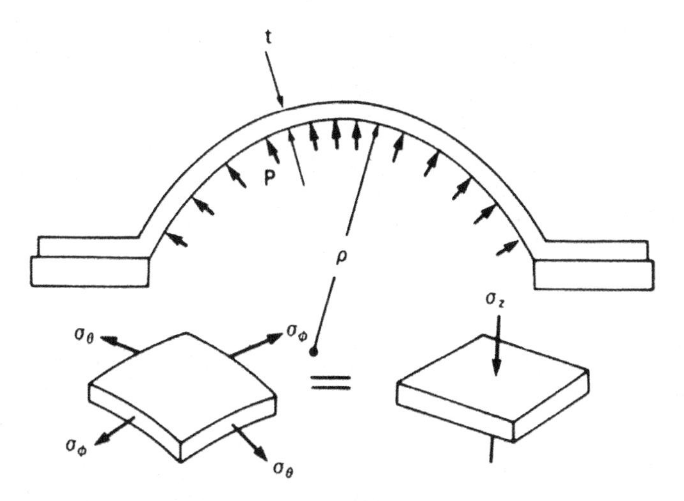

Figure 3.10. Schematic of a bulge test.

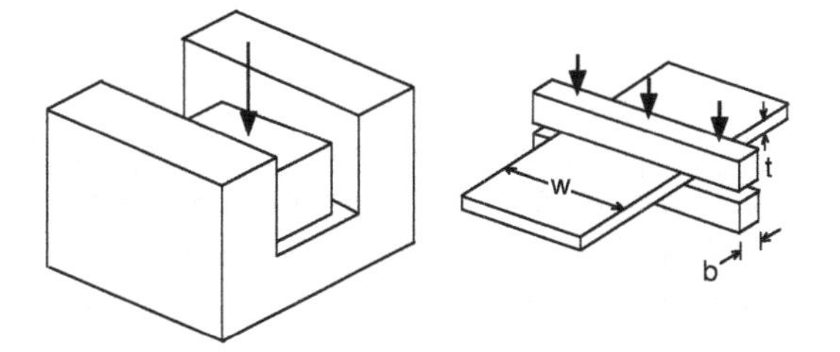

Figure 3.11. Two ways of making plane-strain compression tests.

one side as illustrated in Figure 3.10. As the sheet bulges, the region near the dome becomes nearly spherical. The tensile stresses are

$$\bar{\sigma} = \sigma_\theta = \sigma_\phi = P\rho/2t, \tag{3.19}$$

where P is the pressure, ρ is the radius of curvature and t is the thickness. The effective strain,

$$\bar{\varepsilon} = -2\varepsilon_\theta = -2\varepsilon_\phi = \ln(t/t_0). \tag{3.20}$$

It is necessary to measure the strain to find the thickness t. Duncan* developed a special device for measuring ρ. The advantage of bulge testing is that it can be carried out to much higher strains than in tension testing.

3.9 PLANE-STRAIN COMPRESSION

Figure 3.11 shows two simple ways of making plane-strain compression tests. One is to put a specimen in a well-lubricated channel that prevents lateral spreading. The other

* J. L. Duncan, The hydrostatic bulge test as laboratory experiment, *Bulletin of Mechanical Engineering Education*, v.4, Pergamon Press, (1965). R. F. Young, J. E. Bird and J. L. Duncan, "An Automated Hydraulic Bulge Tester," *J. Applied Metalworking*, 2, (1981).

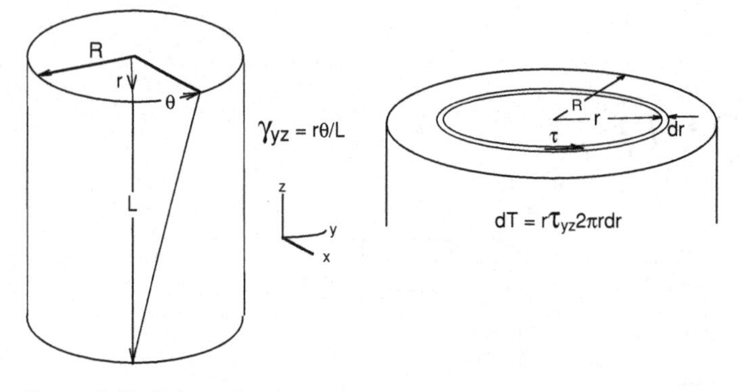

Figure 3.12. Schematic of a torsion test.

is to indent a wide thin specimen with overlapping indenters. The material outside the plastic region will prevent spreading if $w \gg b$. There is friction in both methods. With channel compression there is friction on the sidewalls as well as on the top and bottom.

3.10 TORSION TESTING

Although torsion testing is not widely used, very high strains can be reached without complications from friction or dimensional changes. Figure 3.12 illustrates the essentials. The shear strain in an element is given by

$$\gamma = r\theta/L, \tag{3.21}$$

where r is the radial position of the element.

The torque can be measured and it is related to the shear stress by

$$\mathrm{d}T = 2\pi\tau_{yz}r^2\mathrm{d}r \quad \text{or} \quad T = 2\pi\int_0^R \tau_{yz}r^2\mathrm{d}r. \tag{3.22}$$

The problem is that τ_{yz} varies with r. The solution to this is to test either very thin-wall tubes, which tend to buckle or to test two bars of slightly different diameters and take the difference in torques as the torque that would have been measured on a tube whose thickness equals the difference in diameters.*

NOTE OF INTEREST

J. Herbert Hollomon (1919–1985) was born in Norfolk, Va. He studied at physics and metallurgy at M.I.T where he earned a doctorate in 1946. He worked at the Watertown Arsenal during World War II. There he coauthored a book, *Ferrous Metallurgical Design* with L. D. Jaffe, Wiley, in 1947. Following the war he worked for the General Electric Research Labs where he rose to the position of General Manager. In 1962, President Kennedy appointed him the first assistant secretary for science and technology in the Department of Commerce. In 1960, he had made a speech calling for a national

* D. S. Fields and W. A. Backofen, *Proc. ASTM* 57 (1957), pp. 1259–72.

academy of engineering, which was reprinted in *Science*. The academy was established in 1964, and he became a founding member.

REFERENCES

R. M. Caddell, *Deformation and Fracture of Solids*, Prentice-Hall, 1980.
C. R. Calladine, *Engineering Plasticity*, Pergamon Press, 1969.
P. Han, ed. *Tension Testing*, ASM International, 1992.
W. F. Hosford, *Mechanical Behavior of Materials*, Cambridge University Press, 2005.
W. Johnson and P. M. Mellor, *Engineering Plasticity*, Van Nostrand Reinhold, 1973.
E. G. Thomsen, C. T. Yang and S. Kobayashi, *Mechanics of Plastic Deformation in Metal Processing*, Macmillan, 1965.

PROBLEMS

3.1. When a brass tensile specimen, initially 0.505 in. in diameter, is tested, the maximum load of 12,000 lbs was recorded at an elongation of 40%. What would the load be on an identical tensile specimen when the elongation is 20%?

3.2. During a tension test the tensile strength was found to be 340 MPa. This strength was recorded at an elongation of 30%. Determine n and K if the approximation $\bar{\sigma} = K\bar{\varepsilon}^n$ applies.

3.3. Show that the plastic work per volume is $\sigma_1\varepsilon_1/(n+1)$ for a metal stretched in tension to ε_1 if $\bar{\sigma} = k\bar{\varepsilon}^n$.

3.4. For plane-strain compression (Figure 3.11):

a) Express the incremental work per volume, dw, in terms of $\bar{\sigma}$ and $d\bar{\varepsilon}$ and compare it with $dw = \sigma_1 d\varepsilon_1 + \sigma_2 d\varepsilon_2 + \sigma_3 d\varepsilon_3$.

b) If $\bar{\sigma} = k\bar{\varepsilon}^n$, express the compressive stress, σ_1, as a function of ε_1, K, and n.

3.5. The following data were obtained from a tension test:

Load (kN)	Min. dia. (mm)	Neck radius (mm)	True strain ε	True stress corrected σ (MPa)
0	8.69	∞	0	0
27.0	8.13	∞	0.133	520
34.5	7.62	∞		
40.6	6.86	∞		
38.3	5.55	10.3		
29.2	3.81	1.8		

a) Compute the missing values.

b) Plot both σ and $\bar{\sigma}$ vs. ε on a logarithmic scale and determine K and n.

c) Calculate the strain energy per volume when $\varepsilon = 0.35$.

3.6. Consider a steel plate with a yield strength of 40 ksi, Young's modulus of 30×10^6 psi, and a Poisson's ratio of 0.30 loaded under balanced biaxial tension. What is the volume change, $\Delta V/V$, just before yielding?

3.7. The strain-hardening of a certain alloy is better approximated by $\sigma = A[1 - \exp(-B\varepsilon)]$ than by $\bar{\sigma} = k\bar{\varepsilon}^n$. Determine the true strain at necking in terms of A and B.

3.8. Express the tensile strength, in terms of A and B, for the material in Problem 3.7.

3.9. A metal sheet undergoing plane-strain deformation is loaded to a tensile stress of 300 MPa. What is the major strain if the effective stress–strain relationship is $\bar{\sigma} = 650(0.015 + \bar{\varepsilon})^{0.22}$ MPa?

4 Plastic Instability

Different phenomena limit the extent to which a metal may be deformed. Buckling may occur under compressive loading if the ratio of height-to-diameter is too great. Fracture may occur under tension. The thrust in this chapter is with a different type of phenomenon called plastic instability. When a structure is deformed, there is often a maximum force or maximum pressure after which deformation continues at decreasing loads or pressures. It is assumed throughout this chapter that the strain hardening is described by $\bar{\sigma} = K\bar{\varepsilon}^n$. If other expressions better represent the behavior, they can be used with the same procedures. Solutions for effective strain at instability are functions of n.

4.1 UNIAXIAL TENSION

In a tension test of a ductile metal, the deformation is uniform up to maximum load. After this, localized deformation starts to form a neck. Since $F = \sigma A$, the condition for maximum load can be expressed as

$$\mathrm{d}F = \sigma \mathrm{d}A + A\mathrm{d}\sigma = 0. \tag{4.1}$$

Rearranging,

$$\mathrm{d}\sigma/\sigma = -\mathrm{d}A/A = \mathrm{d}\varepsilon, \tag{4.2}$$

or

$$\mathrm{d}\sigma/\mathrm{d}\varepsilon = \sigma. \tag{4.3}$$

This is illustrated in Figure 4.1.

Deformation occurs in the weakest region. At lower strains the deformation is uniform because the weakest section is the least deformed section. At strains higher than n, the most deformed region is the weakest so deformation will concentrate there forming a neck. A construction attributed to Considére* (Figure 4.2) illustrates this.

$$\text{With } \sigma = K\varepsilon^n, \mathrm{d}\sigma/\mathrm{d}\varepsilon = nK\varepsilon^{n-1} = \sigma, \quad \text{or} \quad \varepsilon = n. \tag{4.4}$$

* A. Considére, *Ann. Ponts et Chausses*, v. 9 (1885), pp. 574–775.

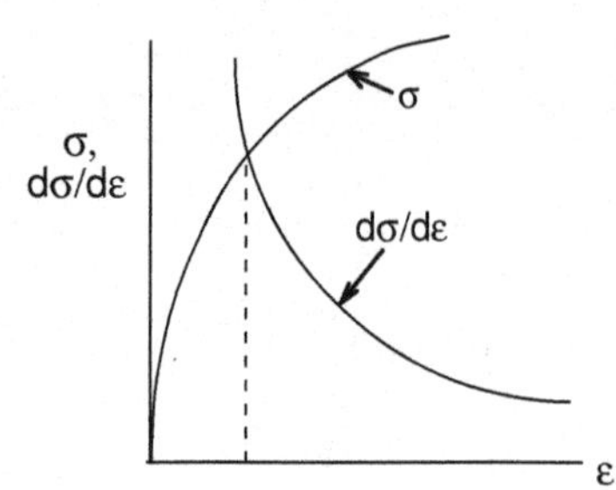

Figure 4.1. The maximum load in tension is reached when $d\sigma/d\varepsilon = \sigma$.

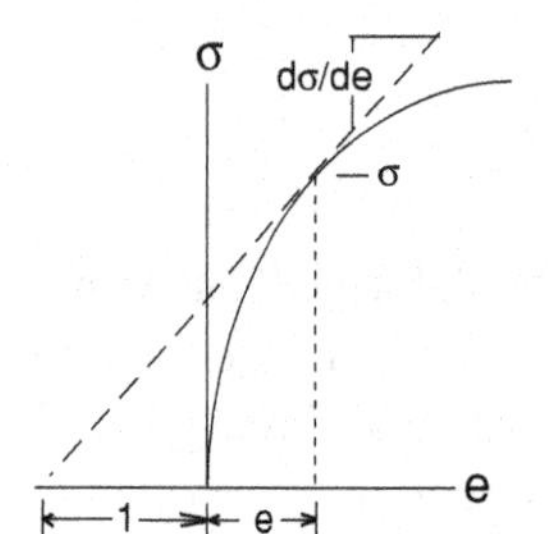

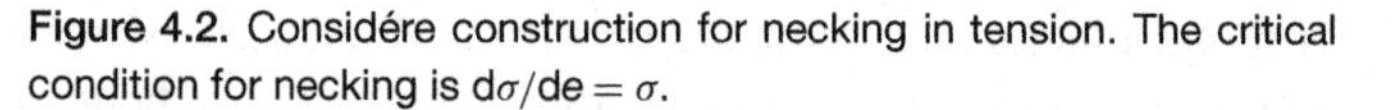

Figure 4.2. Considére construction for necking in tension. The critical condition for necking is $d\sigma/de = \sigma$.

Thus the maximum load, the tensile strength and the onset of necking occur at a strain equal to the strain-hardening exponent.

4.2 EFFECT OF INHOMOGENEITIES

Material properties and dimensions are usually considered to be uniform in analyses. However real materials are inhomogeneous: the cross-sectional diameter or thickness may vary from one place to another; there may also be variations in grain size, composition, and statistical crystal orientation. These latter inhomogeneities are assumed to influence only K in the expression $\bar{\sigma} = K\bar{\varepsilon}^n$, so they will affect the forces in the same manner as dimensional variations.

The effect of inhomogeneity is illustrated by a tensile specimen having homogeneous properties but two regions of different dimensions, a and b as shown in Figure 4.3.

An inhomogeneity factor, f, may be defined as

$$f = A_{ao}/A_{bo} \quad \text{where} \quad A_{ao} < A_{bo}. \tag{4.5}$$

The two regions, a and b, must support the same force so $F_a = F_b$, or

$$A_a\sigma_a = A_b\sigma_b. \tag{4.6}$$

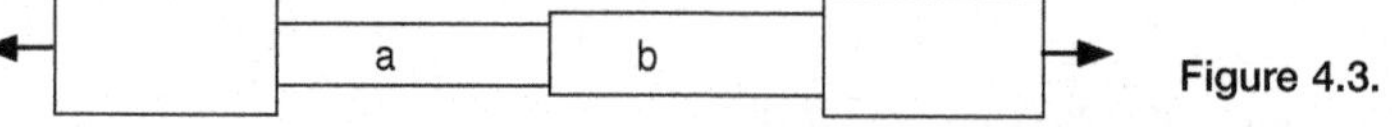

Figure 4.3. Stepped tensile specimen.

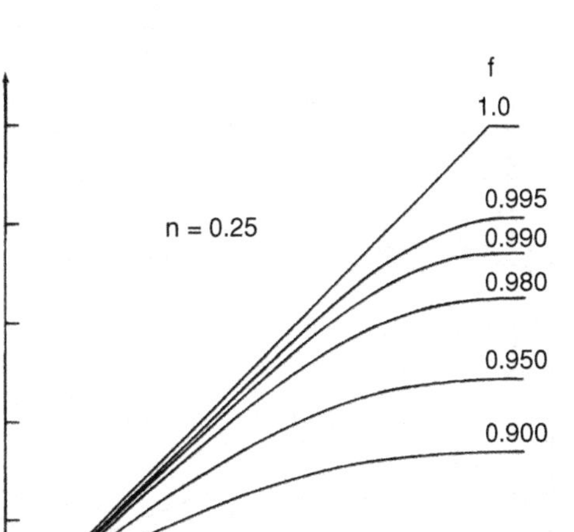

Figure 4.4. Strains induced by various inhomogeneity factors.

With power-law hardening, and substituting $A_i = A_{i0} \exp(-\varepsilon_i)$, where A_{i0} is the original cross-sectional area of region i,

$$A_{ao} \exp(-\varepsilon_a) K \varepsilon_a^n = A_{bo} \exp(-\varepsilon_b) K \varepsilon_b^n. \tag{4.7}$$

Substituting equation 4.5, equation 4.7 simplifies to

$$f \exp(-\varepsilon_a) \varepsilon_a^n = \exp(-\varepsilon_b) \varepsilon_b^n. \tag{4.8}$$

For given values of f and n, equation 4.8 can be solved numerically to give ε_b as a function of ε_a. Figure 4.4 shows this dependence for several values of f with $n = 0.25$. Note that as $f \rightarrow 1$, $\varepsilon_a \rightarrow \varepsilon_b$ up to a limiting strain of 0.25 where necking occurs. This implies that the maximum strain outside the neck equals n if $f = 1$. However if $f < 1$, ε_b lags behind ε_a and saturates at ε^* which may be considerably less than n. Figure 4.5 shows such results.

Note that for a typical tensile specimen with a nominal 0.505-in diameter, a diameter variation of 0.001 in., $f = 0.996$ and $\varepsilon_b^* = 0.0208$ instead of 0.25. Figure 4.6 shows the combined effect of f and n on the uniform elongation.

4.3 BALANCED BIAXIAL TENSION

Figure 4.7 illustrates the biaxial stretching of a sheet with $\sigma_1 = \sigma_2$. Instability will occur when $F_1 = F_2$ equals zero. $F_1 = \sigma_1 A_1$, so $\bar{\varepsilon} = 2n$

$$\mathrm{d}F_1 = \sigma_1 \mathrm{d}A_1 + A_1 \mathrm{d}\sigma_1 \tag{4.9}$$

or

$$\mathrm{d}\sigma_1/\sigma_1 = -\mathrm{d}A_1/A_1 = \mathrm{d}\varepsilon_1. \tag{4.10}$$

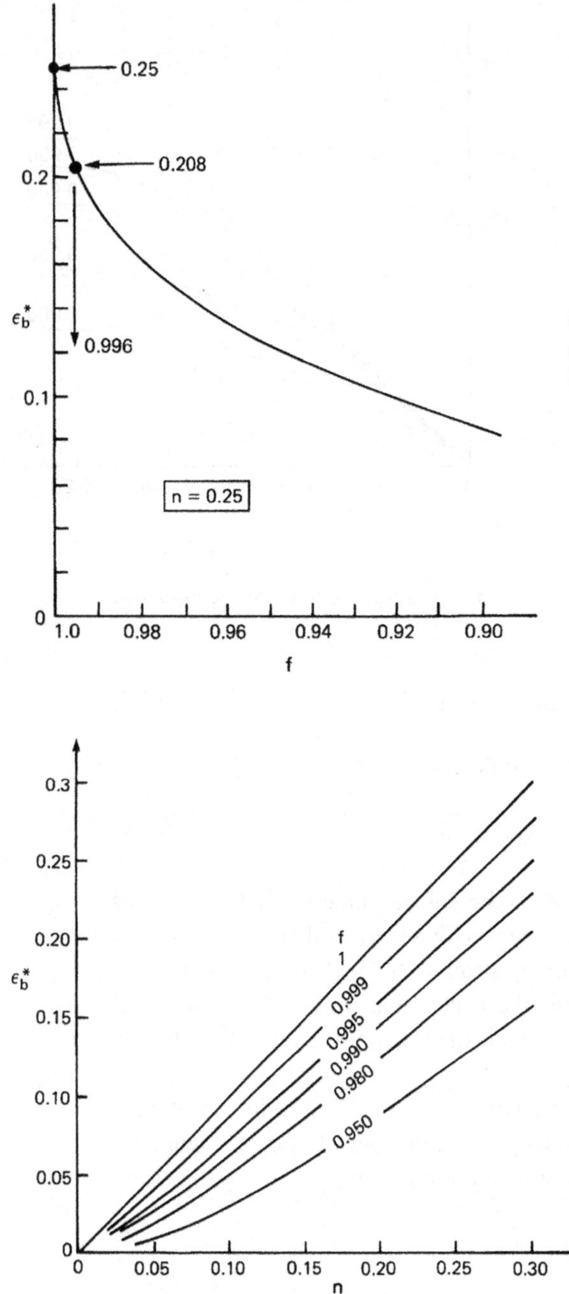

Figure 4.5. Effect of the inhomogeneity factor, *f*, on the limit strain, ε_0^*, outside of the neck. A strain-hardening exponent of 0.25 was assumed.

Figure 4.6. Effects of *f* and *n* on the uniform elongation, ε_b^*.

With $\sigma_1 = \sigma_2$, the effective stress, $\bar{\sigma}$, is

$$\bar{\sigma} = \sigma_1 = \sigma_2. \tag{4.11}$$

According to the flow rules

$$\bar{\varepsilon} = 2\varepsilon_1 = 2\varepsilon_2 = -\varepsilon_3. \tag{4.12}$$

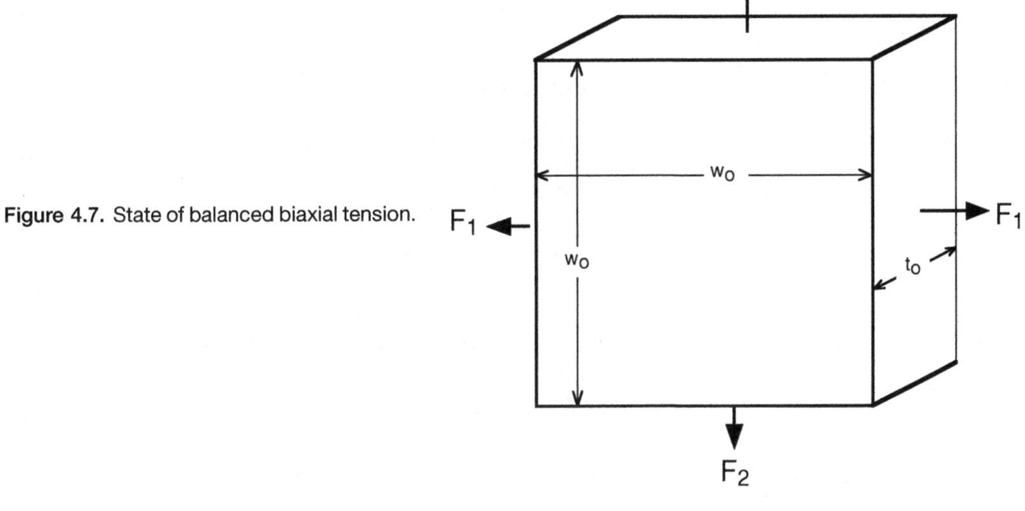

Figure 4.7. State of balanced biaxial tension.

Substituting equations 4.7 and 4.6 into equation 4.5

$$\frac{d\bar{\sigma}}{d\bar{\varepsilon}} = \frac{\bar{\sigma}}{2}, \tag{4.13}$$

so

$$\bar{\varepsilon} = 2n \tag{4.14}$$

at instability. Note that this is twice the instability strain for uniaxial tension.

4.4 PRESSURIZED THIN-WALL SPHERE

Figure 4.8 is a free-body diagram of half of a pressurized thin-wall spherical pressure vessel.

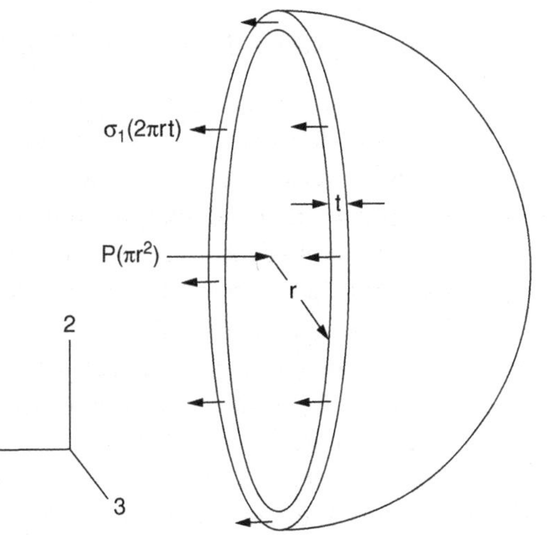

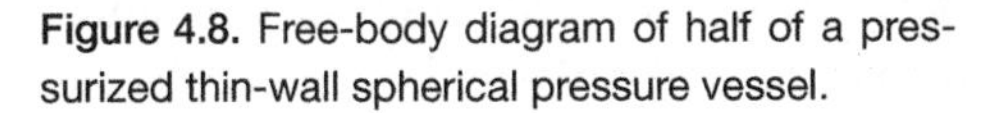
Figure 4.8. Free-body diagram of half of a pressurized thin-wall spherical pressure vessel.

A force balance in the 1-direction indicates that $2\pi rP = 2\pi rt\sigma_1$, where P is the internal pressure, r and t are the radius and wall thickness so

$$P = \frac{2t\sigma_1}{r}. \tag{4.15}$$

The maximum pressure corresponds to

$$\mathrm{d}P = 0 = \frac{2t\mathrm{d}\sigma_1}{r} + \frac{2\sigma_1\mathrm{d}t}{r} + \frac{2t\sigma_1\mathrm{d}r}{r^2} \tag{4.16}$$

or

$$\frac{\mathrm{d}\sigma_1}{\sigma_1} = \frac{\mathrm{d}r}{r} - \frac{\mathrm{d}t}{t}. \tag{4.17}$$

Substituting $\mathrm{d}r/r = \mathrm{d}\varepsilon_1$, $\mathrm{d}t/t = \mathrm{d}\varepsilon_3$, and $\mathrm{d}\varepsilon_1 = \mathrm{d}\varepsilon_2 = -\mathrm{d}\varepsilon_3/2$,

$$\frac{\mathrm{d}\sigma_1}{\sigma_1} = -(3/2)\,\mathrm{d}\varepsilon_3. \tag{4.18}$$

Substituting $\mathrm{d}\bar{\varepsilon} = -\mathrm{d}\varepsilon_3$ and $\bar{\sigma} = \sigma_1$, $\mathrm{d}\bar{\sigma}/\bar{\sigma} = (3/2)\,\mathrm{d}\bar{\varepsilon}$ or

$$\mathrm{d}\bar{\sigma}/\mathrm{d}\bar{\varepsilon} = (3/2)\bar{\sigma}. \tag{4.19}$$

With power-law hardening, the instability occurs at

$$\bar{\varepsilon} = (2/3)n. \tag{4.20}$$

Substituting $t = t_0\mathbf{e}^{\varepsilon_3} = t_0\mathbf{e}^{-\bar{\varepsilon}}$, $r = r_0\mathbf{e}^{\varepsilon_1} = r_0\mathbf{e}^{-\bar{\varepsilon}/2}$ and $\bar{\sigma} = K\bar{\varepsilon}^n$,

$$P = 2K\bar{\varepsilon}^n\frac{t_0}{r_0}\mathbf{e}^{(-3/2)\bar{\varepsilon}}. \tag{4.21}$$

Setting $\mathrm{d}P$ to zero,

$$P_{\max} = 2K(3/2n)^n\mathbf{e}^{-n}. \tag{4.22}$$

Figure 4.9 illustrates the variation of pressure with $\bar{\varepsilon}$ and r.

4.5 SIGNIFICANCE OF INSTABILITY

Instability conditions in this chapter relate to a maximum force or pressure. However uniform deformation does not generally cease at a maximum force or pressure as it does in a tension test. Reconsider the case of the thin-wall tube under internal pressure. Although the maximum pressure occurs when $\bar{\varepsilon} = (2/3)n$, strain localization cannot occur, because localization would cause a decrease of the local radius of curvature, which would decrease the stress in that area below that of the rest of the sphere. Instead the sphere will continue to expand uniformly. The walls are in biaxial tension so the maximum wall force will occur when $\bar{\varepsilon} = 2n$. Strain localization still cannot occur because any localization would decrease the radius of curvature. Figure 4.9 shows these points.

Any strain localization under biaxial tension must occur because of local inhomogeneity. The role of inhomogeneity is developed in Chapter 15.

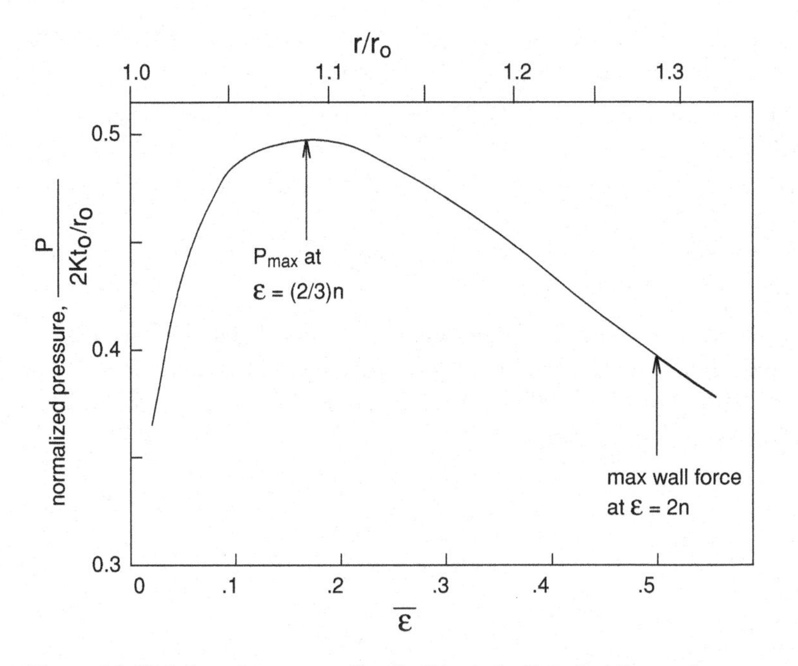

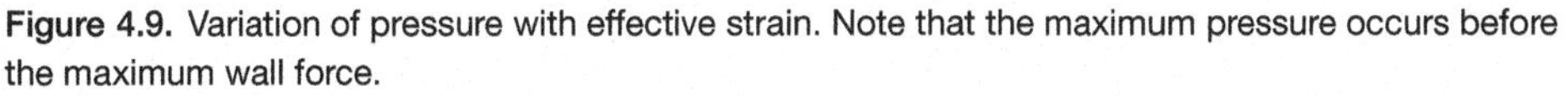

Figure 4.9. Variation of pressure with effective strain. Note that the maximum pressure occurs before the maximum wall force.

NOTE OF INTEREST

Professor Zdzislaw Marciniak is a member of the Polish Academy of Sciences and was head of the Faculty of Production Engineering at the Technical University of Warsaw. For some years he was also acting rector of that university. He has published a number of books (mainly in Polish) on scientific aspects of engineering plasticity and on the technology of metal forming. He developed machines for incremental cold forging including the rocking die and rotary forming processes. His inhomogeneity model of local necking in stretch forming of sheet, published with K. Kuczynski in 1967, has had a profound effect on the understanding of forming limits in sheet metal.

REFERENCES

W. A. Backofen, *Deformation Processing*, Addison-Wesley, 1972.
W. F. Hosford, *Mechanical Behavior of Materials*, Cambridge University Press, 2005.

PROBLEMS

4.1. If $\bar{\sigma} = K\bar{\varepsilon}^n$, the onset of tensile instability occurs when $n = \varepsilon_u$. Determine the instability strain as a function of n if

a) $\bar{\sigma} = A(B + \bar{\varepsilon})^n$.

b) $\bar{\sigma} = Ae^n$, where e is the engineering strain.

4.2. Consider a balloon made of a material that shows linear elastic behavior to fracture and has a Poisson's ratio of ½. If the initial diameter is d_0, find the diameter, d, at the highest pressure.

4.3. Determine the instability strain in terms of n for a material loaded in tension while subjected to a hydrostatic pressure P. Assume $\bar{\sigma} = K\bar{\varepsilon}^n$.

4.4. A thin-wall tube with closed ends is pressurized internally. Assume that $\bar{\sigma} = 150\bar{\varepsilon}^{0.25}$(MPa).

a) At what value of effective strain will instability occur with respect to pressure?

b) Find the pressure at instability if the tube had an initial diameter of 10 mm and a wall thickness of 0.5 mm.

4.5. Figure 4.10 shows an aluminum tube fitted over a steel rod. The steel may be considered rigid, and the friction between the aluminum and the steel may be neglected. If $\bar{\sigma} = 160\bar{\varepsilon}^{0.25}$(MPa) for the tube and it is loaded as indicated, calculate the force, F, at instability.

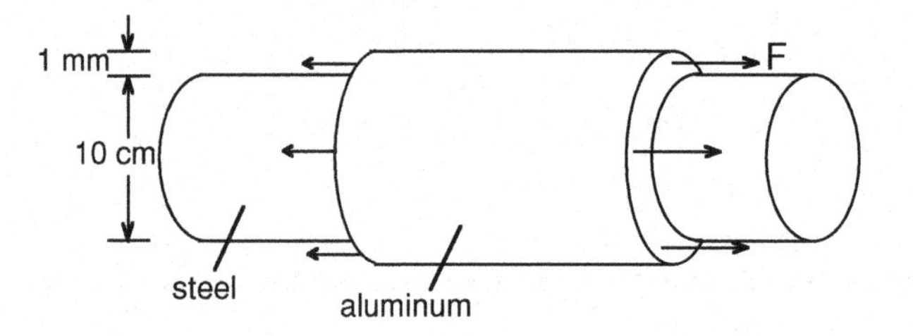

Figure 4.10. Sketch for Problem 4.5.

4.6. A thin-wall tube with closed ends is subjected to an ever-increasing internal pressure. Find the dimensions r and t in terms of the original dimensions r_0 and t_0 at maximum pressure. Assume $\bar{\sigma} = 500\bar{\varepsilon}^{0.20}$ MPa.

4.7. Consider the internal pressurization of a thin-wall sphere by an ideal gas for which $PV =$ constant. One may envision an instability condition for which the decrease of pressure with volume, $(-dP/dV)_{gas}$, due to gas expansion is lower than the rate of decrease in pressure that the sphere can withstand, $(-dP/dV)_{sph}$. For such a condition, catastrophic expansion would occur. If $\bar{\sigma} = K\bar{\varepsilon}^n$, find $\bar{\varepsilon}$ as a function of n.

4.8. For rubber stretched under biaxial tension $\sigma_x = \sigma_y = \sigma$, the stress is given by $\sigma = NkT(\lambda^2 - 1/\lambda^4)$, where λ is the stretch ratio, $L_x/L_{x0} = L_y/L_{y0}$. Consider what this equation predicts about how the pressure in a spherical rubber balloon varies during the inflation. For $t_0 = r_0$, plot P vs. λ and determine the strain, λ, at which the pressure is a maximum.

4.9. For a material that has a stress-strain relationship of the form $\bar{\sigma} = A - B\exp(-C\varepsilon)$, where A, B, and C are constants, find the true strain at the onset of necking and express the tensile strength, s_u, in terms of the constants.

4.10. A tensile bar was machined with a stepped gauge section. The two diameters were 2.0 and 1.9 cm. After some stretching, the diameters were found to be 1.893 and 1.698 cm. Find n in the expression $\bar{\sigma} = K\bar{\varepsilon}^n$, and find $\bar{\varepsilon}$ as a function of n.

4.11. In a rolled sheet, it is not uncommon to find variations of thickness of $\pm 1\%$ from one place to another. Consider a sheet nominally 0.8 mm thick with a $\pm 1\%$

variation of thickness. (Some places are 0.808 mm and others are 0.792 mm thick.) How high would n have to be to ensure that in a tensile specimen every point was strained to at least $\varepsilon = 0.20$ before the thinner section necked?

4.12. A material undergoes linear strain hardening so that $\sigma = Y + 1.35\varepsilon$ is stretched in tension.

a) At what strain will necking begin?

b) A stepped tensile specimen was made from this material with the diameter of region A being 0.990 times the diameter of region B. What would be the strain in region B when region A reached a strain of 0.20?

5 Temperature and Strain-Rate Dependence

The effects of strain hardening on flow stress were treated in Chapter 3. A material's flow stress also depends on strain rate and temperature. It usually increases with increasing strain rate and decreasing with temperature. In this chapter the effect of strain rate at constant temperature will be considered first.

5.1 STRAIN RATE

Increased strain rate normally causes a higher flow stress. The strain-rate effect at constant strain can be approximated by

$$\sigma = C\dot{\varepsilon}^m, \tag{5.1}$$

where C is a strength constant that depends upon the strain, the temperature, and the material, and m is the strain-rate sensitivity of the flow stress. For most metals at room temperature, the magnitude of m is quite low (between 0 and 0.03). The ratio of flow stresses, σ_2 and σ_1, at two strain rates, $\dot{\varepsilon}_2$ and $\dot{\varepsilon}_1$, is

$$\frac{\sigma_2}{\sigma_1} = \left(\frac{\dot{\varepsilon}_2}{\dot{\varepsilon}_1}\right)^m. \tag{5.2}$$

Taking logarithms of both sides, $\ln(\sigma_2/\sigma_1) = m\ln(\dot{\varepsilon}_2/\dot{\varepsilon}_1)$. If, as is likely at low temperatures, σ_2 is not much greater than σ_1, equation 5.2 can be simplified to

$$\frac{\Delta\sigma}{\sigma} \cong m\ \ln\frac{\dot{\varepsilon}_2}{\dot{\varepsilon}_1} = 2.3\,m\log\frac{\dot{\varepsilon}_2}{\dot{\varepsilon}_1}. \tag{5.3}$$

For example, if $m = 0.01$, increasing the strain rate by a factor of 10 would raise the flow stress by only $0.01 \times 2.3 \cong 2\%$. This is why rate effects are often ignored.

Rate effects can be important in some cases, however. For example, if one wishes to predict forming loads in wire drawing or sheet rolling (a process in which the strain rates may be as high as 10^4/sec) from data obtained in a laboratory tension test, in which the strain rates may be as low as 10^{-3}/sec, the flow stress should be corrected unless m is very small. Ratios of (σ_2/σ_1) calculated from equation 5.2 for various levels of $(\dot{\varepsilon}_2/\dot{\varepsilon}_1)$ and m are shown in Figure 5.1.

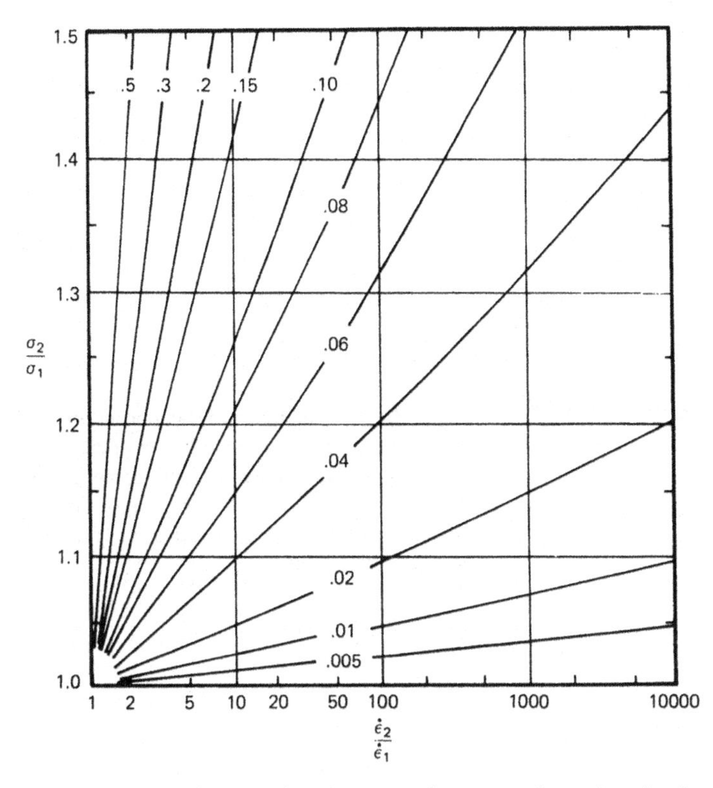

Figure 5.1. The influence of strain rate on flow stress for various levels of strain-rate sensitivity, m, indicated on the curves.

There are two commonly used methods of measuring m. One is to compare the levels of stress, at a fixed strain in tension tests made at different strain rates using equation 5.2. The other is to make abrupt changes of strain rate during a tension test and use the corresponding level of $\Delta\sigma$ in equation 5.3. These are illustrated in Figure 5.2. Increased strain rates cause slightly greater strain hardening, so the use of continuous stress-strain curves yields slightly higher values of m than the second method, which compares the flow stresses for the same structure. The second method has the advantage that several strain-rate changes can be made on a single specimen, whereas continuous stress-strain curves require separate specimens for each strain rate.

A material's strain-rate sensitivity is temperature dependent. At hot-working temperatures, m typically rises to 0.10 or 0.20, making rate effects much larger than at room temperature. Under certain circumstances, m-values of 0.5 or higher have been observed in various metals. Figure 5.3 shows data for a number of metals obtained from continuous constant strain-rate tests. Below $T/T_m = ½$, where T/T_m is the ratio of testing temperature to melting point on an absolute scale, the rate sensitivity is low but it climbs rapidly for $T > T_m/2$.

More detailed data for aluminum alloys are given in Figure 5.4. Although the definition of m in this figure is based upon shear stress-strain rate data, it is equivalent to the definition derived from equation 5.1. For alloys of aluminum and many other metals, there is a minimum in m near room temperature and, as indicated, negative m-values are sometimes found. At low strain rates, solutes segregate to dislocations; this lowers

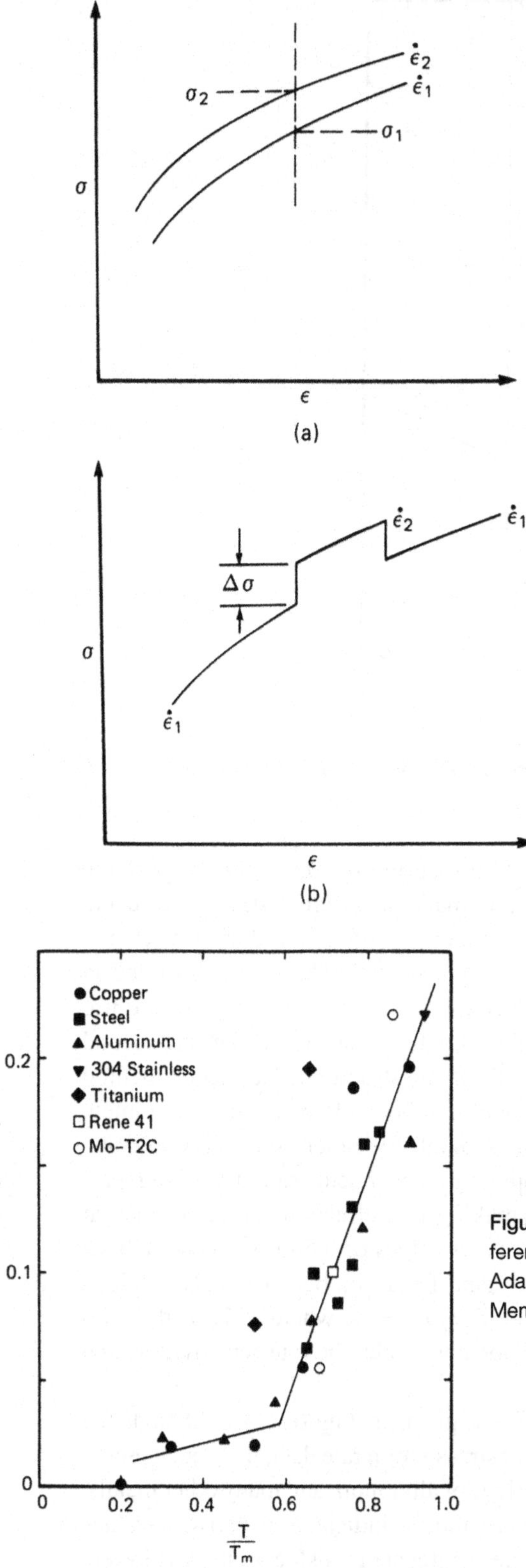

Figure 5.2. Two methods of determining m. (a) Two continuous stress-strain curves at different strain rates are compared at the same strain and $m = \ln(\sigma_2/\sigma_1)/\ln(\dot{\varepsilon}_2/\dot{\varepsilon}_1)$. (b) Abrupt strain-rate changes are made during a tension test and $m = (\Delta\sigma/\sigma)/\ln(\dot{\varepsilon}_2/\dot{\varepsilon}_1)$.

Figure 5.3. Variation of the strain-rate sensitivity of different materials with homologous temperature, T/T_m. Adapted from F. W. Boulger, *DMIC Report 226*, Battelle Mem. Inst. (1966).

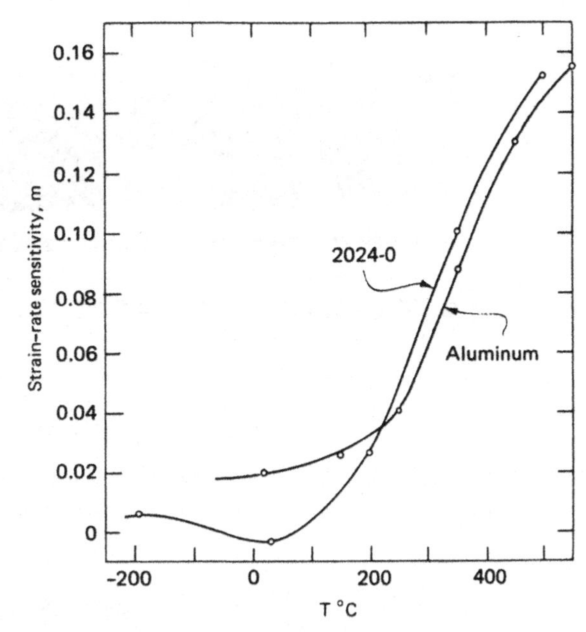

Figure 5.4. Temperature dependence of the strain-rate sensitivities of 2024 and pure aluminum. From D. S. Fields and W. A. Backofen, *Trans. ASM*, 51 (1959).

their energy so that the forces required to move the dislocations are higher than those required for solute-free dislocations. At increased strain rates or lower temperatures, however, dislocations move faster than solute atoms can diffuse, so the dislocations are relatively solute-free and the drag is minimized. A negative rate sensitivity tends to localize flow in a narrow region which propagates along a tensile specimen as a Lüders band. Localization of flow in a narrow band occurs because it allows the deforming material to experience a higher strain rate and therefore a lower flow stress.

The higher values of rate sensitivity at elevated temperatures are attributed to the increased rate of thermally activated processes such as dislocation climb and grain-boundary sliding.

5.2 SUPERPLASTICITY

As already mentioned, rate sensitivities sometimes can be in excess of 0.5. The necessary conditions are:

1. An extremely fine grain size (a few micrometers or less), with generally uniform and equiaxed grain structure,
2. High temperatures ($T > 0.4\ T_m$) and
3. Low strain rates* (10^{-2}/sec or lower).

In addition to the above conditions, the microstructure should be stable, without grain growth during deformation. Under these conditions, extremely high elongations and very low flow stresses are observed in tension tests, so the term *superplasticity*

* The fact that m itself is rate dependent indicates that equation 5.1 is not a true description of the rate effect. Nevertheless, expressed as $m = (\partial \ln \sigma / \partial \ln \dot{\varepsilon})_{\varepsilon T}$, it is a useful index of the strain-rate sensitivity of the flow stress and is a convenient basis for describing and analyzing strain-rate effects.

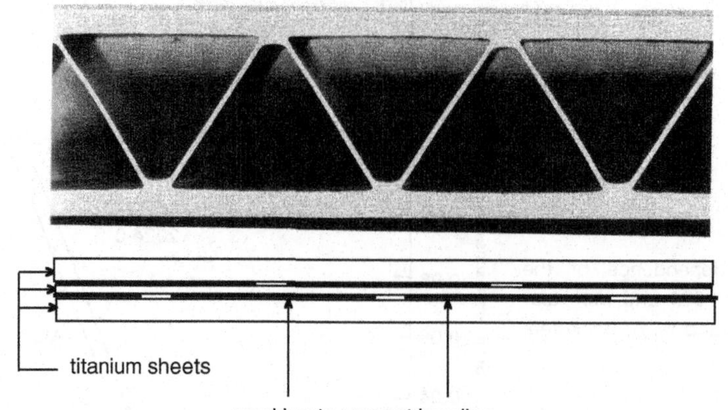

Figure 5.5. Aircraft panel made from a titanium alloy by diffusion bonding and superplastic expansion by internal pressure. Photo courtesy of Rockwell International Corp.

Figure 5.6. Complex sheet-metal part of Zn-22% Al made by superplastic forming. Courtesy of D. S. Fields, IBM Corp.

has been used to describe these effects. There have been several excellent reviews of superplasticity and the mechanisms and characteristics.*

There are two useful aspects of superplasticity. One is that superplasticity is usually accompanied by very low flow stresses at useful working temperatures. This permits creep forging of intricately shaped parts and reproduction of fine detail. The other is the extremely high tensile elongations which permit sheet parts of great depth to be formed with simple tooling.

One application of superplasticity has been in the production of hollow titanium-alloy aircraft panels. An example of such a panel in cross section is shown in Figure 5.5. Using superplastic forming with concurrent diffusion bonding (SPF/DB), three sheets are first diffusion bonded at specific locations, bonding elsewhere being prevented by painting the sheets with an inert ceramic. Then the unbonded channels between the sheets are pressurized internally with argon until the skin pushes against tools, which control the outer shape. Superplastic behavior is required to obtain the necessary

* J. W. Edington, K. N. Melton, and C. P. Cutler, *Progress in Materials Sci.*, 21 (1976). O. D. Sherby and J. Wadsworth, *Superplasticity: Recent Advances and Future Directions* (New York: Oxford, 1990). S. Tang, *Mechanics of Superplasticity; Mathematics for Engineering and Science* (Huntington, New York, 1979).

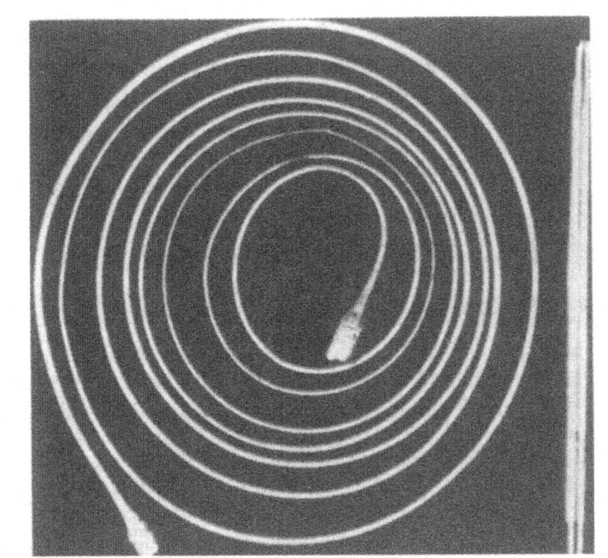

Figure 5.7. Superplastically extended tensile bar of Bi-Sn eutectic. From C. E. Pearson, *J. Inst. Metals*, 54 (1934).

elongation of the interior core sheet ligaments, which is over 100% in the example shown. Figure 5.6 shows an example of a deep part made by superplastic forming of Zn-22%Al sheet. A tensile bar elongated 1950% (a 19-fold increase in length!) is illustrated in Figure 5.7. To appreciate this, one must realize that a tensile elongation over 50% is usually considered large.

The tensile elongations for a number of materials are plotted in Figure 5.8 as a function of m. High m-values promote large elongations by preventing localization of the deformation as a sharp neck. This can be seen clearly in the following example.

Consider a bar that starts to neck. If the cross-sectional area in the neck is A_n and outside the neck is A_u, $F = \sigma_u A_u = \sigma_n A_n$ or $\sigma_u/\sigma_n = A_n/A_u$. From equation 5.2,

$$\frac{\dot{\varepsilon}_u}{\dot{\varepsilon}_n} = \left(\frac{\sigma_u}{\sigma_n}\right)^{1/m} = \left(\frac{A_n}{A_u}\right)^{1/m}. \tag{5.4}$$

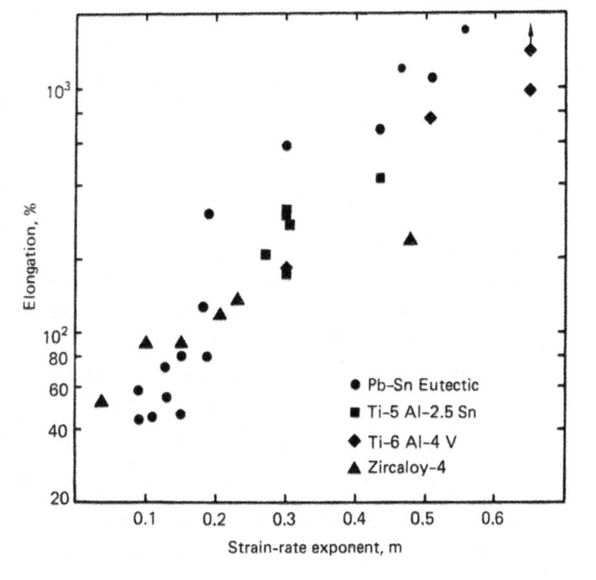

Figure 5.8. Correlation of tensile elongation with strain-rate sensitivity. Data from D. Lee and W. A. Backofen, *TMS-AIME*, **239** (1967), p. 1034; and D. H. Avery and W. A. Backofen, *Trans. Q. ASM*, **58** (1965), pp. 551–62.

Since $A_n < A_u$, if m is low, the strain rate outside the neck will become negligibly low. For example, let the neck region have a cross-sectional area of 90% of that outside the neck. If $m = 0.02$, $\dot{\varepsilon}_u/\dot{\varepsilon}_n = (0.9)^{50} = 5 \times 10^{-3}$. If, however, $m = 0.5$, then $\dot{\varepsilon}_u/\dot{\varepsilon}_n = (0.9)^2 = 0.81$, so that although the unnecked region deforms slower than the neck, its rate of straining is still rather large.

5.3 EFFECT OF INHOMOGENEITIES

Reconsider the tension test in Chapter 4 on a bar on the stepped bar in Figure 4.3. The bar is divided into two regions, one with an initial cross section of A_{b0} and the other $A_{a0} = fA_{b0}$. Neglect strain hardening and assume that $\sigma = C\dot{\varepsilon}^m$. As before, the forces must balance, so $\sigma_b A_b = \sigma_a A_a$. Substituting $A_i = A_{i0}\exp(\varepsilon_i)$ and $\sigma_i = C\dot{\varepsilon}_i^m$, the force balance becomes

$$A_{b0}\exp(-\varepsilon_b)\dot{\varepsilon}_b^m = A_{a_0}\exp(-\varepsilon_a)\dot{\varepsilon}_a^m, \tag{5.5}$$

where $\dot{\varepsilon}_a$ and $\dot{\varepsilon}_b$ are the strains in the reduced and unreduced sections. Expressing $\dot{\varepsilon}$ as $\mathrm{d}\varepsilon/\mathrm{d}t$,

$$\exp(-\varepsilon_b)\left(\frac{\mathrm{d}\varepsilon_b}{\mathrm{d}t}\right)^m = f\exp(-\varepsilon_a)\left(\frac{\mathrm{d}\varepsilon_a}{\mathrm{d}t}\right)^m. \tag{5.6}$$

Raising both sides to the $(1/m)$ power and canceling $\mathrm{d}t$

$$\int_0^{\varepsilon_b}\exp(-\varepsilon_b/m)\mathrm{d}\varepsilon_b = \int_0^{\varepsilon_a} f^{1/m}\exp(-\varepsilon_a/m)\mathrm{d}\varepsilon_a. \tag{5.7}$$

Integration gives

$$\exp(-\varepsilon_b/m) - 1 = f^{1/m}[\exp(-\varepsilon_a/m) - 1]. \tag{5.8}$$

Numerical solutions of ε_b as a function of ε_a for $f = 0.98$ and several levels of m are shown in Figure 5.9. At low levels of m (or low values of f), ε_b tends to saturate early and approaches a limiting strain ε_b^* at moderate levels of ε_a, but with higher m-values, saturation of ε_b is much delayed, that is, localization of strain in the reduced section (or the onset of a sharp neck) is postponed. Thus, the conditions that promote high m-values also promote high failure strains in the necked region. Letting $\varepsilon_a \to \infty$ in equation 5.8 will not cause great error and will provide limiting values for ε_b^*. With this condition,

$$\varepsilon_b^* = -m\ln(1 - f^{1/m}). \tag{5.9}$$

In Figure 5.10, values of ε_b^* calculated from equation 5.9 are plotted against m for various levels of f. The values of tensile elongation corresponding to ε_b^* are indicated on the right margin. It is now clear why large elongations are observed under superplastic conditions. The data in Figure 5.8, replotted here, suggest an inhomogeneity factor of about 0.99 to 0.998 (for a round bar 0.250-in. diameter, a diameter variation of 0.0005 in, corresponds to $f = 0.996$). The general agreement is perhaps fortuitous considering the assumptions and simplifications. The values of ε_b^* are the strains away from the neck, so the total elongation would be even higher than indicated here. On the other hand, strains in the neck, ε_a, are not infinite, so the ε_b^* values corresponding

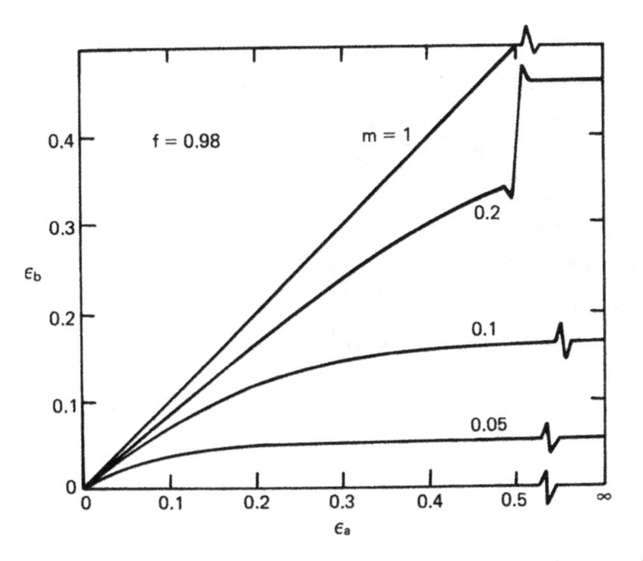

Figure 5.9. Relative strains in unreduced, ε_b, and reduced, ε_a, sections of a stepped tensile specimen for various levels of m, assuming no strain hardening.

to realistic maximum values of ε_a will be lower than those used in Figure 5.10. Also, the experimental values are affected by difficulties in maintaining constant temperature over the length of the bar as well as a constant strain rate in the deforming section. Nevertheless, the agreement between theory and experiments is striking.

Figure 5.11 shows the dependence of flow stress of the Al-Cu eutectic alloy at 520°C upon strain rate. The effect of strain is not important here because strain hardening is

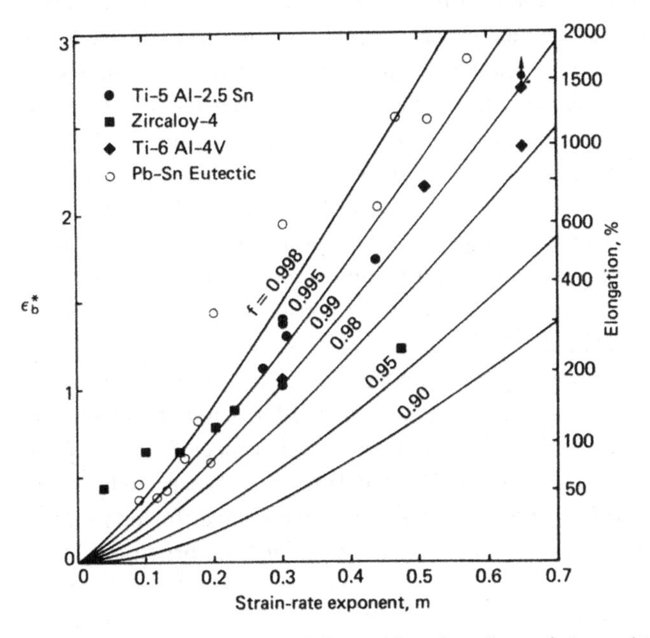

Figure 5.10. Limiting strains, ε_b^*, in unreduced sections of stepped tensile specimens as a function of m and f. Values of percent elongation corresponding to ε_b^* are indicated on the right together with data from Figure 5.9.

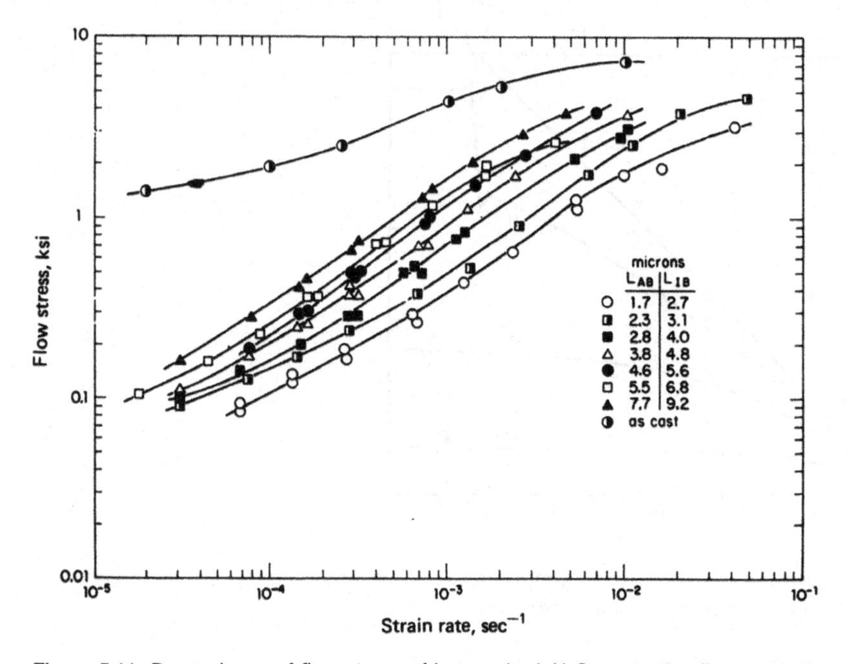

Figure 5.11. Dependence of flow stress of hot-worked Al-Cu eutectic alloy on strain rate at 520°C. The curves are for different grain sizes. L_{AB} is mean free path within the grains of κ and $CuAl_2$ phases and L_{IB} the mean free path between inter-phase boundaries. From D. A. Holt and W. A. Backofen, *Trans. Q. ASM*, 59 (1966).

negligible at this high temperature. Figure 5.12 shows the corresponding value of m as a function of strain rate. At the higher strain rates, $m = 0.2$ is typical of thermally activated slip. At lower strain rates, deformation mechanisms other than slip prevail. Here there are two schools of thought. One school*† maintains that deformation occurs primarily by diffusional creep with *vacancies* migrating from grain boundaries normal to the tensile axis to those parallel to it (i.e., diffusion of *atoms* from boundaries parallel to the tensile axis to boundaries perpendicular to it). Such diffusion causes the grains to elongate in the tensile direction and contract laterally. Whether diffusion is through the lattice or along grain-boundary paths, the strain rate should be proportional to the applied stress and inversely related to the grain size. If diffusion were the only mechanism, m would equal one (Newtonian viscosity) but it is lowered because of the slip contribution to the overall strain. The other school‡ attributes the high rate sensitivity to the role of grain-boundary sliding (shearing on grain boundaries). Although grain-boundary sliding alone would be viscous ($m = 1$), it must be accompanied by another mechanism to accommodate compatibility at triple points where the grains meet. Either slip or diffusion could serve as the accommodating mechanism. Both models explain the need for a very fine grain size, high temperature, and low strain rate, but the diffusional-creep model does not explain why the grains remain equiaxed after large deformations.

* W. A. Backofen, *Deformation Processing* (Addison-Wesley, 1972).

† A. H. Cottrell, *The Mechanical Properties of Matter* (John Wiley, 1964).

‡ M. F. Ashby and R. A. Verrall, *Acta Met.*, 21 (1973).

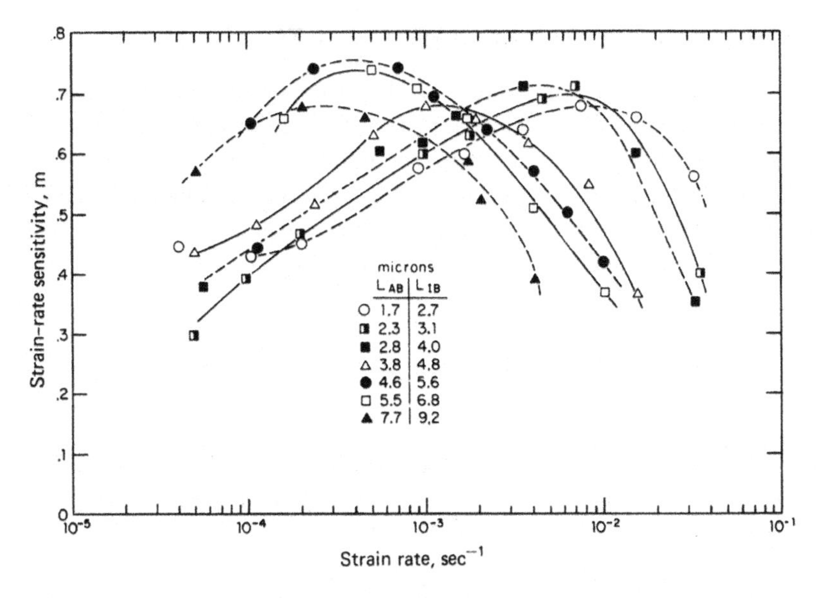

Figure 5.12. Variation of strain-rate sensitivity, $m = (\ln \sigma / \ln \dot{\varepsilon})_{\varepsilon,T}$ for the Al-Cu eutectic at 520°C. Compare with Figure 5.11. Note that the strain rate for peak m increases with decreasing grain size. From D. A. Holt and W. A. Backofen, *ibid.*

One practical problem is the tendency for grain growth during superplastic deformation because of the high temperature and long times dictated by the slow strain rates and because of the deformation itself.* Such grain growth lowers the m-value,† and increases the flow stress, causing the overall superplastic formability to deteriorate. The effects of grain growth are the most important at very slow strain rates (because of the longer times), and it has been suggested that the decrease in m at low strain rates is caused by such grain growth. Because finely distributed second-phase particles markedly retard grain growth, most superplastic alloys have two-phase microstructures. Among the alloys that exhibit superplasticity are Zn-22% Al and a range of steels (eutectoid) and Sn-40% Pb, Sn-5% Bi, Al-33% Cu (eutectics). Other alloys include Cu-10% Al, Zircaloy, several α-β titanium alloys, and some superalloys (γ–γ' with carbides).

Aluminum-base superplastic alloys are of considerable interest in the aerospace industry. It has been found that with controlled size and distribution of inclusions it is possible to generate very fine grain sizes by recrystallization and to prevent excessive grain growth during superplastic forming. Recently it has been found that very fine grain ceramics can be superplastically formed.

Superplasticity of some two-phase ceramics has been studied. These include zirconia-alumina, zirconia-mullite, alumina doped with magnesia, etc. These observations suggest that commercial forming of ceramics is a possibility.

* S. P. Agrawal and E. D. Weisert in *7th North American Metalworking Research Conf.*, Soc. (Dearborn, MI: Man. Engrs., 1979).

† The low m-values often observed at very low strain rates have been attributed in part to grain coarsening during these experiments.

5.4 COMBINED STRAIN AND STRAIN-RATE EFFECTS

Even the low values of m at room temperature can be of importance in determining uniform elongation. In Chapter 4 it was shown that the uniform elongation in a tension test was controlled by the strain-hardening exponent and the inhomogeneity factor f, but strain-rate effects were neglected. Reconsider a tension test on an inhomogeneous specimen with two regions of initial cross-sectional areas A_{b0} and $A_{a0} = fA_{b0}$. Now assume that the material strain hardens and is also rate sensitive, so that the flow stress is given by

$$\sigma = C'\varepsilon^n\dot{\varepsilon}^m. \tag{5.10}$$

Substituting $A_i = A_{i0}\exp(-\varepsilon_i)$ and $\sigma = C'\varepsilon^n\dot{\varepsilon}^m$ into a force balance, $A_b\,\sigma_b = A_a\,\sigma_a$, results in $A_{b0}\exp(-\varepsilon_b)\varepsilon_b^n\dot{\varepsilon}_b^m = A_{a0}\exp(-\varepsilon_a)\varepsilon_a^n\dot{\varepsilon}_a^m$. Following the procedure that produced equation 5.10,

$$\int_0^{\varepsilon_b}\exp(-\varepsilon_b/m)\varepsilon_b^{n/m}d\varepsilon_b = f^{1/m}\int_0^{\varepsilon_a}\exp(-\varepsilon_a/m)\varepsilon_a^{n/m}d\varepsilon_a. \tag{5.11}$$

The results of integration and numerical evaluation are shown in Figure 5.13, where ε_b is plotted as a function of ε_a for $n = 0.2, f = 0.98$, and several levels of m. It is apparent that even quite low levels of m play a significant role in controlling the strains reached in the unnecked region of the bar.

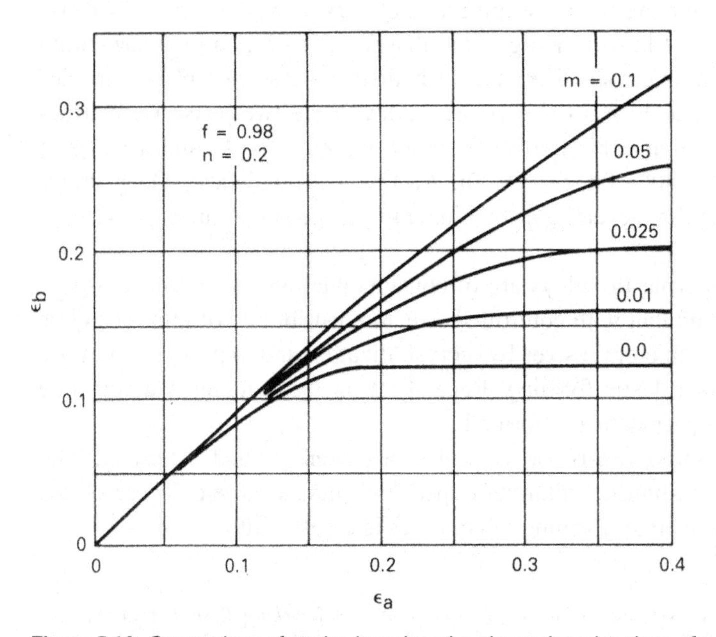

Figure 5.13. Comparison of strains in reduced and unreduced regions of a tensile bar calculated with $f = 0.98$ and $n = 0.20$. Note that even relatively low levels of m influence ε_b.

5.5 ALTERNATIVE DESCRIPTION OF STRAIN-RATE DEPENDENCE

For steel and other bcc metals, strain-rate sensitivity is better described by

$$\sigma = C + m' \ln \dot{\varepsilon}, \tag{5.12}$$

where C is the flow stress at a reference strain rate and m' is the rate sensitivity. Increasing the strain rate from $\dot{\varepsilon}_1$ to $\dot{\varepsilon}_2$ raises the flow stress by $\Delta\sigma = m' \ln(\dot{\varepsilon}_2/\dot{\varepsilon}_1)$.

$$\Delta\sigma = m' \ln(\dot{\varepsilon}_2/\dot{\varepsilon}_1). \tag{5.13}$$

Note the difference between this and equation 5.3, which predicts that $\Delta\sigma = m\sigma \ln(\dot{\varepsilon}_2/\dot{\varepsilon}_1)$. The difference between these two equations is illustrated in Figure 5.14.

Figures 5.15 and 5.16 illustrate this for copper (fcc) and iron (bcc).

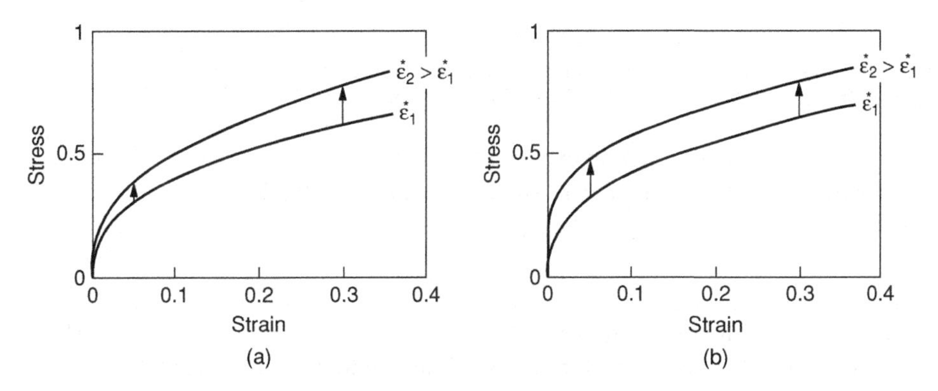

Figure 5.14. Two possible effects of strain rate on flow stress. The strain rates for the two curves differ by a factor of 100. Equation 5.3 predicts that $\Delta\sigma$ is proportional to σ (A) while equation 5.13 predicts that $\Delta\sigma$ is independent of σ (B). From W. F. Hosford, *Mechanical Behavior of Materials, 2nd ed.* Cambridge University Press, 2010.

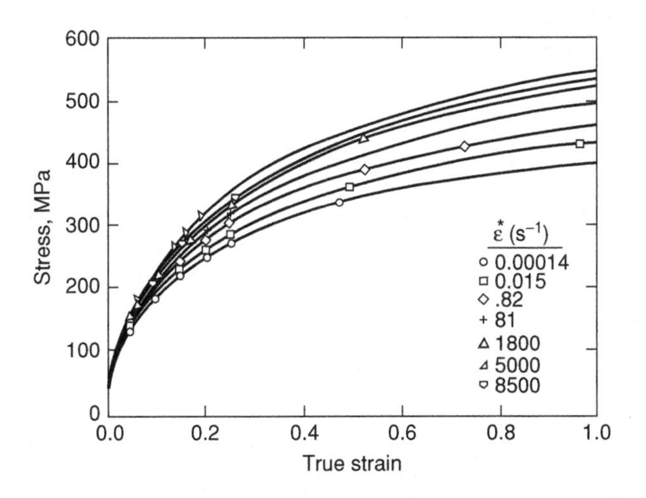

Figure 5.15. Stress-strain curves for pure copper at 25°C. Note that $\Delta\sigma$ between curves is proportional to the stress level. From P. S. Follansbee and U. F. Kocks, *Acta Met.*, v. 36 (1988).

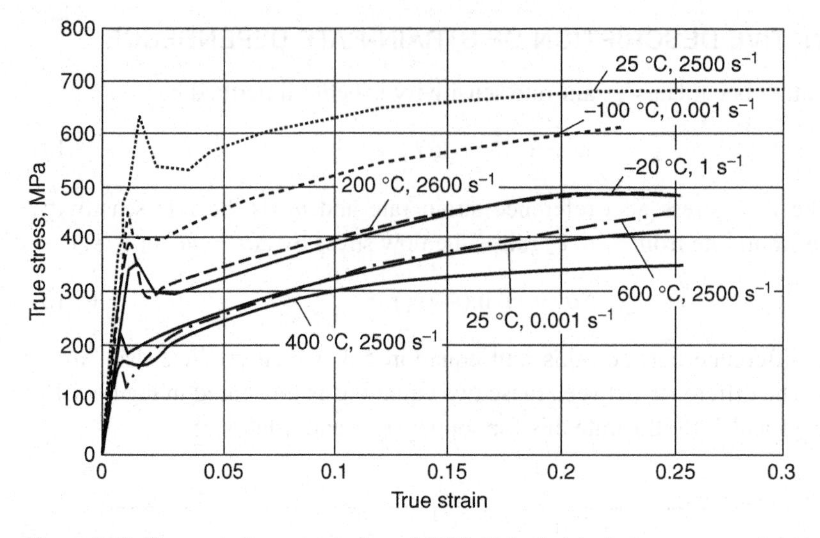

Figure 5.16. Stress-strain curves for iron at 25°C. Note that $\Delta\sigma$ between curves is independent of the stress level. From G. T. Gray in *ASM Metals Handbook*, v. 8, 2000.

For steels the strain-hardening exponent, n, decreases with increasing strain rates. This is because the combined effect of strain and strain-rate hardening is additive. The C in equation 5.12 is $C = K\varepsilon^n$ so the equation becomes

$$\sigma = K\varepsilon^n + m' \ln \dot{\varepsilon}. \tag{5.14}$$

Figure 5.17 is a plot of equation 5.14 for $K = 520$ MPa, $n = 0.22$ and $m' = 10$ MPa. Figure 5.18 is a logarithmic plot for the same data. Note that the slope, n decreases

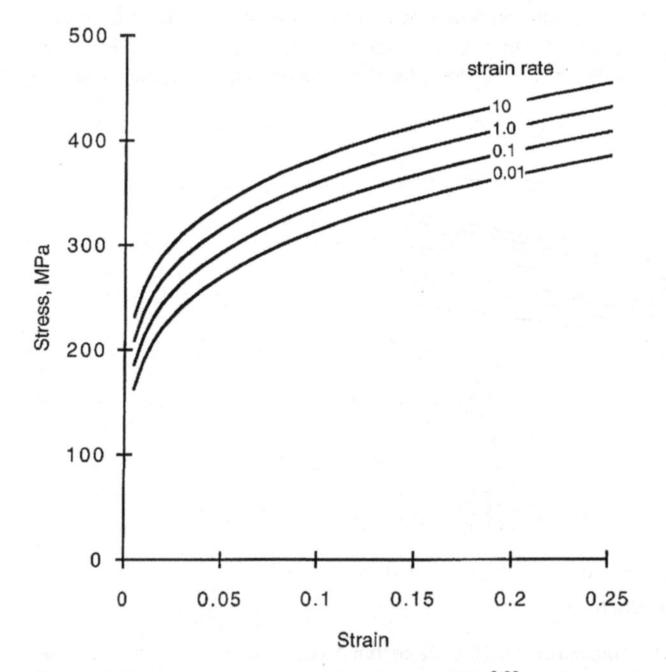

Figure 5.17. True stress-strain curves for $\sigma = 520\varepsilon^{0.22} + 10 \ln \dot{\varepsilon}$ with several strain rates. From Hosford, *ibid*.

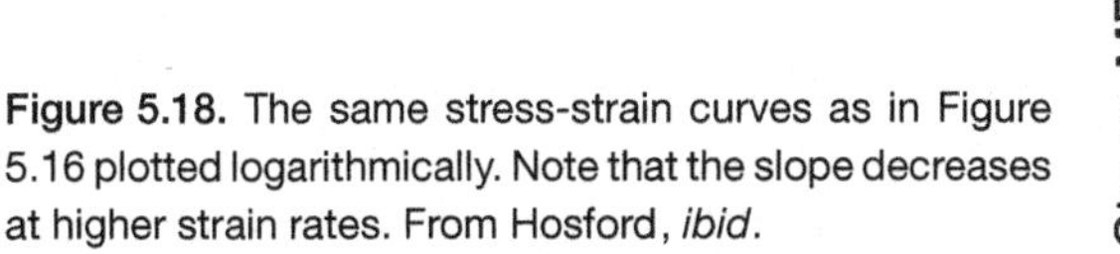

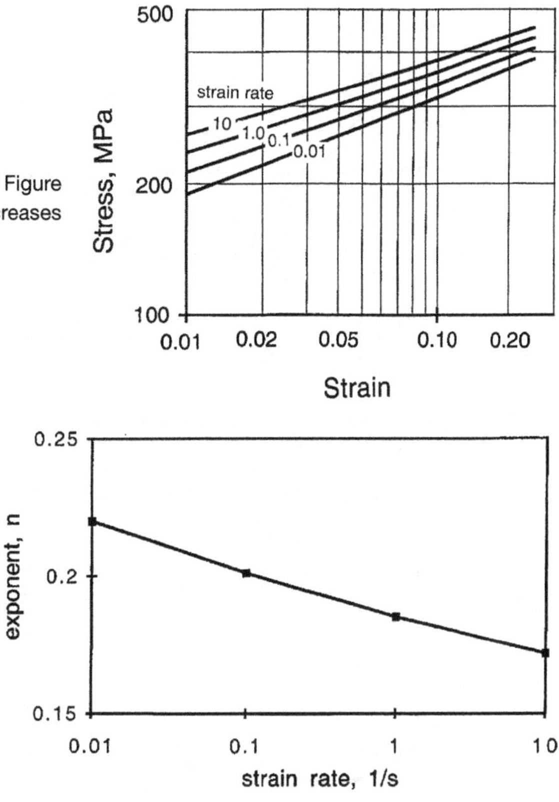

Figure 5.18. The same stress-strain curves as in Figure 5.16 plotted logarithmically. Note that the slope decreases at higher strain rates. From Hosford, *ibid*.

Figure 5.19. The decrease of n in $\sigma = K\varepsilon^n$ calculated from Figure 5.17. From Hosford, *ibid*.

with increasing strain rate indicating that n decreases. The average slope, n, is plotted against strain rate in Figure 5.19.

5.6 TEMPERATURE DEPENDENCE OF FLOW STRESS

At elevated temperatures, the rate of strain hardening falls rapidly in most metals with an increase in temperature, as shown in Figure 5.20. The flow stress and tensile strength, measured at constant strain and strain rate, also drop with increasing temperature as

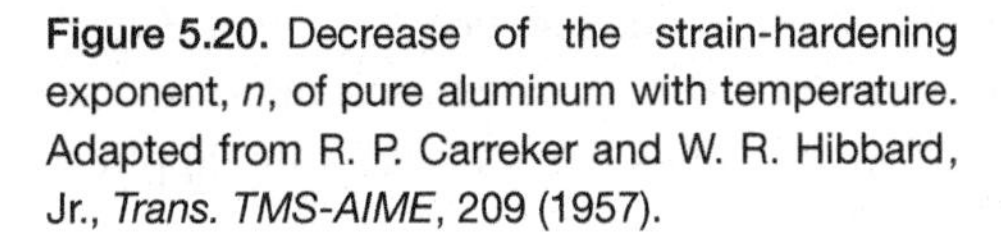

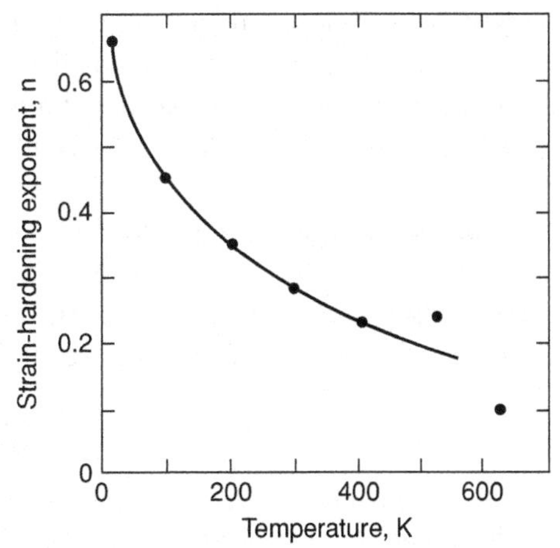

Figure 5.20. Decrease of the strain-hardening exponent, n, of pure aluminum with temperature. Adapted from R. P. Carreker and W. R. Hibbard, Jr., *Trans. TMS-AIME*, 209 (1957).

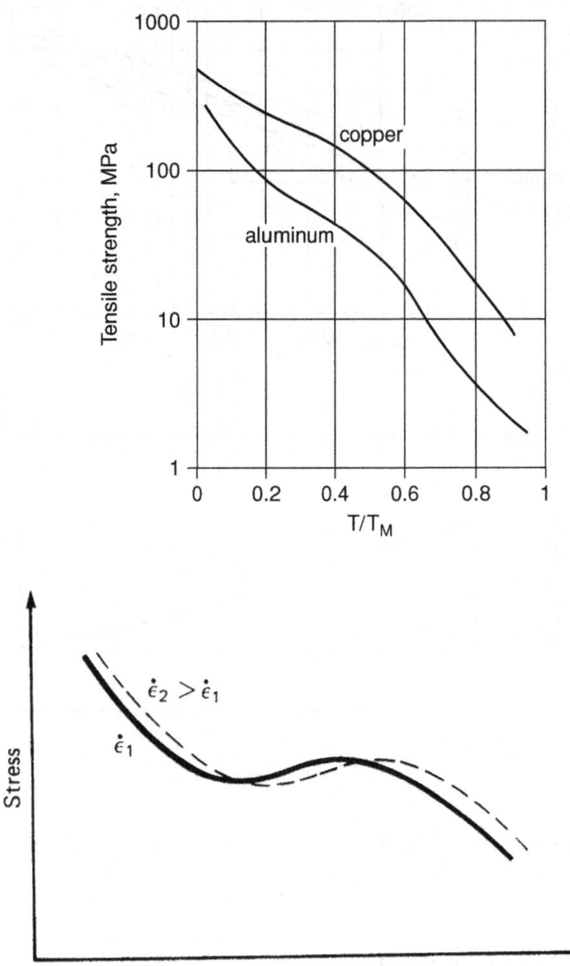

Figure 5.21. Decrease of tensile strength of pure copper, silver, and aluminum with homologous temperature. Adapted from R. P. Carreker and W. R. Hibbard, Jr., *ibid.*

Figure 5.22. Schematic plot showing the temperature dependence of flow stress for some alloys. In the temperature region where flow stress increases with temperature, the strain-rate sensitivity is negative.

illustrated in Figure 5.21. However, the drop is not always continuous; often there is a temperature range over which the flow stress is only slightly temperature dependent or in some cases even increases slightly with temperature. The temperature dependence of flow stress is closely related to its strain-rate dependence. Decreasing the strain rate has the same effect on flow stress as raising the temperature, as indicated schematically in Figure 5.22. Here it is clear that at a given temperature the strain-rate dependence is related to the slope of the σ-versus-T curve; where σ increases with T, m must be negative.

The simplest quantitative treatment of temperature dependence is that of Zener and Hollomon* who argued that plastic straining could be treated as a rate process using Arrhenius rate law,† rate $\propto \exp(-Q/RT)$, which has been successfully applied to many rate processes. They proposed that

$$\dot{\varepsilon} = A\exp(-Q/RT), \tag{5.15}$$

* C. Zener and J. H. Hollomon, *J. Appl. Phys.*, 15 (1994).

† S. Arrhenius, *Z. Phys. Chem.*, 4 (1889).

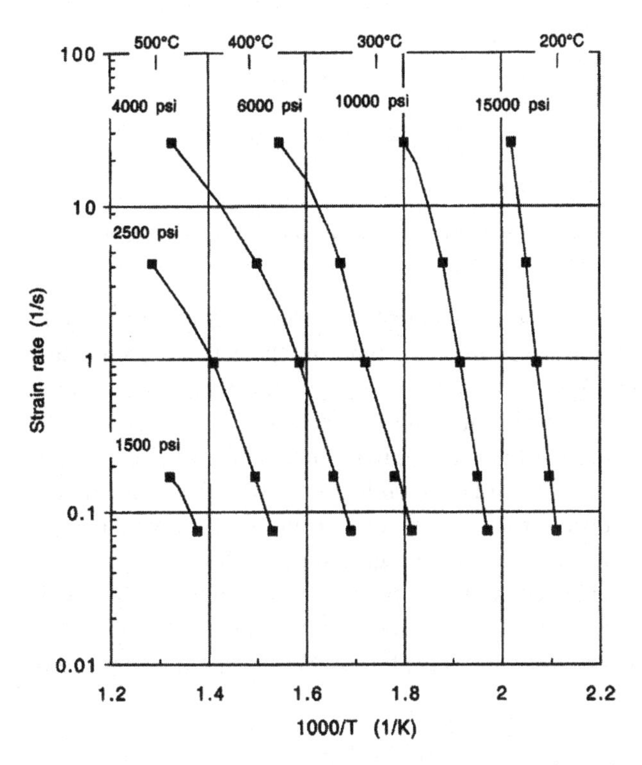

Figure 5.23. Strain rate and temperature combinations for various levels of stress. The data are for aluminum alloy 2024 and stresses are taken at an effective strain of 1.0. From D. S. Fields and W. A. Backofen, *ibid.*

where Q is an activation energy, T the absolute temperature, and R the gas constant. Here the constant of proportionality, A, is both stress and strain dependent. At constant strain, A is a function of stress alone, $A = A(\sigma)$, so equation 5.15 can be written as

$$A(\sigma) = \dot{\varepsilon}\exp(-Q/RT), \tag{5.16}$$

or more simply as

$$\sigma = f(z), \tag{5.17}$$

where the Zener-Hollomon parameter $Z = \dot{\varepsilon}\exp(-Q/RT)$. This development predicts that if the strain rate to produce a given stress at a given temperature is plotted on a logarithmic scale against $1/T$, a straight line should result with a slope of $-Q/R$. Figure 5.23 shows such a plot for 2024-O aluminum.

Correlations of this type are very useful in relating temperature and strain-rate effects, particularly in the high-temperature range. However, such correlations may break down if applied over too large a range of temperatures, strains, or strain rates as apparent in Figure 5.23. One reason is that the rate-controlling process, and hence Q, may change with temperature or strain. Another is connected with the original formulation of the Arrhenius rate law in which it was supposed that thermal fluctuations, alone, overcome an activation barrier, whereas in plastic deformation, the applied stress acts together with thermal fluctuations in overcoming the barriers as indicated in the following development.

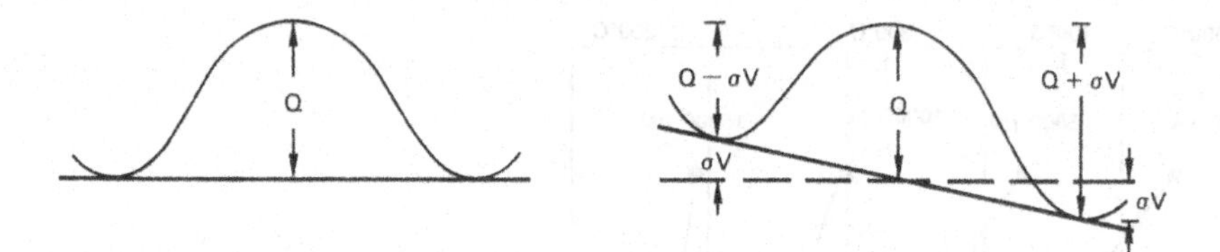

Figure 5.24. Schematic illustration of an activation barrier for slip and the effect of applied stress on skewing the barrier.

Consider an activation barrier for the rate-controlling process, as in Figure 5.24. The process may be cross slip, dislocation climb, etc. Ignoring the details, assume that the dislocation moves from left to right. In the absence of applied stress, the activation barrier has a height Q and the rate of overcoming this barrier would be proportional to $\exp(-Q/RT)$. However, unless the position at the right is more stable, i.e., has a lower energy than the position on the left, the rate of overcoming the barrier from right to left would be exactly equal to that in overcoming it from left to right, so there would be no net dislocation movement. With an applied stress, σ, the energy on the left is raised by σV, where V is a constant with dimensions of volume, and on the right the energy is lowered by σV. Thus the rate from left to right is proportional to $\exp[-(Q - \sigma V)/RT]$ and from right to left the rate is proportional to $\exp[-(Q + \sigma V)/RT]$. The net strain rate then is

$$\begin{aligned}\dot{\varepsilon} &= C\{\exp[-(Q - \sigma V)/RT] - \exp[-(Q + \sigma V)/RT]\}\\ &= C\exp(-Q/RT)[\exp(\sigma V/RT) - \exp(-\sigma V/RT)]\\ &= 2C\exp(-Q/RT)\sinh(\sigma V/RT). \end{aligned} \tag{5.18}$$

To accommodate data better, and for some theoretical reasons, a modification of equation 5.15 has been suggested.*† It is:

$$\dot{\varepsilon} = A[\sinh(\alpha\sigma)]^{1/m}\exp(-Q/RT) \tag{5.19}$$

Steady-state creep data over many orders of magnitude of strain rate correlate very well with equation 5.16, as shown in Figure 5.25.

It should be noted that if $\alpha\sigma \ll 1$, $\sinh(\alpha\sigma) \approx \alpha\sigma$, so equation 5.16 reduces to $\dot{\varepsilon} = A\exp(-Q/RT)\cdot(\alpha\sigma)^{1/m}$ or

$$\sigma = A'\dot{\varepsilon}^m\exp(mQ/RT), \tag{5.20}$$

or $\sigma = A'Z^m$, which is consistent with both the Zener-Hollomon development, equation 5.14 and the power-law expression, equation 5.1. Since $\sinh(x) \to \exp(x)/2$ for $x \gg 1$, at low temperatures and high stresses equation 5.16 reduces to

$$\dot{\varepsilon} = C\exp(\alpha'\sigma - Q/RT). \tag{5.21}$$

But now strain hardening becomes important so C and α' are both strain and temperature dependent. Equation 5.18 reduces to

$$\sigma = C + m'\ln\dot{\varepsilon}, \tag{5.22}$$

* F. Garofalo, *TMS-AIME*, 227 (1963).

† J. J. Jonas, C. M. Sellers, and W. J. McG. Tegart, *Met. Rev.* (1969).

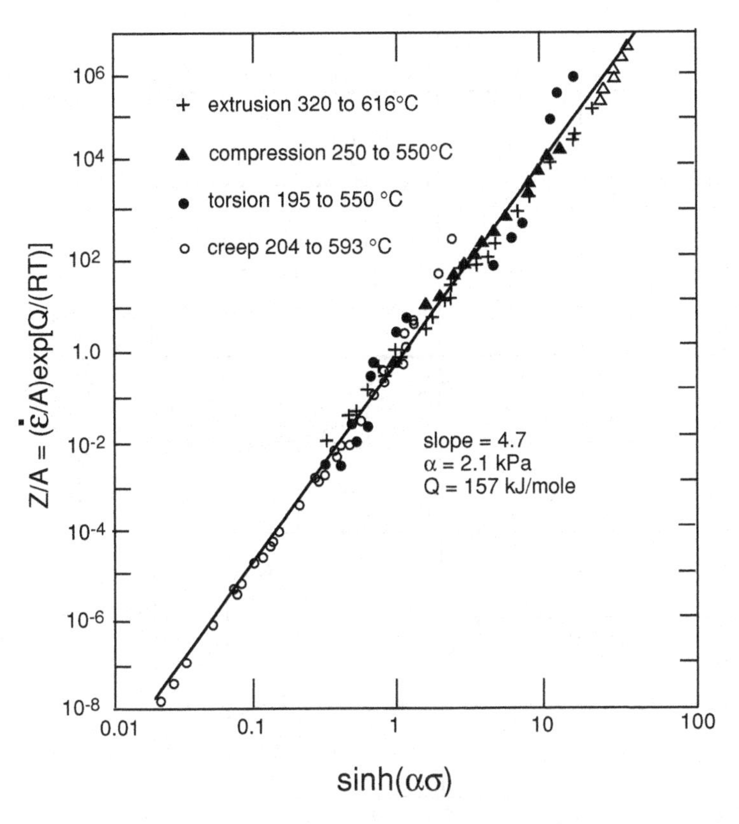

Figure 5.25. Plot of the Hollomon-Zener parameter versus flow stress data showing the validity of the hyperbolic sine relation (equation 5.16). Adapted from J. J. Jonas, *Trans. Q. ASM*, v. 62 (1969).

which is consistent with equation 5.10 and explains the often observed breakdown in the power-law strain-rate dependence at low temperatures and high strain rates.

5.7 DEFORMATION MECHANISM MAPS

The controlling mechanisms of deformation change with temperature and strain rate. A typical deformation mechanism map is shown in Figure 5.26. The dominant mechanisms change with temperature and stress. At high stresses slip by dislocation motion predominates. At lower stresses different creep mechanisms are important. The boundaries between the dominant mechanism shift with grain size. In particular, with decreased grain size, grain-boundary diffusion (Coble creep) becomes important at higher temperatures.

5.8 HOT WORKING

The decrease in flow stress at high temperatures permits forming with lower tool forces and, consequently, smaller equipment and lower power. *Hot working* is often defined as working above the recrystallization temperature so that the work metal recrystallizes as it deforms. However, this is an oversimplified view. The strain rates of many metal working processes are so high that there is not time for recrystallization to occur during deformation. Rather, recrystallization may occur in the time period between

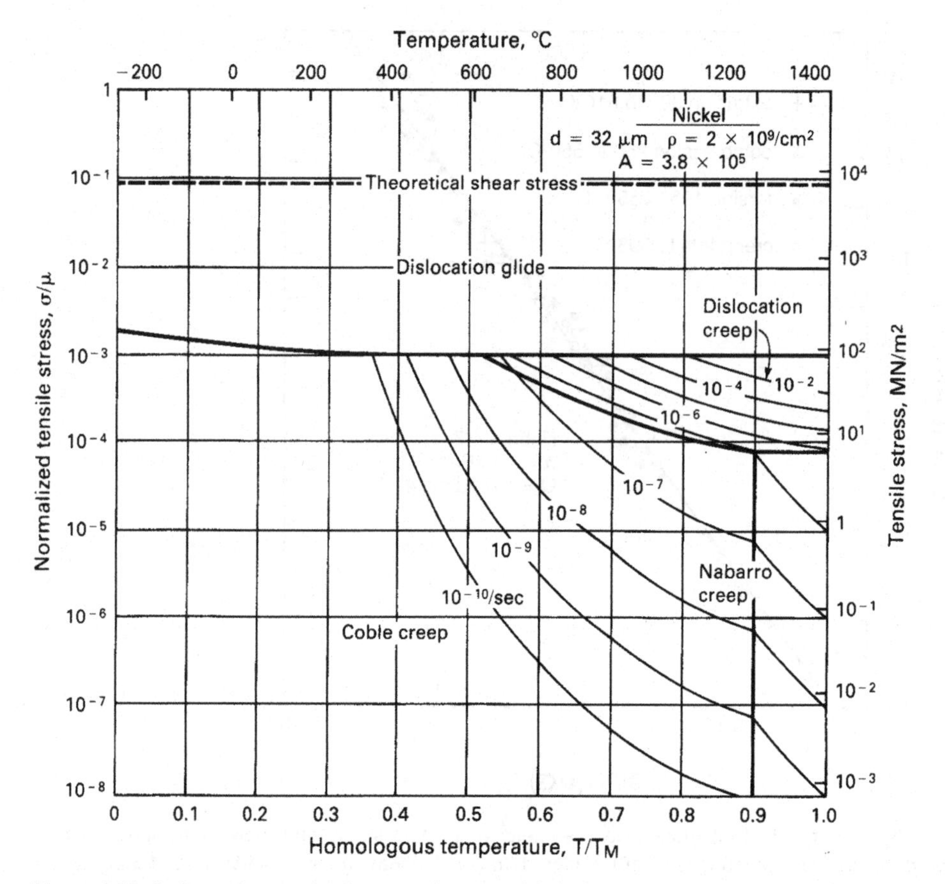

Figure 5.26. Deformation mechanism map for pure nickel with a grain diameter d = 32 μm, showing the strain rate as a function of stress and temperature. Different mechanisms are dominant in different regimes. Coble creep is controlled by grain-boundary diffusion, and Nabarro creep by lattice diffusion. Reprinted from M. F. Ashby, "A First Report on Deformation Mechanism Maps," *Acta Met.* v. 20 (1972), with permission from Pergamon Press Ltd.

repeated operations, as in forging and multiple-stand rolling, or after the deformation is complete, while the material is cooling to room temperature. High temperatures do, however, lower the flow stress whether recrystallization occurs during the deformation or not. Furthermore, the resultant product is in an annealed state.

In addition to lowering the flow stress, the elevated temperature during hot working has several undesirable effects. Among them are:

1. Lubrication is more difficult. Although viscous glasses are often used in hot extrusions, hot working often is done without lubrication.
2. The work metal tends to oxidize. Scaling of steel and copper alloys causes loss of metal and roughened surfaces. While processing under inert atmosphere is possible, it is prohibitively expensive and is avoided except in the case of very reactive metals, such as titanium.
3. Tool life is shortened because of heating, the presence of abrasive scales, and the lack of lubrication. Sometimes scale breakers are employed and rolls are cooled by water spray to minimize tool damage.

4. Poor surface finish and loss of precise gauge control result from the lack of adequate lubrication, oxide scales, and roughened tools.
5. The lack of work hardening is undesirable where the strength level of a cold-worked product is needed.

Because of these limitations, it is common to hot roll steel to a thickness of about 0.10 inches to take advantage of the decreased flow stress at high temperature. The hot-rolled product is then pickled to remove scale, and further rolling is done cold to ensure good surface finish and optimum mechanical properties. The term *cold-rolled* steel refers to the surface finish of the steel. Almost all *cold-rolled* steels have been annealed after cold rolling.

5.9 TEMPERATURE RISE DURING DEFORMATION

The temperature of the metal rises during plastic deformation because of the heat generated by mechanical work. The mechanical energy per volume, w, expended in deformation is equal to the area under the stress-strain curve

$$w = \int_o^{\bar{\varepsilon}} \bar{\sigma}\, d\bar{\varepsilon} \tag{5.23}$$

Only a small fraction of this energy is stored (principally as dislocations and vacancies). This fraction drops from about 5% initially to 1 or 2% at high strains. The rest is released as heat. If the deformation is adiabatic, that is, no heat transfer to the surroundings, the temperature rise is given by

$$\Delta T = \frac{\alpha \int \bar{\sigma}\, d\bar{\varepsilon}}{\rho C} = \frac{\alpha \bar{\sigma}_a \bar{\varepsilon}}{\rho C} \tag{5.24}$$

where $\bar{\sigma}_a$ is the average value of σ over the strain interval 0 to ε, ρ is the density, C is the mass heat capacity, and α is the fraction of energy stored ($\sim$0.98).

EXAMPLE 5.1: Calculate the temperature rise in a high-strength steel that is adiabatically deformed to a strain of 1.0. Pertinent data are: $\rho = 7.87 \times 10^3$ kg/m^3, $\sigma_a =$ 800 MPa, $C = 0.46 \times 10^3$ J/kg°C.

SOLUTION: Substituting in equation 5.21 and taking $\alpha = 1$,

$$\Delta T = \frac{800 \times 10^6 \times 1.0}{7.87 \times 10^3 \times 0.46 \times 10^3} = 221^\circ\text{C}$$

Although both σ_a and ε were high in this example, it illustrates the possibility of large temperature rises during plastic deformation. Any temperature increase causes the flow stress to drop. One effect is that at low strain rates, where heat can be transferred to surroundings, there is less thermal softening than at high strain rates. This partially compensates for the strain-rate effect on flow stress and can lead to an apparent decrease in the strain-rate sensitivity at high strains, when m is derived by comparing continuous stress-strain curves.

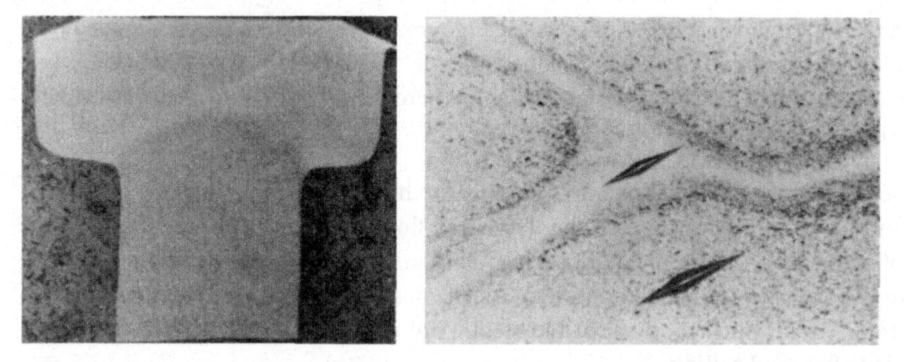

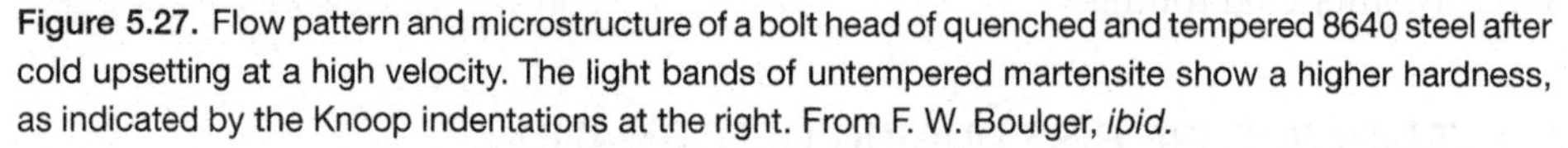

Figure 5.27. Flow pattern and microstructure of a bolt head of quenched and tempered 8640 steel after cold upsetting at a high velocity. The light bands of untempered martensite show a higher hardness, as indicated by the Knoop indentations at the right. From F. W. Boulger, *ibid.*

Another effect of the thermal softening is that it can act to localize flow in narrow bands. If one region or band deforms more than another, the greater heating may lower the flow stress in this region, causing even more concentration of flow and more local heating in this region. An example of this is shown in Figure 5.27, where in the upsetting of a steel bolt head, localized flow along narrow bands raised the temperature sufficiently to cause the bands to transform to austenite. After the deformation, these bands were quenched to martensite by the surrounding material. Similar localized heating, reported in punching holes in armor plate, can lead to sudden drops in the punching force. Such extreme localization is encouraged by conditions of high flow stresses and strains which increases ΔT, low rates of work hardening and work-piece tool geometry that encourages deformation along certain discrete planes.

NOTES OF INTEREST

Svante August Arrhenius (1859–1927) was a Swedish physical chemist. He studied at Uppsala where he obtained his doctorate in 1884. It is noteworthy that his thesis on electrolytes was given a fourth (lowest) level pass because his committee was skeptical of its validity. From 1886 to 1890 he worked with several noted scientists in Germany who did appreciate his work. In 1887, he suggested that a very wide range of rate processes follow what is now known as the Arrhenius equation. For years his work was recognized throughout the world, except in his native Sweden.

Count Rumford (Benjamin Thompson) was the first person to measure the mechanical equivalent of heat. He was born in Woburn, Massachusetts in 1753, and studied at Harvard. At the outbreak of the American Revolution, after being denied a commission by Washington, he was commissioned by the British. When the British evacuated Boston in 1776, he left for England, where he made a number of experiments on heat. After being suspected of selling British naval secrets to the French, he went to Bavaria. In the Bavarian army he eventually became Minister of War and eventually Prime Minister. While inspecting a canon factory, he observed a large increase in temperature during the machining of bronze canons. He measured the temperature rise and with the known heat capacity of the bronze, he calculated the heat generated by machining. By equating this to the mechanical work done in machining, he was able to deduce the mechanical equivalent of heat.

REFERENCES

W. A. Backofen, *Deformation Processing*, Addison-Wesley, 1972.
W. F. Hosford, *Mechanical Behavior of Materials, 2nd ed.* Cambridge University Press, 2010.

PROBLEMS

5.1. Low-carbon steel is being replaced by HSLA steels in automobiles to save weight because the higher strengths of HSLA steels permit use of thinner gauges. In laboratory tests at a strain rate of about 10^{-3} s^{-1}, one grade of HSLA steel has a yield strength of 420 MPa with a strain-rate exponent of $m = 0.005$ while for the low-carbon steel, $Y = 240$ MPa and $m = 0.015$. Calculate the percent weight saving possible for the same panel strength assuming
a) a strain rate of 10^{-3} s^{-1},
b) crash conditions with a strain rate of 10^{+4} s^{-1}.

5.2. The thickness of a sheet varies from 8.00 mm to 8.01 mm depending on location so a tensile specimen cut from a sheet has different thicknesses in different locations
a) For a material with $n = 0.15$ and $m = 0$, what will be the strain in the thicker region when the thinner region necks?
b) If $n = 0$ and $m = 0.05$, find the strain in the thicker region when the strain in the thinner region is 0.5 and ∞.

5.3. **a)** Find the % elongation in the diagonal ligaments in Figure 5.6, assuming that the ligaments make an angle of 75° with the horizontal.
b) Assuming that $f = 0.98$ and $n = 0$, what value of m is required for the variation of thickness along the ligaments to be held to 20%? (The thickness of the thinnest region is 0.80 times the thickness of the thickest region.)

5.4. Find the value of m' in equation 5.12 that best fits the data in Figure 5.28.

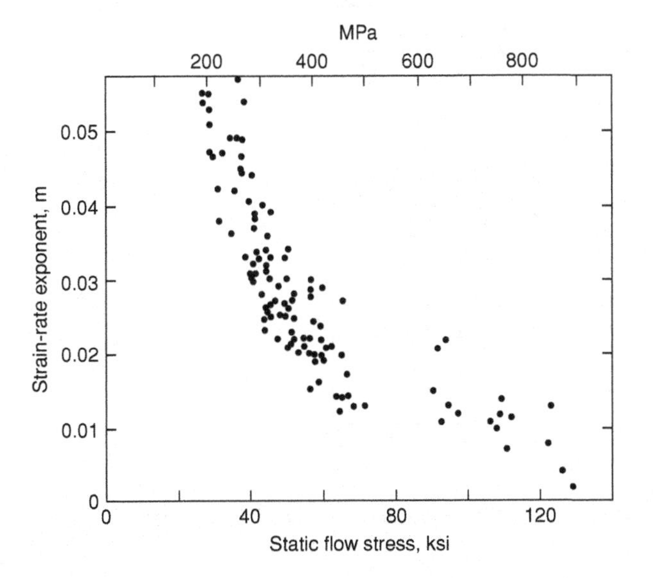

Figure 5.28. Effect of stress level on the strain-rate sensitivity of steel. Adapted from A. Saxena and D. A. Chatfield, *SAE Paper 760209* (1976).

5.5. From the data in Figure 5.23, estimate Q in equation 5.12 and m in equation 5.1 for aluminum at 400°C.

5.6. Estimate the total elongation in a tensile bar if
a) $f = 0.98$, $m = 0.5$ and $n = 0$
b) $f = 0.75$, $m = 0.8$ and $n = 0$.

5.7. Estimate the shear strain necessary in the shear bands of Figure 5.27 necessary to explain the formation of untempered martensite if the tensile strength level was 1.75 GPa, $n = 0$, and adiabatic conditions prevailed.

5.8. During superplastic forming it is often necessary to maintain a constant strain rate.
a) Describe qualitatively how the gas pressure should be varied to form a hemispherical dome by bulging a sheet clamped over a circular hole with gas pressure.
b) Compare the gas pressure required to form a hemispherical dome of 5 cm diameter with the pressure for a 0.5 m diameter dome.

5.9. **a)** During a creep experiment under constant stress, the strain rate was found to double when the temperature was suddenly increased from 290°C to 300°C. What is the apparent activation energy for creep?
b) The stress level in a tension test increased by 1.8% when the strain rate was increased by a factor of 8. Find the value of m.

5.10. Figure 5.29 gives data for high-temperature creep of α-zirconium. In this range of temperatures, the strain rate is independent of strain.
a) Determine the value of m that best describes the data at 780°C.

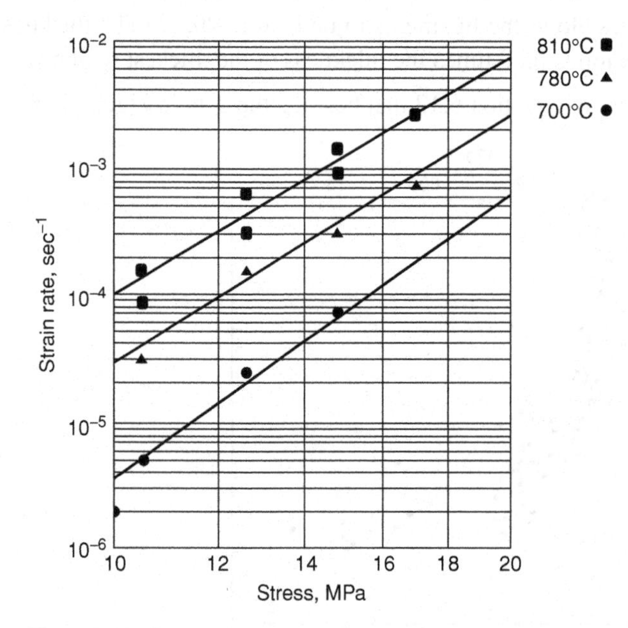

Figure 5.29. Strain rate vs. stress for α-zirconium at several temperatures.

b) Determine the activation energy, Q, in the temperature range 700°C to 810°C at about 14 MPa.

5.11. Tension tests were made in two different labs on two different materials. In both the strain-hardening exponent was found to be 0.20, but the post-uniform elongations were quite different. Offer two plausible explanations.

6 Work Balance

The work or energy balance is a very simple method of estimating the forces and energy involved in some metal forming operations. It does not, however, permit predictions of the resulting properties. The energy to complete an operation can be divided into the ideal work, w_i, that would be required for the shape change in the absence of friction and inhomogeneous flow, the work against friction, w_f, and the redundant work, w_r.

6.1 IDEAL WORK

To calculate the ideal work for a shape change, it is necessary to envision an ideal process for achieving the desired shape change. It is not necessary that the ideal process be physically possible. For example, the axially symmetric deformation in the extrusion or wire drawing of a circular rod or wire can be simulated by tension test. The fact that necking would occur in a tension test can be ignored. The ideal work is

$$w_i = \int_0^{\varepsilon} \sigma \, d\varepsilon, \tag{6.1}$$

where $\varepsilon = \ln(A_0/A_f)$. With power-law hardening,

$$w_i = K\varepsilon^{n+1}/(n+1). \tag{6.2}$$

Other expressions of work hardening, if more appropriate, can be used with equation 6.1.

EXAMPLE 6.1: The strain-hardening behavior of a metal is approximated by $\bar{\sigma} = 140\bar{\varepsilon}^{0.25}$ MPa. Find the ideal work per volume if a bar of the material is reduced from 12.7 to 11.5 mm diameter in tension.

SOLUTION: This is an ideal process. Using equation 6.2, $\bar{\varepsilon} = 2\ln(12.7/11.5) = 0.199$.

$$w_i = 140 \times 10^6 (0.199)^{1.25}/1.25 = 14.7 \text{ MJ/m}^3.$$

A frictionless compression test could serve as the ideal process for forging. Flat rolling can be simulated by frictionless plane-strain compression.

If a mean flow stress, Y_m, is known w_i can be taken as

$$w_i = Y_m \bar{\varepsilon}. \tag{6.3}$$

6.2 EXTRUSION AND DRAWING

Figure 6.1 illustrates direct or forward extrusion. A billet of diameter, D_0, is extruded through a die of diameter, D_1. Except for the very first and last material to be extruded, this is a steady-state operation. The volume of metal exiting the die, $A_1\Delta\ell_1$, must equal the material entering the die. The total external work, W_a is $W_a = F_e\Delta\ell$, where F_e is the extrusion force. Substituting the work per volume as $w_a = W_a/(A_0\Delta\ell_0)$,

$$w_a = \frac{F_e\Delta\ell}{A_0\Delta\ell}. \tag{6.4}$$

Therefore extrusion pressure, P_e must equal w_a, so

$$w_a = F_e/A_0 = P_e. \tag{6.5}$$

Although $w_a = w_i$ for an *ideal* process, for an actual process $w_a > w_i$. Therefore

$$P_e > \int \sigma \, d\varepsilon, \tag{6.6}$$

so equation 6.5 underestimates the extrusion work and is a lower bound to the actual value.

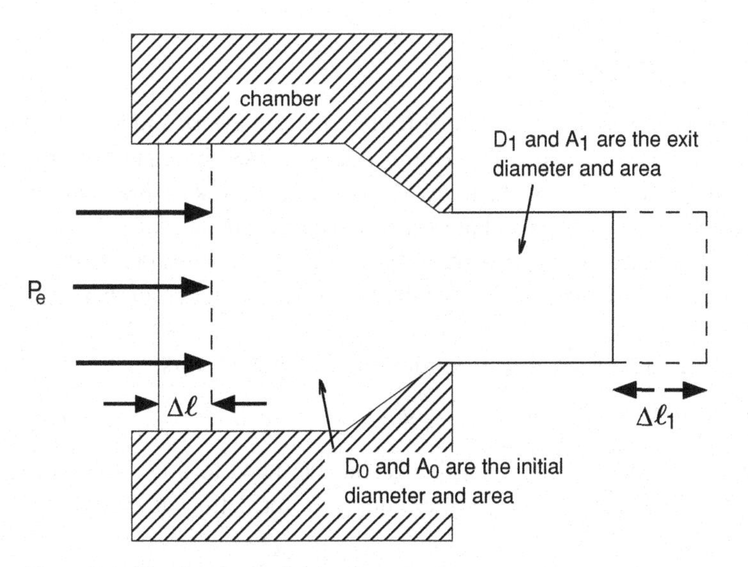

Figure 6.1. Direct or forward extrusion.

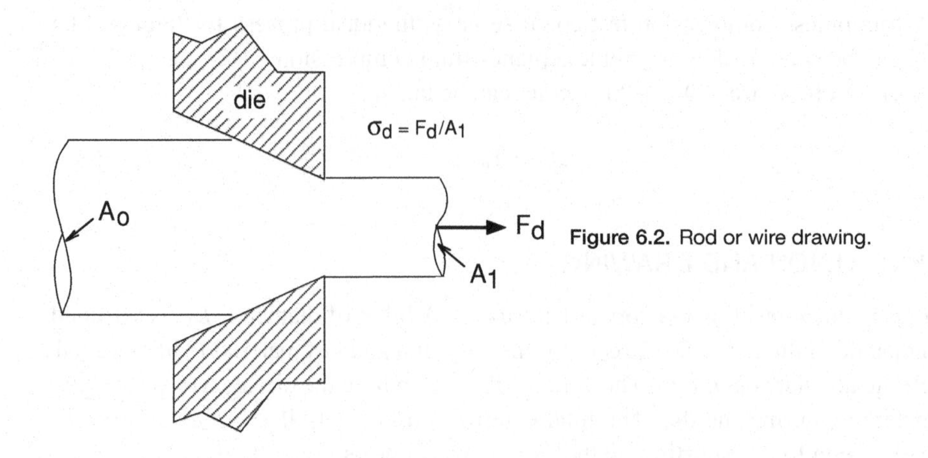

Figure 6.2. Rod or wire drawing.

Wire and rod drawing (Figure 6.2) can be analyzed in a similar way. The drawing force, F_{d}, pulls the metal through the die. The actual work, W_a, in drawing a length, $\Delta\ell$, is $W_a = F_{\mathrm{d}}\Delta\ell$, so the work per volume is

$$w_a = F_{\mathrm{d}}/A_1 = \sigma_{\mathrm{d}}, \tag{6.7}$$

where σ_{d} is the drawing stress. Again because $w_a > w_i$,

$$\sigma_{\mathrm{d}} > \int \sigma \mathrm{d}\varepsilon. \tag{6.8}$$

EXAMPLE 6.2: Find the drawing stress necessary to reduce the bar in Example 6.1 from 12.7 to 11.5 mm, diameter by drawing it through a die assuming an ideal process.

SOLUTION: For an ideal process, $\sigma_{\mathrm{d}} = w_i = 14.7$ MPa.

6.3 DEFORMATION EFFICIENCY

In addition to the ideal work, there is work against friction between work and tools, W_f and work to do redundant or unwanted deformation, W_r. Expressed on a per volume basis these are w_f and w_r. Figure 6.3 illustrates the redundant work in drawing or extrusion. If the deformation were ideal, plane sections would remain plane. In real processes, the surface layers are sheared relative to the center. The material undergoes more strain than required for the diameter reduction and consequently strain hardens more and is less ductile.

The actual work is the sum of the ideal, frictional and redundant works

$$w_a = w_i + w_f + w_r. \tag{6.9}$$

It is often difficult to find w_f and w_r explicitly, but the need to do this can be avoided by lumping the inefficient terms and defining a deformation efficiency, η, where

$$\eta \equiv w_i/w_a. \tag{6.10}$$

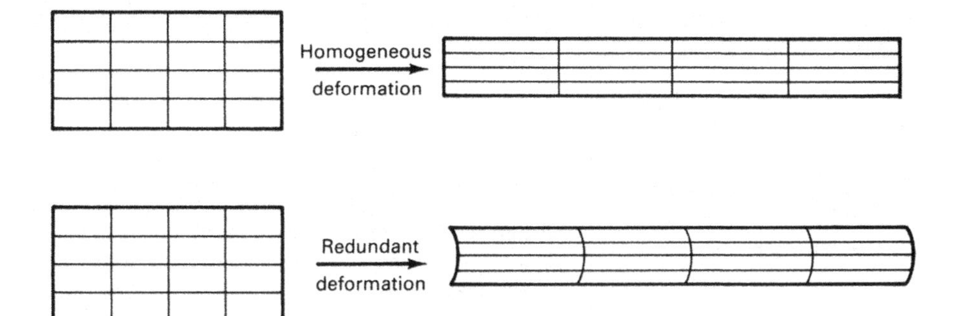

Figure 6.3. Comparison of ideal and actual deformation to illustrate the meaning of redundant work.

Experience allows one to make reasonable estimates of η. For wire and rod drawing, η is often between 0.5 and 0.65 depending on lubrication, reduction and die angle. Using an efficiency, the extrusion pressure or drawing stress can be expressed as

$$P_{\text{ext}} = \frac{1}{\eta} \int \bar{\sigma}\, \mathrm{d}\bar{\varepsilon}. \tag{6.11}$$

If strain hardening is small

$$P_{\text{ext}} = \frac{1}{\eta} \int \bar{\sigma}\, \mathrm{d}\bar{\varepsilon} = \frac{Y_m \varepsilon}{\eta}. \tag{6.12}$$

EXAMPLE 6.3: Calculate the extrusion pressure to extrude the bar in Example 6.1 through a die, reducing its diameter from 12.7 mm to 11.5 mm, if the efficiency is 70%.

SOLUTION: $P_{\text{ext}} = \frac{1}{\eta} \int \bar{\sigma}\, \mathrm{d}\bar{\varepsilon} = w_i/\mu = 14.7/0.7 = 21$ MPa.

6.4 MAXIMUM DRAWING REDUCTION

For drawing, there is a maximum possible reduction per pass because the drawing stress cannot exceed the strength of the drawn wire or rod. Once drawing has started, it is a steady-state process. The maximum reduction corresponds to $\sigma_d = \bar{\sigma}$. If the effect of hardening caused by redundant strain is neglected, the limiting strain, ε^* can be found by equating expressions for σ_d and $\bar{\sigma}$. For power-law hardening, $\bar{\sigma} = K\varepsilon^{*n}$ and $\sigma_d = K\varepsilon^{*n+1}/(n+1)$ so

$$\varepsilon^* = \eta(n+1). \tag{6.13}$$

For an ideally plastic metal ($n = 0$) and perfect efficiency ($\eta = 1$), the maximum strain would be $\varepsilon^* = 1$ which corresponds to a reduction, r^*, of 63%. With a more reasonable value of $\eta = 0.65$, $\varepsilon^* = 0.65$ and $r^* = 48\%$. In practice, multiple passes are used for wire drawing and after the first or second pass additional strain hardening can be neglected ($n \to 0$).

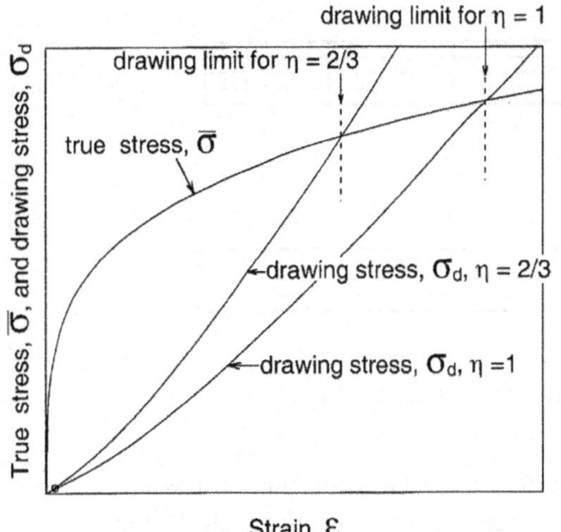

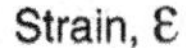

Figure 6.4. The drawing limit, ε^*, corresponds to the intersection of the plots of drawing stress with the stress-strain curve.

For all conditions, (strain-hardening rules and efficiencies), the maximum drawing strain corresponds to $\sigma_\mathrm{d} = \bar{\sigma}/\eta$ as illustrated by Figure 6.4.

The conditions are different in the initial start-up. Undrawn material must be fed through the die. If it is annealed, the maximum drawing force will correspond to the tensile strength, or

$$\sigma_{\mathrm{d}_{(\max)}} = Kn^n, \tag{6.14}$$

and the maximum strain is

$$\varepsilon^* = [\eta(n+1)n^n]^{1/(n+1)}. \tag{6.15}$$

More likely, however, the end of the wire to be fed through the die will be reduced by swaging (See Section 6.6). In this case it will have been strain hardened at least as much as if it had been drawn.

6.5 EFFECTS OF DIE ANGLE AND REDUCTION

Figure 6.5 shows schematically the dependence of the friction and redundant work terms on die angle. For a given reduction, the contact area between the die and material decreases with increasing die angle. The *pressure* the die exerts on the material in the die gap is almost independent of the die angle, so the *force* between the die and workpiece increases with greater area of contact at die angles. With a constant coefficient of friction, w_f, increases as α decreases. The redundant work term, w_f, increases with die angle. The ideal work term, w_i, doesn't depend on the die angle. For each reduction, there is an optimum die angle, α^*, for which the work is a minimum.

The efficiency and optimum die angle are functions of the reduction. In general, the efficiency increases with reduction. Figures 6.6 and 6.7 are adapted from the measurements of Wistreich.* The optimum die angle increases with reduction.

* J. Wistreich, *Metals Rev.* (1958), pp. 97–142.

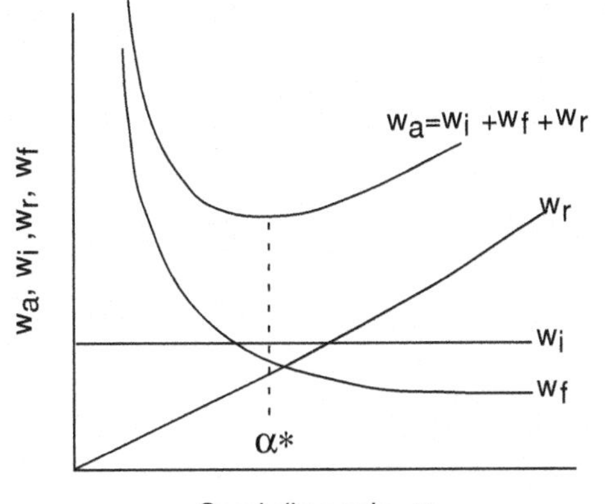

Figure 6.5. Dependence of the various work terms on the die angle.

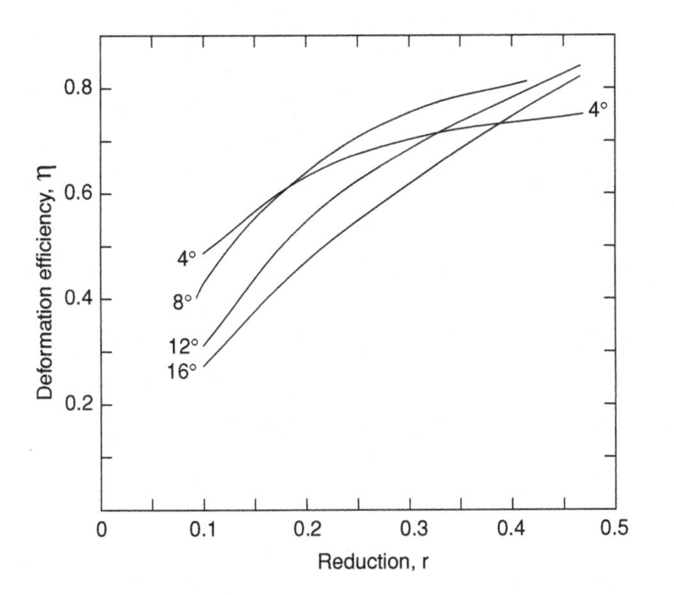

Figure 6.6. Wistreich's data show an increase of efficiency with increasing reduction.

In extrusion die angle and reduction have similar effects on efficiency apply. However, there is no theoretical limiting reduction, though excessive extrusion forces may be beyond the capacity of the extrusion machinery. Therefore the reductions and die angles in extrusion tend to be much larger than in drawing. Area reductions of 8 : 1 ($r = 0.875$) are not uncommon.

6.6 SWAGING

The diameters of wires and tubes may be reduced by rotary swaging. Shaped dies are rotated around the wire or tube and hammer the work piece as they rotate. This causes a swirled microstructure. In a wire drawing plant, swaging is used to reduce the ends of wires so they can be fed through drawing dies in the initial set up. Figure 6.8 is a schematic of the tooling used in swaging.

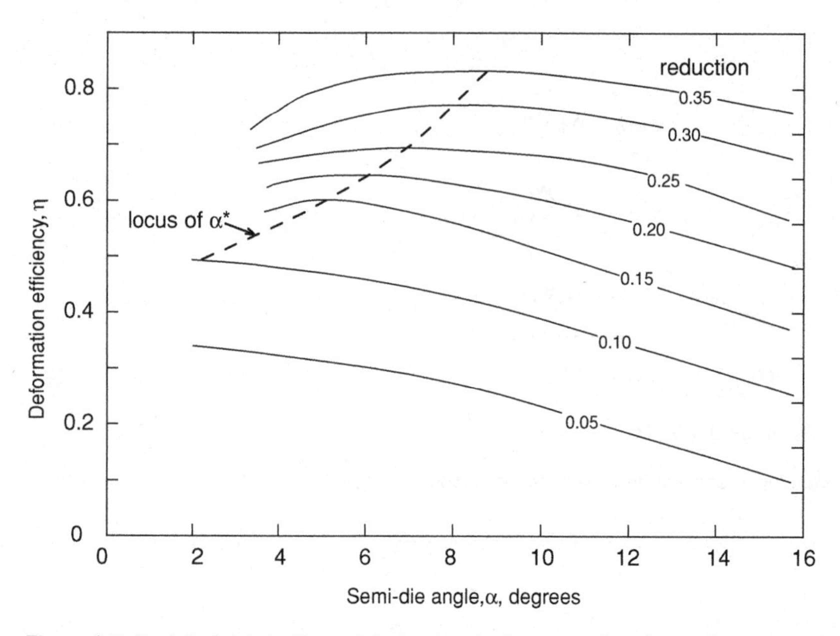

Figure 6.7. Replot of data in Figure 6.6 showing the increase of optimum die angle with reduction.

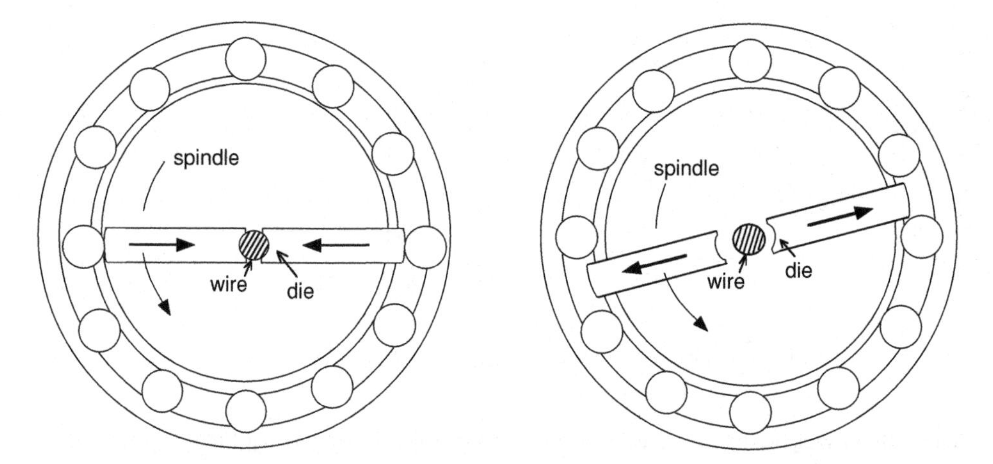

Figure 6.8. Tooling for swaging. As the spindle rotates it causes the die to hammer the work at different angles (in this case 30°).

REFERENCES

W. A. Backofen, *Deformation Processing*, Addison Wesley, 1972.
Metals Handbook: Forging and Forming, vol 4. 9th ed. ASM, 1988.

PROBLEMS

6.1. The diameter, D_0, of a round rod can be reduced to D_1 either by a tensile force of F_1 or by drawing through a die with a force, F_d, as sketched in Figure 6.9. Assuming ideal work in drawing, compare F_1 and F_d (or σ_1 and σ_d) to achieve the same reduction.

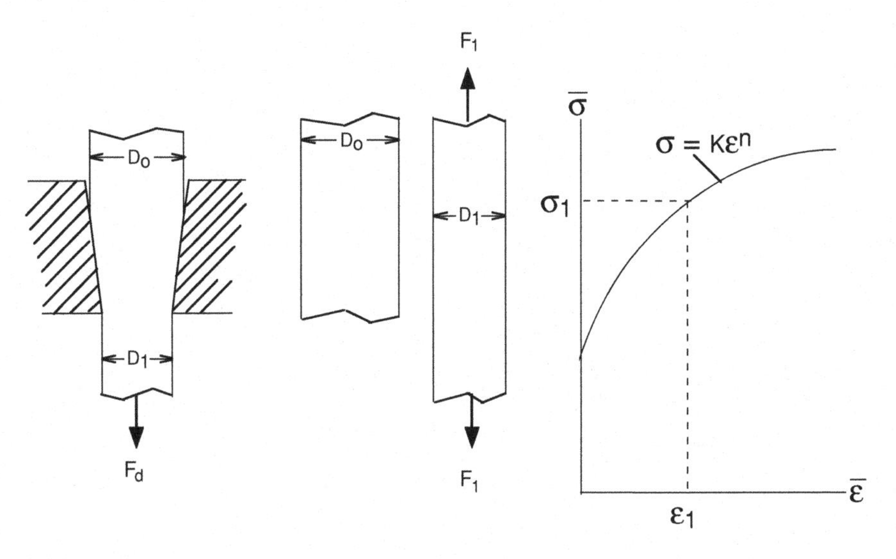

Figure 6.9. Sketch for Problem 6.1.

6.2. Calculate the maximum possible reduction, r, in wire drawing for a material whose stress-strain curve is approximated by $\bar{\sigma} = 200\bar{\varepsilon}^{0.18}$ MPa. Assume an efficiency of 65%.

6.3. An aluminum alloy billet is being hot extruded from 20 cm diameter to 5 cm diameter as sketched in Figure 6.10. The flow stress at the extrusion temperature is 40 MPa. Assume $\eta = 0.5$.

a) What extrusion pressure is required?

b) Calculate the lateral pressure on the die walls.

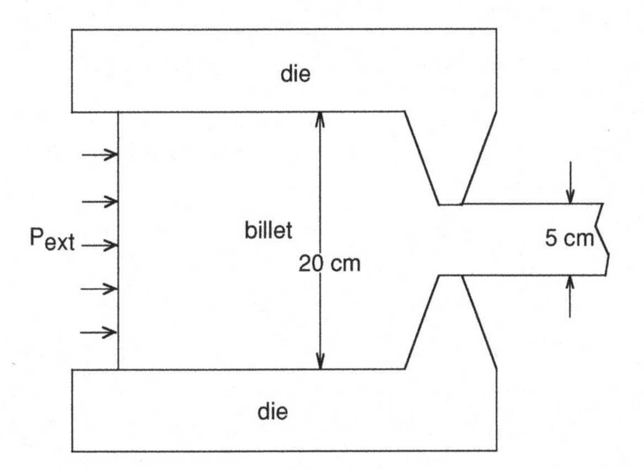

Figure 6.10. Aluminum billet being extruded.

6.4. An unsupported extrusion process (Figure 6.11) has been proposed to reduce the diameter of a bar from D_0 to D_1. The material does not strain harden. What is the largest reduction, $\Delta D/D_0$, that can be made without the material yielding before it enters the die? Neglect the possibility of buckling and assume $\eta = 60\%$.

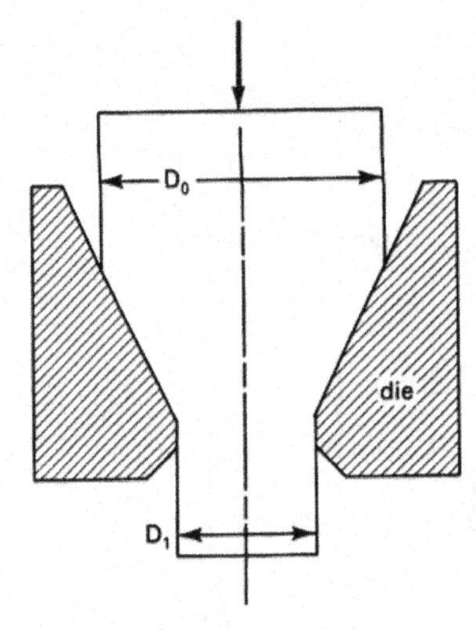

Figure 6.11. Unsupported extrusion.

6.5. A sheet, 1-m wide and 8-mm thick, is to be rolled to a thickness of 6 mm in a single pass. The strain-hardening expression for the material is $\bar{\sigma} = 200\bar{\varepsilon}^{0.18}$ MPa. A deformation efficiency of 80% can be assumed. The von Mises yield criterion is applicable. The exit speed from the rolls is 5 m/s. Calculate the power required.

6.6. The strains in a material for which $\bar{\sigma} = 350\bar{\varepsilon}^{0.20}$ MPa are $\varepsilon_1 = 0.200$ and $\varepsilon_2 = -0.125$. Calculate the work per volume assuming $\eta = 1$.

6.7. You are asked to plan a wire-drawing schedule to reduce copper wire from 1 mm to 0.4 mm diameter. How many wire-drawing passes would be required if, to be sure of no failures, the drawing stress never exceeds 80% of the flow stress and the efficiency is assumed to be 60%?

6.8. Derive an expression for ε^* at the initiation of drawing when the outlet diameter is produced by machining.

6.9. For a material with a stress-strain relation, $\sigma = A + B\varepsilon$, find the maximum strain per wire-drawing pass if $\mu = 0.75$.

7 Slab Analysis

The slab analysis is based on making a force balance on a differentially thick slab of material. It is useful in estimating the role of friction on forces required in drawing, extrusion and rolling. The important assumptions are:

1. The principal axes are in the directions of the applied loads.
2. The effects of friction do not alter the directions of the principal axes or cause internal distortion. The deformation is homogeneous so plane sections remain plane.

7.1 SHEET DRAWING

Figure 7.1 illustrates *sheet* or *strip drawing*. A force, F, pulls a strip through a pair of wedges. The strip width, w, is much greater than the thickness, t, so the width doesn't change and plane strain prevails. The drawing stress, $\sigma_\mathrm{d} = F/wt_e$.

Figure 7.2 shows the differential slab element. Pressure normal to the die, P, acts on two areas of $w\,\mathrm{d}x/\cos\alpha$, and its component in the x-direction is $2Pw\,\mathrm{d}x(\sin\alpha/\cos\alpha)$. The normal pressure also causes a frictional force on the work metal, $2\mu w\,\mathrm{d}x/\cos\theta$, with a horizontal component $2\mu w\,\mathrm{d}x(\cos\alpha/\cos\alpha)$, where μ is the coefficient of friction. Both of these forces act in the negative x-direction. The drawing force, $\sigma_x wt$, acts in the negative x-direction while the force in the positive x-direction is $(\sigma_x + \mathrm{d}\sigma_x)(t + \mathrm{d}t)w$. An x-direction force balance gives

$$(\sigma_x + \mathrm{d}\sigma_x)(t + \mathrm{d}t)w + 2\mu w(\tan\alpha)P\mathrm{d}x + 2Pw\,\mathrm{d}x = \sigma_x wt. \qquad (7.1)$$

Simplifying and neglecting second order differentials,

$$\sigma_x \mathrm{d}t + t\mathrm{d}\sigma_x + 2\mu P\mathrm{d}_x + 2(\tan\alpha)P\mathrm{d}x = \sigma_x t. \qquad (7.2)$$

Substituting $2\mathrm{d}x = \mathrm{d}t/\tan\alpha$ and $B = \mu\cot\alpha$, $P\mathrm{d}t + \sigma_x\mathrm{d}t + t\mathrm{d}\sigma_x + \mu\cot\alpha P\mathrm{d}t = 0$, or

$$t\mathrm{d}\sigma_x + [\sigma_x + P(1 + B)]\,\mathrm{d}t = 0. \qquad (7.3)$$

or

$$\frac{\mathrm{d}\sigma_x}{\sigma_x + P(1 + B)} = -\frac{\mathrm{d}t}{t} \qquad (7.4)$$

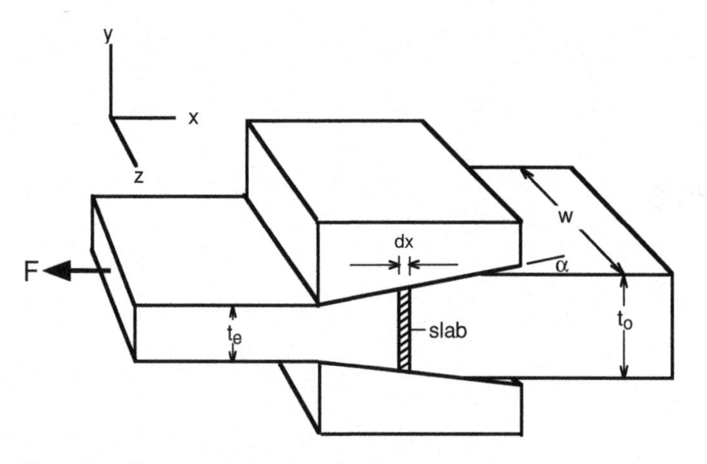

Figure 7.1. Plane-strain drawing of a sheet.

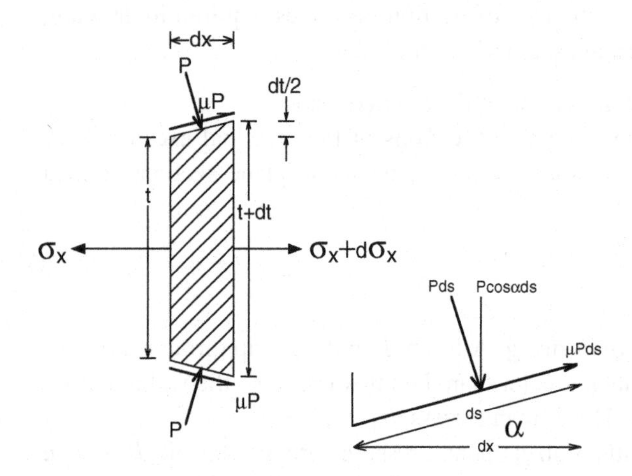

Figure 7.2. Slab used for force balance in sheet drawing.

To integrate equation 7.4, the relation between P and σ_x must be found and substituted. The fact that P does not act in a direction normal to x will be ignored and P will be taken as a principal stress, $P = -\sigma_y$. For yielding in plane strain, $\sigma_x - \sigma_y = 2k$, where k is the yield strength in shear. For Tresca $2k = Y$ and for von Mises $2k = (2/\sqrt{3})Y$. Substituting $P = 2k - \sigma_x$ into equation 7.4

$$\frac{\mathrm{d}\sigma_x}{B\sigma_x - 2k(1+B)} = \frac{\mathrm{d}t}{t}. \tag{7.5}$$

Integrating, between $\sigma_x = 0$ and $\sigma_x = \sigma_d$ and between $t = t_0$ and t_e, and solving for σ_d,

$$\frac{\sigma_\mathrm{d}}{2k} = \frac{(1+B)}{B}\left[1 - \left(\frac{t_e}{t_0}\right)^B\right]. \tag{7.6}$$

Finally substituting $\varepsilon_h = \ln(t_0/t_e)$ as the homogeneous strain,

$$\frac{\sigma_d}{2k} = \frac{(1+B)}{B}[1 - \exp(-B\varepsilon_h)]. \tag{7.7}$$

Several assumptions are involved in the derivation of equation 7.7. First, it is assumed that $2k$ is constant which isn't true if the material work hardens. The effect of work hardening can be approximated by using an average value of $2k$. Another is that the friction coefficient is constant. The assumption that P is a principal stress is reasonable for low die angles and low coefficients of friction, but the assumption gets progressively worse at high die angles and high friction coefficients.

EXAMPLE 7.1: A 2.5 mm thick metal sheet 25 cm wide is drawn to a thickness of 2.25 mm through a die of included angle 30°. The flow stress is 200 MPa and the friction coefficient is 0.08. Calculate the drawing force.

SOLUTION: Using equation 7.7, and taking $B = \mu \cot\alpha = 0.08\cot(15°) = 0.299$ and $\varepsilon_h = \ln(2.5/2.25) = 0.1054$.

From equation 7.7, $\sigma_d = 1.15(200)(1.299/.299)[1 - \exp(-0.299 \times 0.1054)] = 31$ MPa. $F_d = 31\,\text{MPa}\,(.25\text{m})(0.0225) = 175$ kN.

Note that if $\mu = 0$ (i.e., $B = 0$) were substituted into equation 7.7, it can be shown that with L'Hospital's rule

$$\frac{\sigma_d}{2k} = \varepsilon_h, \tag{7.8}$$

which is exactly what would be predicted by the work balance.

7.2 WIRE AND ROD DRAWING

Sachs* analyzed wire or rod drawing by an analogous procedure. The basic differential equation is

$$\frac{d\sigma}{B\sigma - (1+B)\bar{\sigma}} = 2\frac{dD}{D}, \tag{7.9}$$

where D is the diameter of the rod or wire. Realizing that $2dD/D = -d\varepsilon$,

$$\int_0^{\sigma_d} \frac{d\sigma}{B\sigma - (1+B)\bar{\sigma}} = -\int_0^{\varepsilon_h} d\varepsilon. \tag{7.10}$$

Integrating,

$$\sigma_d = \sigma_a\left(\frac{1+B}{B}\right)[1 - \exp(-B\varepsilon_h)], \tag{7.11}$$

where σ_a is the average flow stress of the material in the die. This analysis neglects redundant strain and has the same limitations as the plane-strain drawing in Section 7.1 and becomes unrealistic at high die angles and low reductions.

* G. Sachs, *Z. Angew. Math. Mech.*, v. **7**, p. 235 (1927).

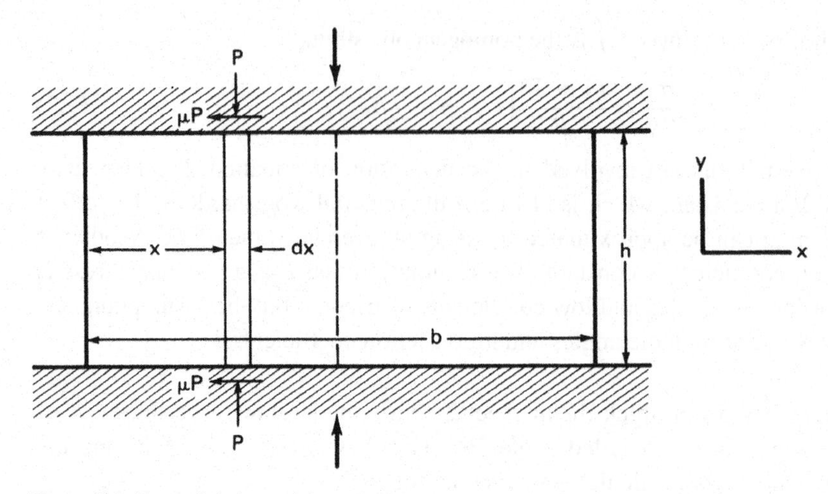

Figure 7.3. Essentials for a slab analysis.

7.3 FRICTION IN PLANE-STRAIN COMPRESSION

The slab analysis can also be used for plane-strain compression. Figure 7.3 shows a specimen with $h < b$ being compressed with a constant coefficient of friction, μ. Making a force balance on the differential slab, $\sigma_x h + 2\mu P \mathrm{d}x - (\sigma_x + \mathrm{d}\sigma_x)h = 0$. This simplifies to

$$2\mu P \mathrm{d}_x = h \mathrm{d}\sigma_x. \tag{7.12}$$

Again taking $\sigma_y = -P$ and σ_x as principal stresses and realizing that for plane strain, $\sigma_x - \sigma_y = 2k$, so $\mathrm{d}\sigma_x = -\mathrm{d}P$. Equation 7.12 becomes $2\mu P\mathrm{d}x = h\mathrm{d}P$ or

$$\frac{\mathrm{d}P}{P} = \frac{2\mu \mathrm{d}P}{h}. \tag{7.13}$$

Integrating from $P = 2k$ at $x = 0$ to P at x, $\ln(P/2k) = 2\mu x/h$ or

$$\frac{P}{2k} = \exp\left(\frac{2\mu x}{h}\right). \tag{7.14}$$

This predicts a friction hill, which is illustrated in Figure 7.4.

This is valid from the edge ($x = 0$) to the centerline ($x = b/2$). The average pressure, P_a, can be found by integrating P over half of the block to find the force and dividing that by the area of the half-block.

$$F_y = \int_0^{b/2} P\mathrm{d}x = \int_0^{b/2} 2k \exp\left(\frac{2\mu x}{h}\right)\mathrm{d}x = 2k\left[\left(\frac{h}{2\mu}\right)\exp\left(\frac{\mu b}{h}\right) - 1\right] \tag{7.15}$$

$$\frac{P_{av}}{2k} = \left(\frac{h}{\mu b}\right)\left[\exp\left(\frac{\mu b}{h}\right) - 1\right] \tag{7.16}$$

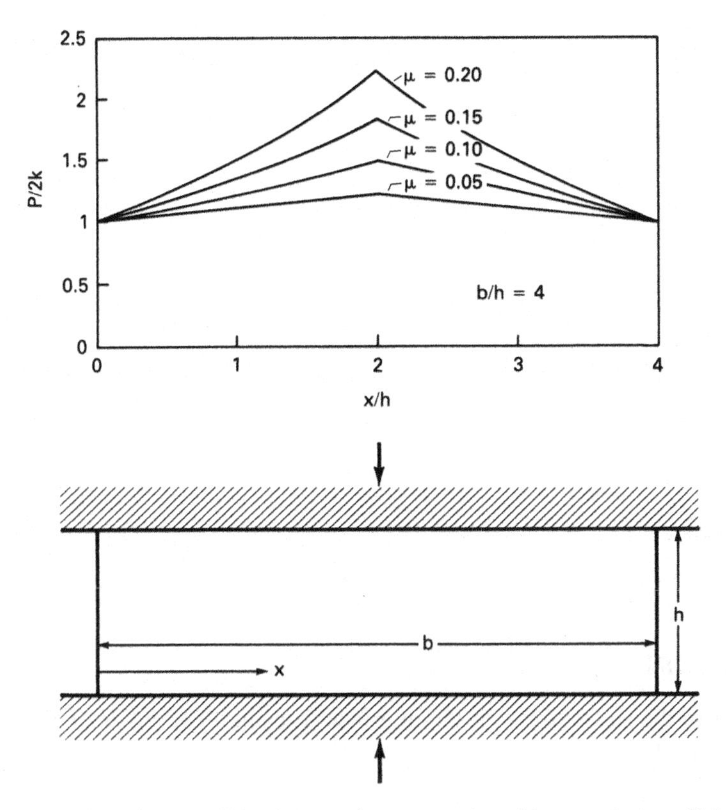

Figure 7.4. Friction hill in plane-strain compression with a constant coefficient of friction.

A simple approximate solution can be found by expanding $\exp(\mu b/h) - 1 = 1 + (\mu b/h) + (\mu b/h)^2/2 + \cdots$. For small values of $(\mu b/h)$,

$$\frac{P_{av}}{2k} \approx 1 + \frac{\mu b}{2h} \tag{7.17}$$

EXAMPLE 7.2: Plane-strain compression is conducted on a slab of metal 20 cm wide and 2.5 cm high. with a yield strength in shear of $k = 100$ MPa. Assuming a coefficient of friction of $\mu = 0.10$,

a) Estimate the maximum pressure at the onset of plastic flow;
b) Estimate the average pressure at the onset of plastic flow.

SOLUTION:

a) From equation 7.14, $P_{\max} = 2k \exp(\mu b/h) = 200 \exp(0.1 \times 20/2.5) = 445$ GPa.
b) Using the exact solution (equation 7.16) $P_{av} = (200)(0.25)/(0.1 \times 20))[\exp(.25) - 1] = 306$ MPa. The approximate solution (equation 7.17) gives

$$P_{av} = 200[1 + (0.1)(20)/5] = 280 \text{ MPa}$$

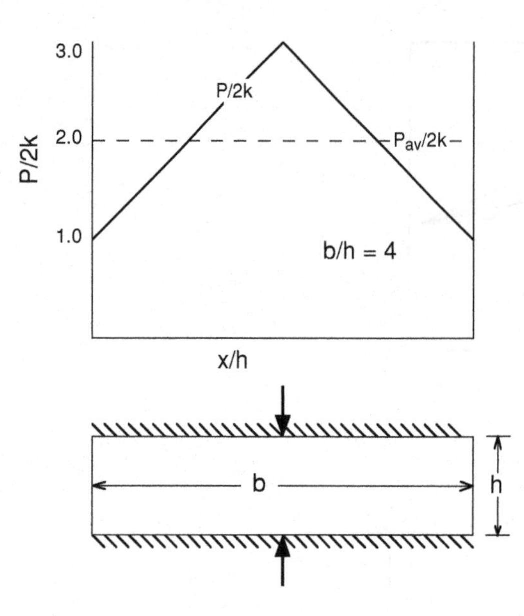

Figure 7.5. Friction hill in plane-strain with sticking friction.

7.4 STICKING FRICTION

There is an upper limit to the shear stress on the interface. It cannot exceed $2k$. This limits equations 7.16 and 7.17 to

$$x \leq \frac{-h}{2\mu}\ln(2\mu) \quad \text{and} \quad \frac{b}{h} \leq -\ln(2\mu)/\mu. \tag{7.18}$$

Otherwise the tool and work piece will stick at the interface, and the work piece will be sheared. If sticking occurs, the shear stress will be k instead of μP and equation 7.14 will become

$$P/2k = 1 + x/h. \tag{7.19}$$

The pressure distribution is shown in Figure 7.5.

The average pressure is given by

$$\frac{P_{av}}{2k} = 1 + \frac{b}{4h}. \tag{7.20}$$

EXAMPLE 7.3: Repeat Example 7.2 for sticking friction.

SOLUTION:

a) Using equation 7.19, $P_{max} = 200(1 + 20/5) = 1{,}000$ MPa
b) Using equation 7.20, $P_{av} = 200(1 + 20/10) = 600$ MPa

7.5 MIXED STICKING-SLIDING CONDITIONS

If $b/h \geq -\ln(2\mu)/\mu$, sticking is predicted at the center and sliding at the edges. Equation 7.14 predicts P for $x \leq x^*$ where $x^* = -h\ln(2\mu)/(2\mu)$. From x^* to the

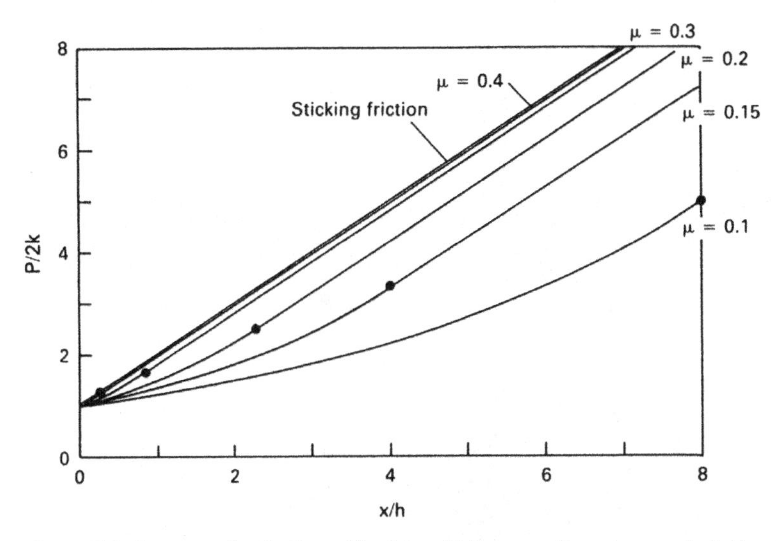

Figure 7.6. Pressure distribution with sliding friction near the edges and sticking friction in the center for several values of μ. Points mark the boundaries between sticking and sliding friction.

centerline, $h\,dP = -2k\,dx$. Integrating between $P = 2k\exp(2\mu x^*/h)$ at $x = x^*$ and P at x,

$$\frac{P}{2k} = \frac{x - x^*}{h} + \exp\left(\frac{2\mu x^*}{h}\right). \tag{7.21}$$

This is plotted in Figure 7.6.

The average pressure can be found by integrating equation 7.21.

$$\frac{P_{av}}{2k} = \left(\frac{2}{b}\right)\int_{x^*}^{b/2}\left[\frac{x - x^*}{h} + \exp\left(\frac{2\mu x^*}{h}\right)dx + \left(\frac{2}{b}\right)\int_0^{x^*}\exp\left(\frac{2\mu x}{h}\right)dx\right]$$

$$= \frac{b}{4h} - \frac{x^*}{h} + \frac{h}{2b\mu}\exp\left(\frac{\mu b}{h}\right) + \left(1 + \frac{x^*}{2b} - \frac{h}{2b\mu}\right)\exp\left(\frac{2\mu x^*}{h}\right). \tag{7.22}$$

7.6 CONSTANT SHEAR STRESS INTERFACE

Films of soft materials such as lead or a polymer are sometimes used as lubricants. In this case there will be a constant shear stress, $\tau = mk$, in the interface, where m is the ratio of the shear strength of the film to that of the work piece. The local pressure is now given by

$$P/2k = 1 + mx/h \tag{7.23}$$

and the average pressure by

$$\frac{P_{av}}{2k} = 1 + \frac{mb}{4h}. \tag{7.24}$$

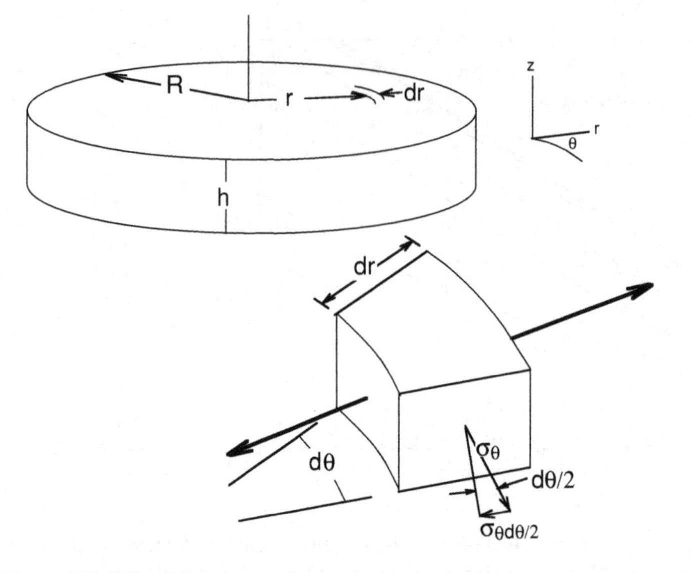

Figure 7.7. Differential element for analysis of axially symmetric compression.

7.7 AXIALLY SYMMETRIC COMPRESSION

An analysis similar to that in Section 7.3 can be used for axially symmetric compression. Making a force balance on a differential element (Figure 7.7), $\sigma_r hr\,\mathrm{d}\theta + 2\mu\,\mathrm{Pr}\,\mathrm{d}\theta\mathrm{d}r + (2\sigma_\theta h\,\mathrm{d}r\,\mathrm{d}\theta)/2 - (\sigma_r + \mathrm{d}\sigma_r)(r + \mathrm{d}r)h\mathrm{d}\theta = 0$. Simplifying,

$$2\mu\,\mathrm{Pr}\,\mathrm{d}r + h\sigma_\theta\,\mathrm{d}r = h\sigma_r\,\mathrm{d}r + hr\,\mathrm{d}\sigma_r. \tag{7.25}$$

For axially symmetric flow, $\varepsilon_\theta = \varepsilon_r$ so $\sigma_\theta = \sigma_r$. At yielding $\sigma_r + P = Y$ or $\mathrm{d}\sigma_r = -\mathrm{d}P$. Substituting into equation 7.25 gives $2\mu\mathrm{Pr}\,\mathrm{d}r = -hr\,\mathrm{d}P$ or

$$\int_Y^P \frac{\mathrm{d}P}{P} = -\int_0^R \frac{2\mu}{h}\mathrm{d}r. \tag{7.26}$$

Letting $P = -Y$, $\mathrm{d}\sigma_r = -\mathrm{d}P$ and integrating,

$$P = Y\exp\left[\left(\frac{2\mu}{h}\right)(R - r)\right]. \tag{7.27}$$

The average pressure is $P_{av} = (1/\pi R^2)\int_0^R 2P\pi r\,\mathrm{d}r$.

$$\begin{aligned} P_{av} &= \left(\frac{2Y}{R^2}\right)\int_0^R r\exp\left[\left(\frac{2\mu}{h}\right)(R - r)\right]\mathrm{d}r \\ &= \left(\frac{1}{2}\right)\left(\frac{h}{\mu R}\right)^2 Y\left[\exp\left(\frac{2\mu R}{h}\right) - \left(\frac{2\mu R}{h}\right) - 1\right]. \end{aligned} \tag{7.28}$$

For small values of $\frac{\mu R}{h}$, this reduces to

$$P_{av} = Y[1 + (2\mu R/h)/3 + (2\mu R/h)^2/12 + \cdots] \tag{7.29}$$

Here $Y = 2k$ for Tresca and $Y = \sqrt{3}k$ for von Mises. In this analysis, it was tacitly assumed that $\mu P \leq k$ so $Y\exp[(2\mu/h)(R - r)] \leq k/\mu$. Therefore the limiting radius, r^*, is $r^* = R - (h/2\mu)\ln(k/\mu Y)$. For sliding to prevail over the entire surface, $R \leq (h/2\mu)\ln(k/\mu Y)$.

For sticking friction over the entire interface, the shear term in equation 7.25 is $2kr\,\mathrm{d}r$ instead of $2\mu Pr\mathrm{d}r$, so $2k\mathrm{d}r = -h\mathrm{d}P$ and

$$\int_R^r 2k\mathrm{d}r = -\int_Y^P h\mathrm{d}P. \tag{7.30}$$

Integrating

$$P = Y + (2k/h)(R - r). \tag{7.31}$$

The average pressure is

$$P_{av} = Y + 2kR/(3h). \tag{7.32}$$

EXAMPLE 7.4: A solid 10 cm diameter, 2.5 cm high disc is compressed. The tensile and shear yield strengths are 300 and 150 MPa. Estimate the force needed to deform the disc assuming sticking friction.

SOLUTION: Using equation 7.32, $P_{av} = 300 + 5\text{cm}(150 \times 2)/[3(2.5\text{cm})] = 500$ MPa.

$$F = \pi(.005)^2\,\text{m}^2\,500 \times 10^6\,\text{Pa} = 3.9\,\text{kN}.$$

7.8 SAND-PILE ANALOGY

The analyses for axially symmetric and plane-strain compression with sticking friction can be interpreted in terms of the shape of a sand-pile. Dry sand piled on a flat surface will form a hill with a constant slope. This slope is analogous to the linear increase of P with distance from the edge of the work piece. The effect of sticking friction can be analyzed using this effect. Sand can be piled onto cardboard or other flat material cut to the shape of the work piece. The volume of sand, found by pouring into a calibrated vessel or by weighing, is proportional to the integral of ($P - Y$) over the compressed surface, and thus to the total compressive force minus Y times the area. This method can be used to analyze complex shapes.

7.9 FLAT ROLLING

Flat rolling of plates and sheets is essentially a plane-strain compression because the length of contact between rolls and work piece, L, is usually much smaller than the width of the sheet, w, (Figure 7.8). As the plastic region is thinned by the compressive stress, σ_z, it is free to expand in the rolling direction, x. However lateral expansion in the y-direction is constrained by the undeforming material on both sides of the roll gap. The net effect is a condition of plane strain, $\varepsilon_y = 0$ and $\varepsilon_z = -\varepsilon_x$, except at the edges.

On the inlet side of the gap, the roll surface is moving faster than the work material, whereas on the outlet side material moves faster than the roll surface (Figure 7.9). This causes friction to act toward the neutral point, N, creating a friction hill.

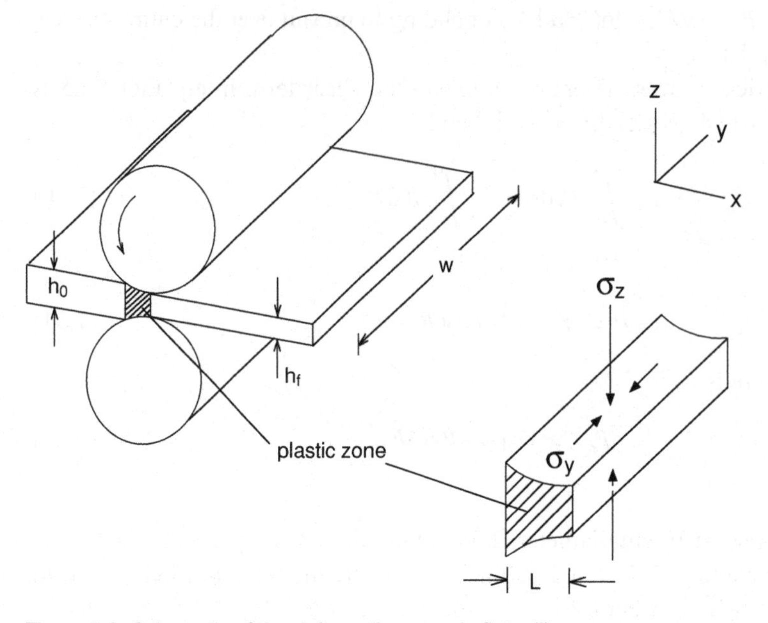

Figure 7.8. Schematic of the deformation zone in flat rolling.

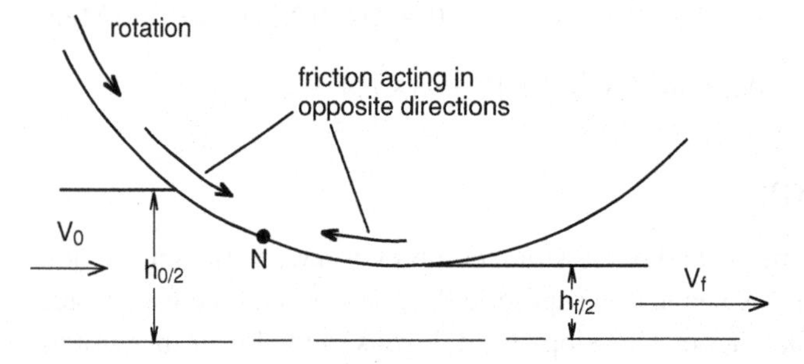

Figure 7.9. On the inlet side, the surface of the roll moves faster than the work piece and on the outlet side the work piece moves faster. This causes friction to act on the work piece toward the neutral point, N.

Figure 7.10 shows the roll-gap geometry, where R is the roll radius, $\Delta h = h_0 - h_f$ and L is the projected contact length. It can be seen that

$$L^2 = R^2 - \left(R - \frac{\Delta h}{2}\right)^2 = R\Delta h - \left(\frac{\Delta h}{2}\right)^2. \tag{7.33}$$

Neglecting the last term,

$$L = \sqrt{R\Delta h} = \sqrt{Rrh_0}, \tag{7.34}$$

where the reduction, $r = \Delta h / h_0$.

The frictional effects are similar to those in plane-strain compression. If the curvature of the roll contact area is neglected, equation 7.16 with L substituted for b and $(h_0 + h_f)/2$ substituted for h can be used to find the average pressure, so

$$P_{av} = \frac{h}{\mu L}\left(\exp\frac{\mu L}{h} - 1\right)\sigma_0, \tag{7.35}$$

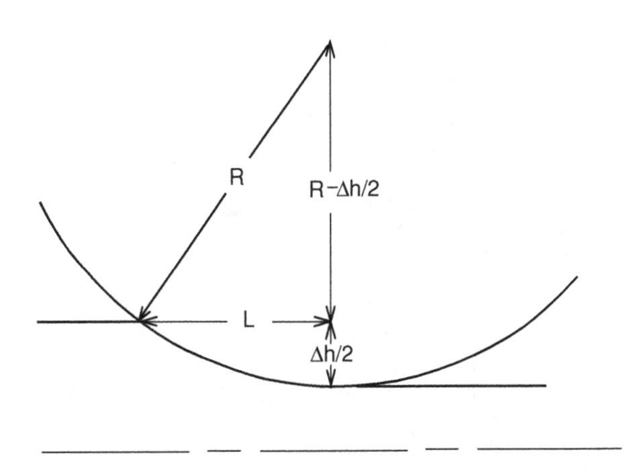

Figure 7.10. Geometry of the roll gap.

where σ_0 is the average plane-strain flow stress in the roll gap. If the material strain hardens, a simple approximation is $\sigma_0 = (\sigma_1 + \sigma_2)/2$ where σ_1 and σ_2 are the flow stresses of the material at the entrance and exit if the roll gap.

If front tension or back tension is applied, this has the effect of lowering σ_0 so equation 7.35 becomes

$$P_{av} = \frac{h}{\mu L}\left(\exp\frac{\mu L}{h} - 1\right)[\sigma_0 - (\sigma_{bt} + \sigma_{ft})/2], \tag{7.36}$$

where σ_{bt} and σ_{ft} are the back and front tensile stresses. Figure 7.11 illustrates the effects of back and front tension. The position of the neutral point shifts with front or back tension.

EXAMPLE 7.5: The plane-strain flow stress, σ_0, of a metal is 200 MPa. A sheet 0.60 m wide and 3 mm thick is to be cold rolled to 2.4 mm in a single pass using 30 cm diameter rolls. Assuming a coefficient of friction is 0.075,

a) Compute the roll pressure.
b) If front tension of 75 MPa were applied, what would be the average roll pressure?

SOLUTION:

a) Substituting $h = (3 + 2.4)/2 = 2.7\,\text{mm}$, $L = \sqrt{(150 \times 0.6)} = 9.487\,\text{mm}$, into equation 7.36, $P_{av} = [2.7(9.487)/0.075][\exp(0.075)(9.487)/2.7](200) = 988\,\text{MPa}$.
b) $P_{av} = [2.7(9.487)/0.075][\exp(0.075)(9.487)/2.5](200 - 37.5) = 803\,\text{MPa}$.

7.10 ROLL FLATTENING

With thin sheets and large roll diameters, the pressure from the friction hill can be very large causing P_{av} to be very high. The roll separating force per width, $F_s = P_{av}L$, increases even more rapidly. The high separating force causes the roll surfaces to elastically flatten much as an automobile tire flattens under the weight of a car. The

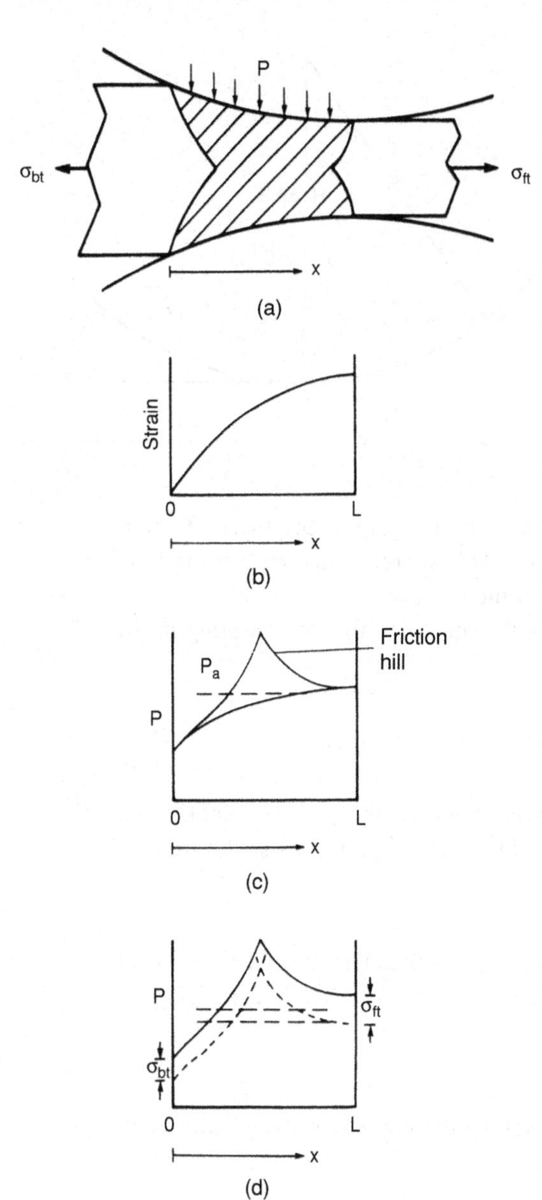

Figure 7.11. Roll gap (a) showing how the strain (b) and roll pressure (c) vary across the gap. The effect of front and back tension (d).

actual radius of contact, R', is larger than R as illustrated in Figure 7.12. Hitchcock* derived an approximate expression for R'

$$R' = R\left(1 + \frac{16F_s}{\pi E'\Delta h}\right), \tag{7.37}$$

where $E' = E/(1 - \nu^2)$. With $L = \sqrt{R'\Delta h}$ the roll separating force becomes

$$F_s = P_{av}\sqrt{R'\Delta h}, \tag{7.38}$$

* J. Hitchcock, "Roll neck bearings," *App. I ASME* (1935), pp. 286–96.

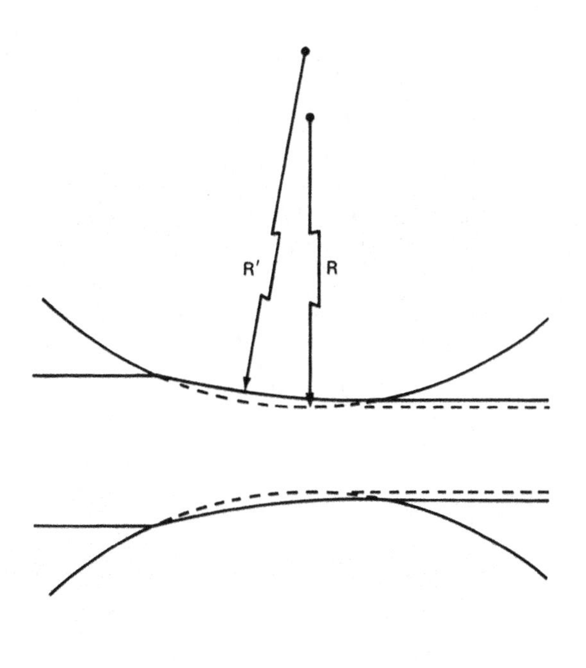

Figure 7.12. Roll flattening.

where

$$P_{av} = \frac{h}{\mu\sqrt{R'\Delta h}}\left[\exp(\mu\sqrt{R'\Delta h}/h) - 1\right](\sigma_0 - \sigma_t) \tag{7.39}$$

and $\sigma_t = (\sigma_{ft} + \sigma_{bt})/2$. The roll separating force, $F_s = P_{av}L$, may be written

$$F_s = \frac{h}{\mu}\left[\exp(\mu\sqrt{R'\Delta h}/h) - 1\right](\sigma_0 - \sigma_t) \tag{7.40}$$

The effect of roll flattening is to increase the roll separating force because both P_{av} and L increase. Both F_s and R' can be found by solving equations 7.38 and 7.40. Figure 7.13 is a plot of both equations for $\sigma_0 = 100{,}000$ lbs, $E' = 33\times106$ psi, $r = 5\%$, $R = 5$ in. and $\mu = 0.2$. Initial thicknesses of $h_0 = 0.100$, 0.040 and 0.020 in. were assumed. The intersections give the appropriate values of F_s and R'. There is no intersection for $h_0 = 0.02$ because the roll flattening is so severe that that thickness cannot be achieved. There is a minimum thickness, h_0, that can be rolled.

$$h_{\text{min}} = \frac{C\mu R}{E'}(\sigma_0 - \sigma_t), \tag{7.41}$$

where C is between 6 and 7.

With $C = 7$, and the conditions cited above, $h_{\text{min}} = 0.021$ in, which explains why there is no solution for $h_0 = 0.020$ in Figure 7.13.

Methods of achieving thinner sheets and foils include better lubrication (lower μ), application of back and front tension (σ_{bt} and σ_{ft}), lower σ_0 (achieved by annealing), and use of smaller diameter rolls. Small diameter rolls will bend under high separating forces. The use of back-up rolls lessens this effect. An example is the Sendzimir mill shown in Figure 7.14. Use of carbide rolls instead of steel rolls increases E'.

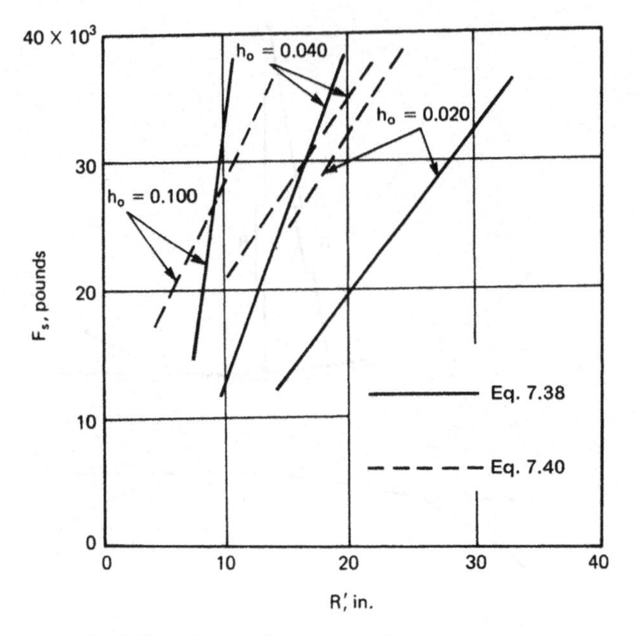

Figure 7.13. Variation of roll separating force vs. R' (solid line) and dependence of flattened radius on roll separating force (dashed line). A flow stress of 100 ksi was assumed. The intersections satisfy both conditions. Note that there is no solution for $h_0 = 0.02$ in.

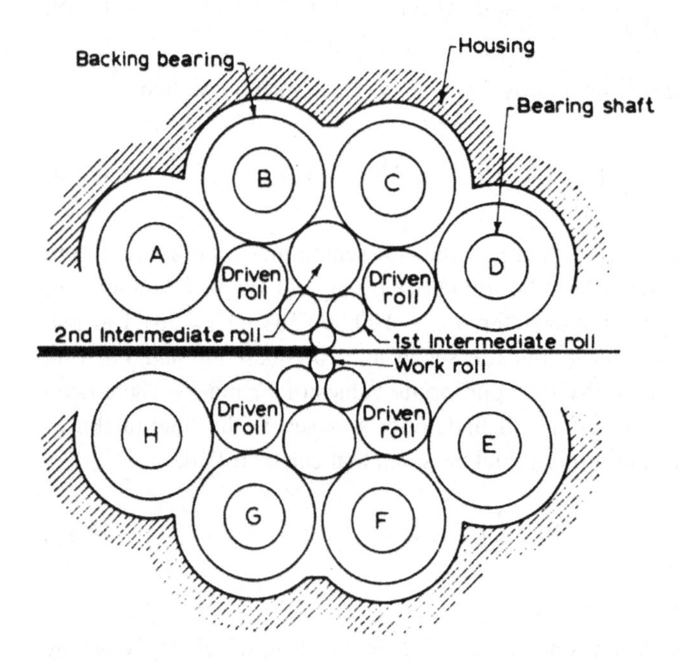

Figure 7.14. Sendzimir mill.

EXAMPLE 7.6: A sheet of steel with a plane-strain yield strength of 500 MPa, is cold rolled between 25 cm diameter rolls to a reduction of 5%. The plane-strain modulus, $E' = 225$ GPa and the initial sheet thickness is 2.5 mm. The coefficient of friction is 0.15.

a) Find the roll separating force of length of roll using equation 7.38.
b) Repeat using equation 7.40.
c) Compare with Figure 7.13.
d) If σ_0 were 50 ksi, to what minimum thickness could the sheet be rolled on this mill?

SOLUTION:

a) $R' = R\left(1 + \frac{16F_s}{\pi E'\Delta h}\right)$ so $F_s = (R'/R - 1)\frac{\pi E'\Delta h}{16}$.

$$F_s = (2 - 1)\pi(33 \times 10^6 \times 0.005)/16 = 32{,}400 \text{ lbs.}$$

b) $F_s = 100{,}000(0.0975/0.15)\{\exp[0.15(10 \times 0.005)^{1/2}/0.975] - 1\} = 27{,}000$ lbs.
c) The left-hand pair of lines in Figure 7.13 intersect at $R' = 9$ in. and $F_s = 26{,}000$ lb. Equation 7.38 predicts $F_s = 32{,}000$ lbs at $R' = 10$ in. as found in (a). Equation 7.37 predicts $F_s = 26{,}000$ lb at $R' = 9$ in. as found in (b).
d) Using Equation 7.41 with $C = 7.5$, $h_{\min} = (7.5)(0.15)(50{,}000)/33 \times 10^6 = 0.0085$ in.

7.11 ROLL BENDING

Roll bending would produce sheets with varying thickness. To counter this effect, rolls are usually cambered (crowned) as shown in Figure 7.15. The degree of cambering varies with the width of the sheet, the flow stress and reduction per pass. The results of insufficient camber are shown in Figure 7.16. The thicker center requires the edges to be elongated more. This can cause edge wrinkling, or warping of a plate. The center is left in residual tension and center cracking can occur.

If the rolls are over-cambered, as shown in Figure 7.17, the residual stress pattern is the opposite. Centerline compression and edge tension may cause edge cracking, lengthwise splitting, and a wavy center.

There are large economic incentives for proper cambering in addition to assuring flatness and freedom from cracks. A variation of only ±0.001 in. in a sheet of 0.32 in. thickness between center and edge is 3%. If a minimum thickness is required, some

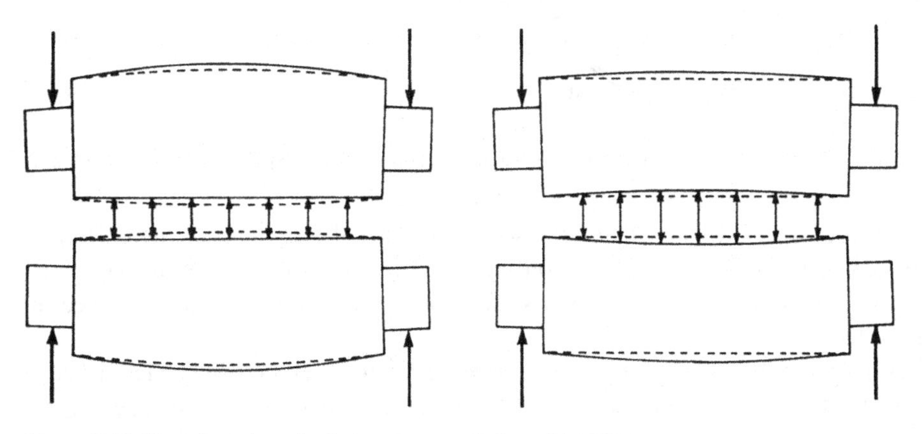

Figure 7.15. Use of cambered rolls to compensate for roll bending.

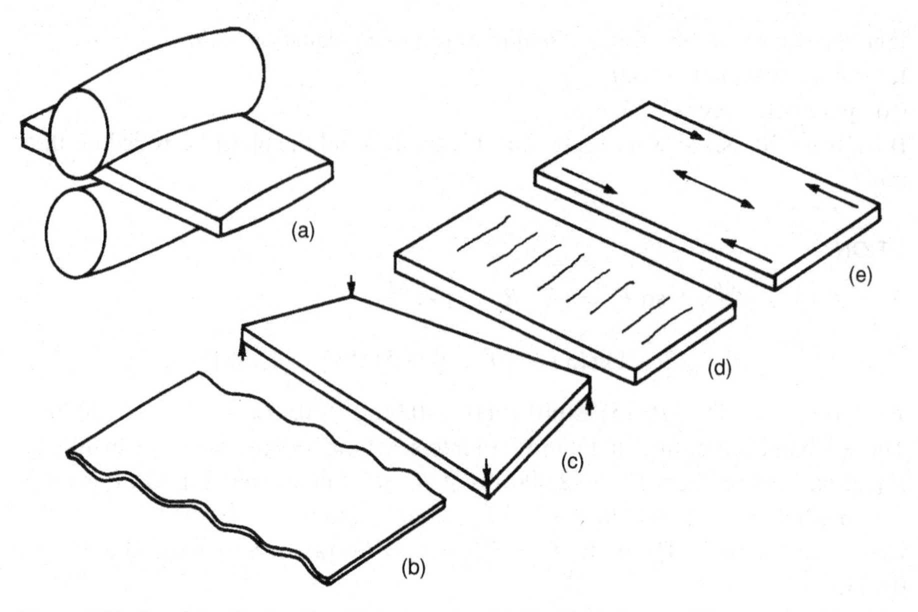

Figure 7.16. Possible effects of insufficient camber (a). Residual stresses (b) center cracking (c) warping (d) and edge wrinkling (e).

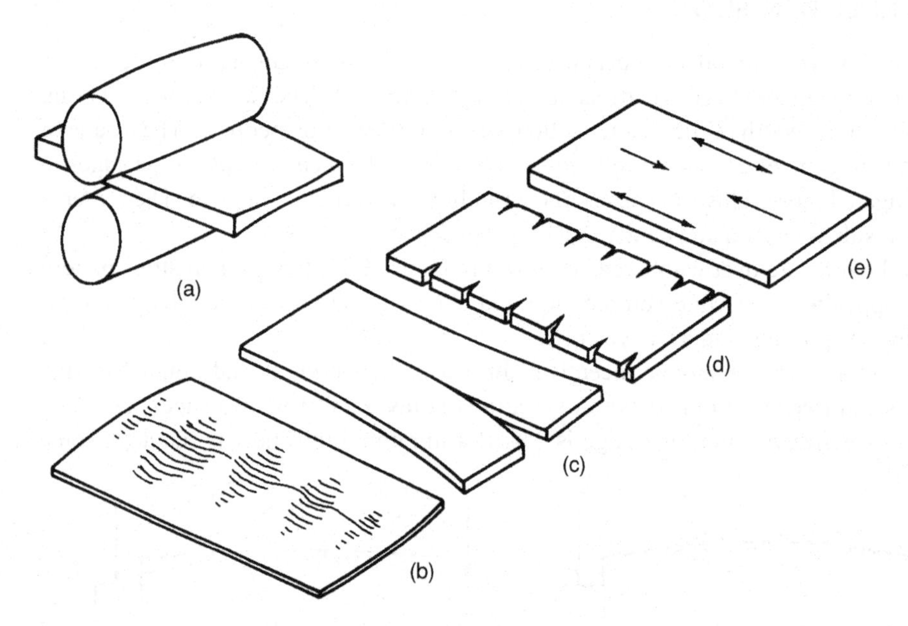

Figure 7.17. Effects of over-cambering include edge cracking, centerline splitting and wavy center.

of the sheet will be thicker than necessary and either the supplier or the customer (depending on whether the sheet is sold by weight or area) will suffer an economic loss. Furthermore the formability will suffer from variable thickness as discussed in Chapter 16.

Even with proper cambering there is a tendency for edge cracking. Material just outside the roll gap constrains flow to plane-strain ($\varepsilon_y = 0$), except at the edges where plane-stress ($\sigma_y = 0$) prevails. Uniaxial compression at the edge would cause only half as much elongation as in the center but this isn't possible. Instead compatibility with

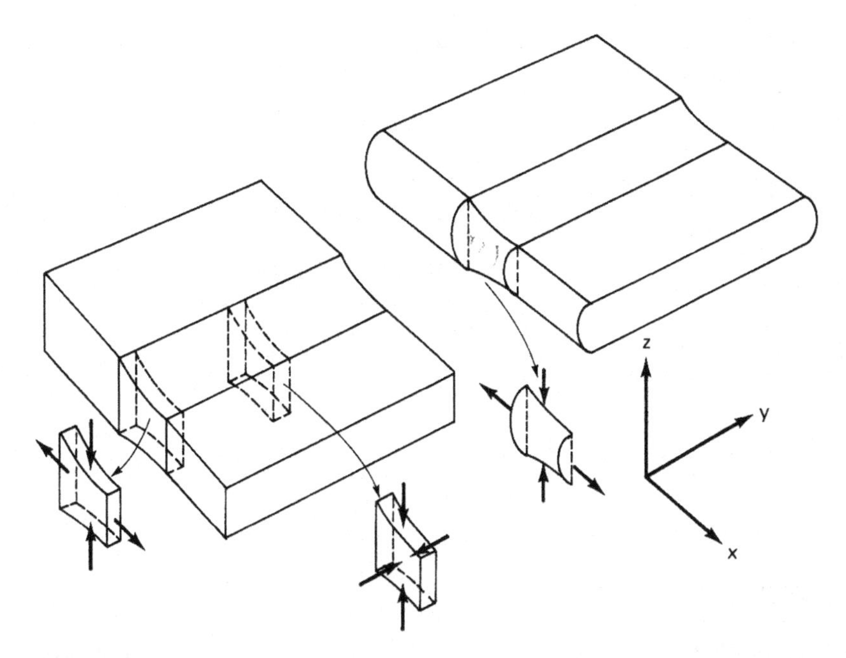

Figure 7.18. Stress states at the edge of a rolled strip. Edge cracking is more likely with rounded edges.

the center requires that the edges experience tension in the rolling direction. This can cause edge cracking. The situation is aggravated if the edges become rounded as in Figure 7.18. With a bulged edge the material at the mid-plane experiences even less compression so the tensile stresses necessary for compatibility are even larger. In multiple-pass rolling it is common to use small edge rollers to maintain square edges. Figure 12.1 in Chapter 12 illustrates the greater formability when square edges are maintained.

7.12 COINING

Coining is a compression operation that embosses the compressed surface with a design. For the design to be embossed, the entire surface must be at its yield strength although the amount of reduction may be small. The pressure must be at least as high as the predictions of equations in Section 7.7 with sticking friction. Forming a sharp detail is similar to making a hardness indentation. Since the pressure in hardness indentation is about three times the yield strength, the local pressure near a sharp detail in coining must be that high.

7.13 REDUCING THE AREA OF CONTACT

One simple way to lower the forces required for forging is to decrease the area of contact by compressing one region at a time. Marciniak and Chodakowski* developed a machine whereby the effect of friction on the force required for forging can be reduced by periodically moving a reduced area of contact between tool and work piece. Figure 7.19 is a schematic illustration of orbital forging.

* Z. Marciniak and A. Chodakowski, *Stahl und Eisen* v. 90, (1970).

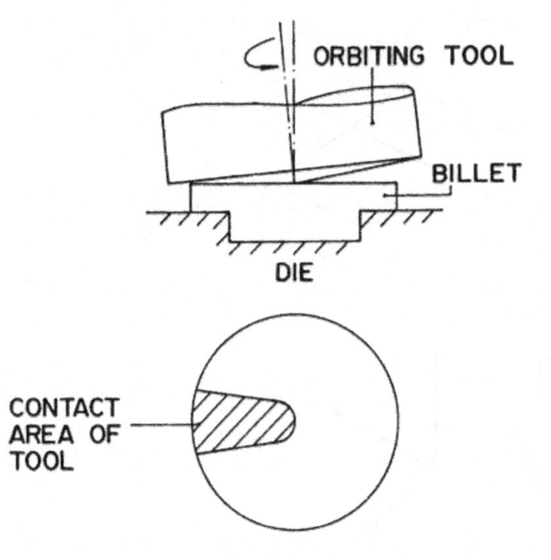

Figure 7.19. Schematic of forging with a rotating tool to reduce the area of contact. From R. Sowerby, *Sci. Prog. Oxf* v. 64 (1977), p. 1077.

NOTES OF INTEREST

One way of circumventing equation 7.41 is to roll two sheets at the same time, artificially increasing h_{min}. Aluminum foil is made by rolling two thin sheets together that are separated after the rolling. The shiny surface was in contact with the rolls and the matte surface was in contact with the other half of the foil.

Georg Sachs (1896–1962) was born in Moscow of German parents. He taught at Frankfurt University (1930–35) and later at Case Institute of Technology.

REFERENCES

J. Hitchcock, "Roll Neck Bearings" *App. I ASME* (1935) pp. 286–96.
W. Johnson and P. B. Mellor, *Plasticity for Mechanical Engineers*, Van Nostrand, 1973.
G. T. van Rooyen and W. A. Backofen, *J. Mech. Phys. Solids*, 7 (1959).

PROBLEMS

7.1. A coil of steel, 252-mm wide and 3-mm thick, is drawn though a pair of dies of semi-angle 8° to a final thickness of 2.4 mm in a single pass. The outlet speed is 3.5 m/s. The average yield strength is 700 MPa, and the friction coefficient is 0.06. Calculate the power in kw consumed.

7.2. An efficiency of 65% was found in a rod-drawing experiment with a reduction of 0.2 and a semi-die angle of 6°.

a) Using Sachs' analysis, find the coefficient of friction.

b) Using the value of η found in (a) what value of efficiency should be predicted from the Sachs' analysis for $a = 6$ and $r = 0.4$?

c) The actual value of η found for the conditions in (b) was 0.80. Explain.

7.3. Estimate the force required to coin a U.S. 25¢ piece. Assume that the mean flow stress is 30,000 psi, the diameter is 0.95 in., and the thickness after forming is 0.060 in.

7.4. Figure 7.20 shows a billet before and after hot forging from an initial size of 2.5 mm × 2.5 mm × 25 mm to 5 mm × 1.25 mm × 10 mm. This is accomplished by using a flat-face drop hammer. Sticking friction can be assumed. For the rate of deformation and the temperature, a flow stress of 18 MPa can be assumed.

a) Find the *force* necessary.

b) Find the *work* required. (Remember that work $= \int F dL$ and that F changes with L.)

c) From what height would the hammer of 3 kg have to be dropped?

d) Compute the efficiency, η.

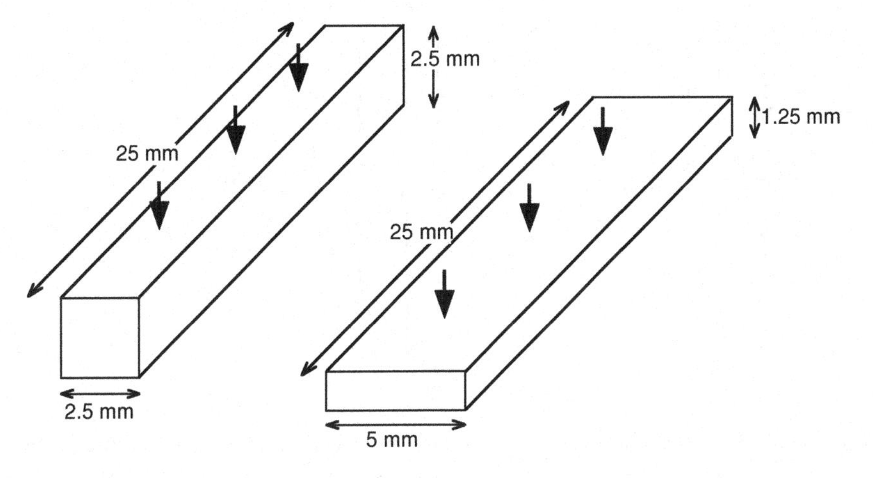

Figure 7.20. Compression in Problem 7.4.

7.5. Two steel plates are brazed as shown in Figure 7.21 The steel has a tensile yield strength of 70 MPa and the filler material has a tensile yield strength of 7 MPa.

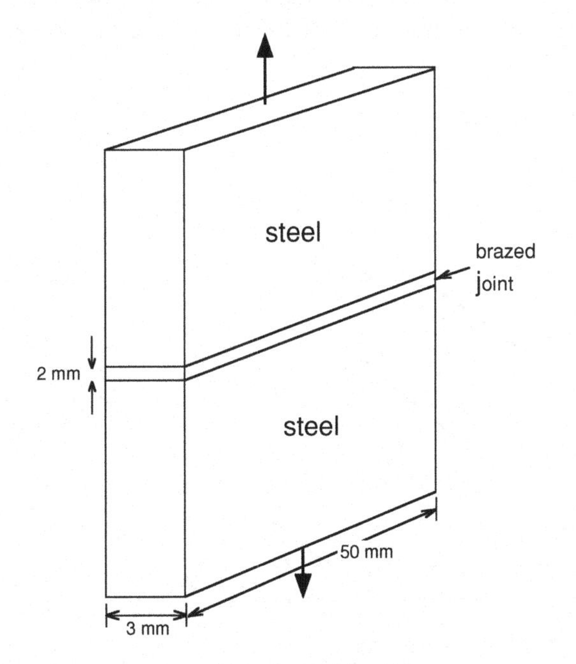

Figure 7.21. Brazed joint for Problem 7.5.

Assuming that the bonds between the filler and the steel do not break, determine the force necessary to cause yielding of the joint.

7.6. Figure 7.22 shows a thin lead ring being used as a gasket. To ensure an acceptable seal, the gasket must be compressed to a thickness of 0.25 mm. Assume that the flow stress of lead is 15 MPa and strain hardening is negligible. Find the required force.

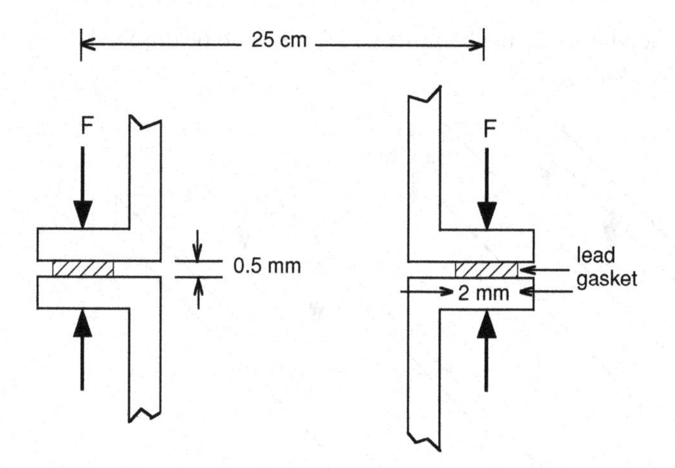

Figure 7.22. Lead gasket for Problem 7.6.

7.7. Magnetic permalloy tape is produced by roll flattening of drawn wire. The final cross section is 0.2 mm × 0.025 mm. It is physically possible to achieve this cross section with different rolling schedules. However, it has been found that the best magnetic properties result with a maximum amount of lateral spreading. For production, the rolling direction must be parallel to the wire axis. Describe how you would vary each of the parameters below to achieve the maximum spreading.

a) roll diameter
b) reduction per pass
c) the friction
d) back and front tension

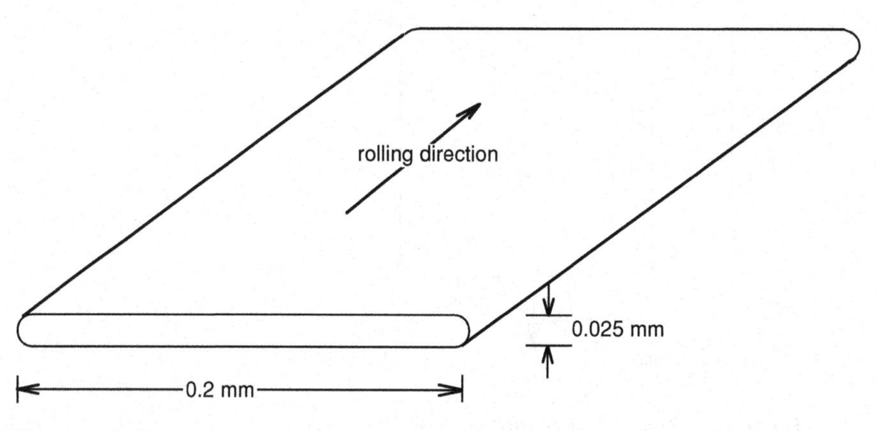

Figure 7.23. Sketch for Problem 7.7.

7.8. A metal with a flow stress of 35 MPa is to be drawn from a diameter of 25 mm to 20 mm through a die of 15 semi-angle. Calculate the necessary drawing stress if:

a) The conditions are frictionless
b) There is sticking friction

7.9. Consider the rolling of a sheet 15-cm wide from a thickness of 1.8 mm to 1.2 mm in a single pass by steel rolls 20 cm in diameter. Assume a friction coefficient of 0.10 and a flow stress of 125 MPa.

a) Calculate the roll pressure if roll flattening is neglected.
b) Calculate the roll pressure taking into account roll flattening.
c) Estimate the minimum thickness that could be achieved.

7.10. Use equation 7.14 to predict how the ratio of w_f/w_i depends on μ, α, and ε_h. (Realize that equation 7.14 neglects redundant work. Expand the exponential term after simplifying, and assume that ε_h is small enough so that higher order terms can be neglected.) Describe in words how w_i depends on μ, α, and ε_h.

7.11. In the force balance in the slab analysis for frictional effects in plane-strain compression, P was assumed to be a principle stress, even though with finite friction it can't be. Examine this assumption by assuming a constant shear stress interface with $\tau = mk$. Also, derive an expression for the angle, θ, between the principal axis, 1, and the x-axis. Express your answer in terms of m, x, h, L, and $2k$. (Not all of these need be in the final expression.) See Figure 7.24.

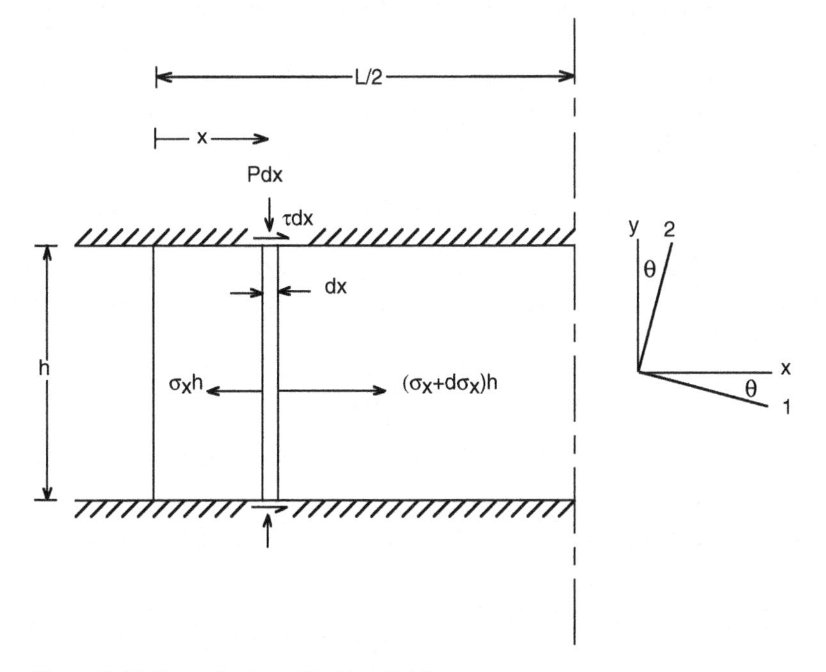

Figure 7.24. Permalloy tape (Problem 7.11).

8 Friction and Lubrication

8.1 GENERAL

Friction in metal working is a result of both tools and work pieces having microscopically rough surfaces as illustrated in Figure 8.1. There are several lubrication regimes, depending on film thickness. With thick films there is no contact between tools and work piece. In the thin film regime, there is contact between tools and work piece and most of the load is carried by this contact. In the mixed film regime a significant fraction of the load is carried by contact between surface asperities. Finally in the boundary regime, almost all the load is carried by contact between asperities of the tools and work piece. Solid lubricants form an easily sheared layer between tools and work piece.

Hydrodynamic lubrication involves a thick film that has a thickness, h, greater than the roughness so that the friction is completely due to the viscosity of the lubricant, so the frictional stress, τ_f, is

$$\tau_f = \mu(U - V)/h, \tag{8.1}$$

where μ is the fluid viscosity, h is the film thickness, and U and V are the velocities of the tool and the work piece. However, h may vary through the deformation zone. For ironing (Figure 8.2),

$$h_1 = 3\mu U_1/(\sigma \tan\theta), \tag{8.2}$$

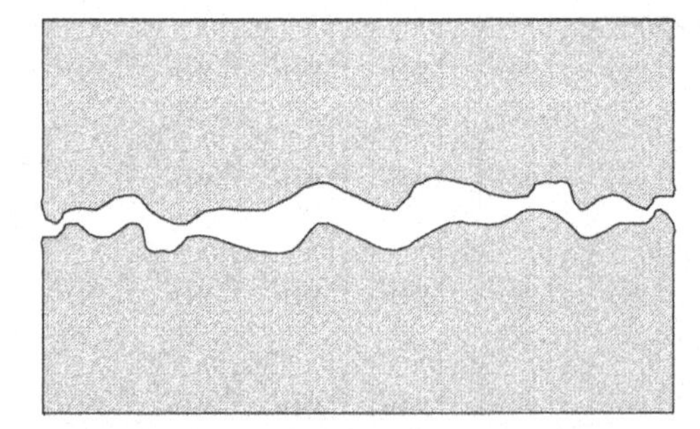

Figure 8.1. Asperities on tool and work-piece surfaces.

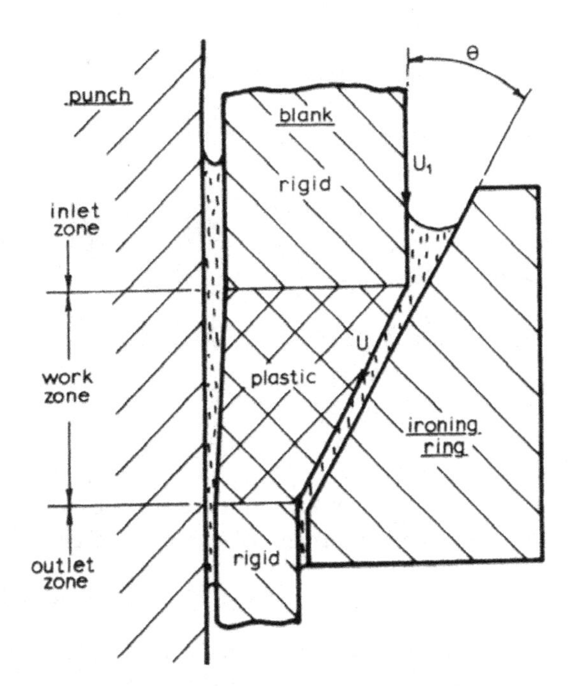

Figure 8.2. Lubrication during ironing.

where h_1 is the film thickness at the inlet, U_1 is the inlet velocity, σ is the material flow stress, and θ is the ironing-ring angle. Into the deformation zone,

$$h = h_1 U_1 / U. \tag{8.3}$$

Hydrodynamic lubrication requires sufficient speed so that the film thickness, h, is greater than the surface roughness. At high speeds, however, the fluid is heated with a resulting decrease of viscosity (i.e., increased speed lowers the viscosity); the result is that the film thickness, h, decreases. The effect of speed is illustrated schematically in Figure 8.3.

With thin films, there is some contact between tools and the work piece, but most of the load is still carried by the lubricant film.

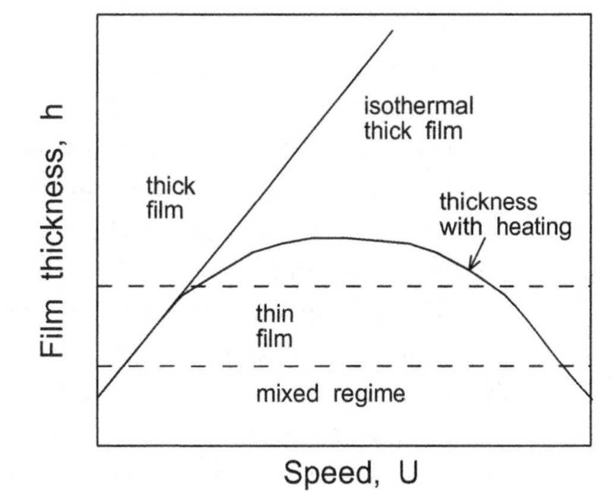

Figure 8.3. Change of film thickness with sliding speed, showing the effect of heating.

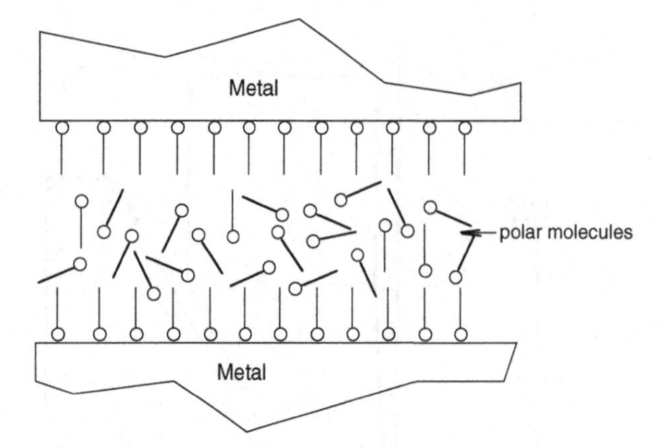

Figure 8.4. Polar molecules bonding with metal surfaces.

In the mixed regime, a significant fraction of the load is carried by contact between surface asperities. The frictional shear stress, $\bar{\tau}_f$, is given by

$$\bar{\tau}_f = a\tau_b + (1 - a)\bar{\tau}_t, \tag{8.4}$$

where a is the fractional contact area, τ_b is the shear strength of the work piece, and $\bar{\tau}_t$ is the mean shear stress in the lubricant.

Loads for boundary regime are entirely supported by contact between asperities.

The mean pressure, $\bar{p} = ap_i$, where p_i is the indentation pressure of the asperity of the work piece. Because p_i is usually approximately 1/5 of the shear strength, $\bar{\tau}_s$, of the work piece,

$$\mu = a\bar{\tau}_s/p = \tau_s/p_i \cong 0.2. \tag{8.5}$$

Most boundary lubricants are compounds that contain polar molecules, perhaps longer than 20 μm, that react with metal surfaces producing strongly bonded films. With these,

$$\mu = a\tau_b/\bar{p} = \tau_b/p_i, \tag{8.6}$$

where τ_b is the shear strength of the boundary film. Other boundary lubricants include fatty acids and organic compounds of sulfur, chlorine, and phosphorus. One end of these molecules reacts with the surface of the work piece, preventing metal-to-metal contact as shown in Figure 8.4.

Solid lubricants form easily sheared layers between tool and work piece. Graphite and molybdenum disulfide (MoS_2) are the most commonly used. They both have lamellar structures that are easily sheared. Sliding aligns the lamellae parallel to the surface in the direction of motion. The lamellae prevent contact even under high loads. The shear stress of the interface is the shear strength of the solid lubricant. Large particles perform best on relatively rough surfaces at low speed, and finer particles perform best on relatively smooth surfaces and at higher speeds.

Other useful solid lubricants include boron nitride, polytetrafluorethylene (Teflon), polyethylene, talc, calcium fluoride, cerium fluoride, and tungsten disulfide. Steel surfaces may be phosphate coated to provide a better surface for lubricants. Molten glass may be used as a lubricant for hot extrusion.

Extreme pressure lubricants react with the surface of the work piece where other protective films have been broken. Chlorinated and fluorinated hydrocarbons are particularly effective. The use of these, however, is severely limited by the Occupational Safety and Health Administration (OSHA).

Lubricant contamination has been a problem in the food and beverage packaging industry. The U.S. Food and Drug Administration (FDA) allows zero amounts a nonfood-grade lubricants and has specified criteria for the acceptable components used in food-grade lubricants.

Water-based lubricants tend to be less expensive and are easier to clean. Lubricants may be applied by rollers or with a spray mist. Recycling of lubricants avoids environmental problems and saves money.

8.2 EXPERIMENTAL FINDINGS

The experimental findings with respect to disc compression are in direct contrast to the predictions of Sections 7.8 and 7.10. It was predicted that there would be sliding friction at the periphery and sticking friction in the center. Measurements of local pressure made by embedding pressure-sensitive pins into the compression platens indicate sticking at the edges and sliding in the center. Figure 8.5 shows that P does increase with distance from the edge, but the slope, $\mathrm{d}P/\mathrm{d}x$ decreases as sliding friction would predict or even remains constant as sticking friction would predict.

The explanation for this is that, early in the compression, lubricant at the edges runs out and the edge of the work piece makes contact with the platen and sticks to it. Lubricant is trapped in the central region so that the frictional shear forces are lower

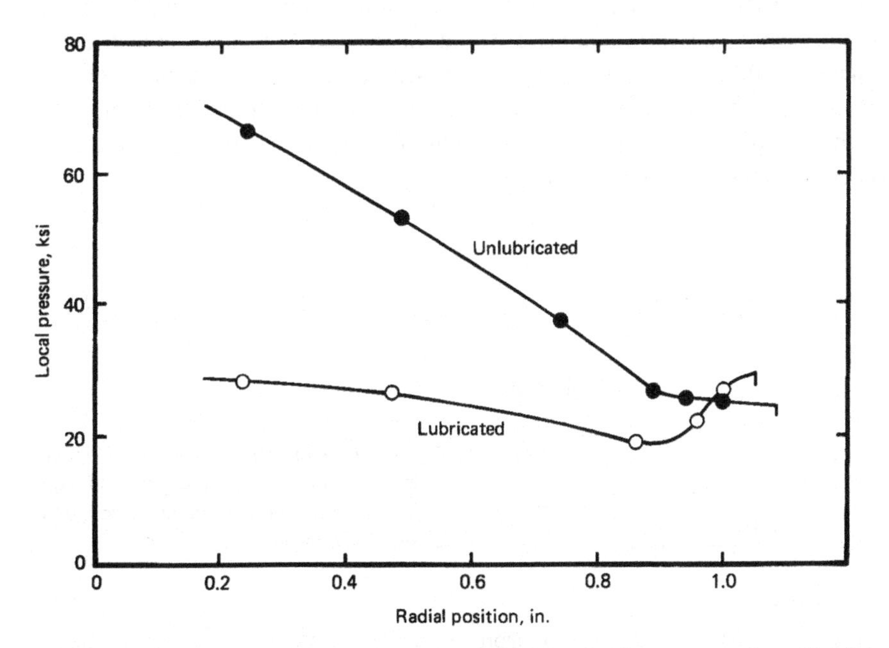

Figure 8.5. Variation of local pressure with radial position in disc compression with lubricated and unlubricated conditions at 3% compression. The data were taken from G. W. Pearsall and W. A. Backofen, *Trans. ASME*, v. 85B (1963).

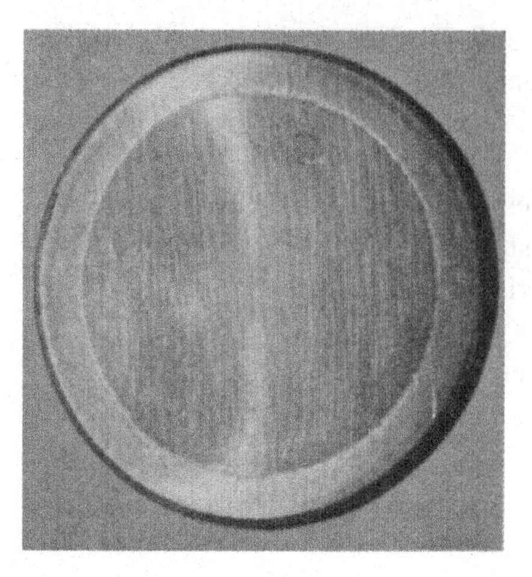

Figure 8.6. Surface appearance of an aluminum disc after compression. The outer ring was originally part of the side walls. Sticking friction occurred there. From G. W. Pearsall and W. A. Backofen, *ibid*.

in the center than at the edges. As compression progresses, the side walls fold up onto the compression platen. Figure 8.6 clearly shows this.

Even with sheets of plastic or soft metal used to create a low shear stress interface, the edges of the work piece cut through the film at low strains. Thus, for axially symmetric compression, neither the assumption of a constant coefficient of friction nor constant shear stress interface is correct.

8.3 RING FRICTION TEST

A simple test for friction in compression involves compressing a ring (Figure 8.7). If there were no friction, the inner diameter would increase by the same percentage as the outer diameter. With high friction, there is a no-slip location between the inner and outer diameters, so the inner diameter must decrease during compression. Figure 8.8 shows the changes of inner diameter as a function of the friction coefficient for a ring with an outer diameter twice the inner diameter.

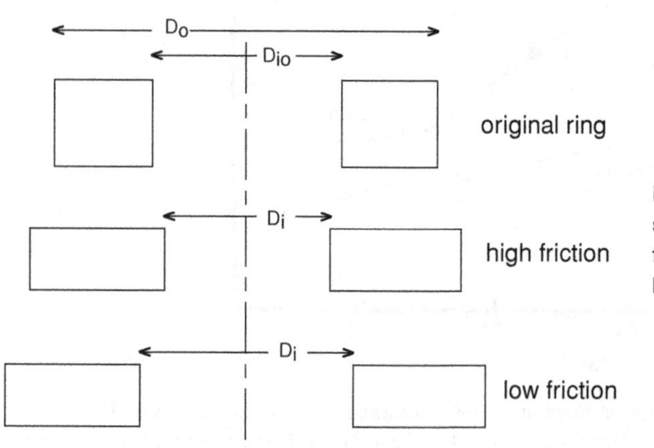

Figure 8.7. Ring compression test. Original specimen (top). After compression with high friction (middle) and after compression with low friction (bottom).

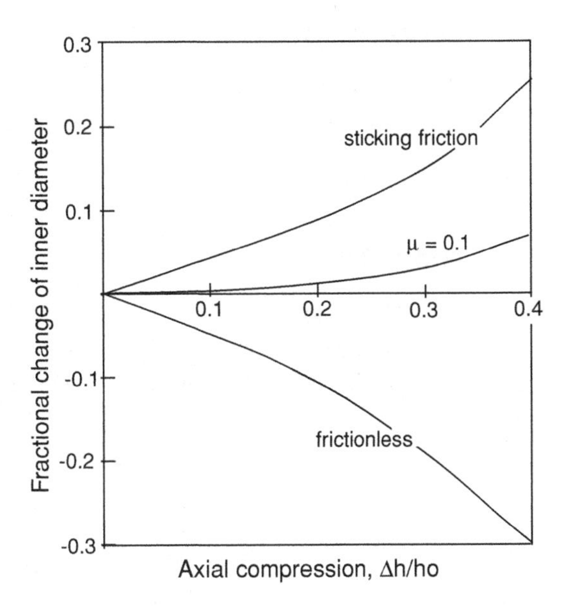

Figure 8.8. Inner diameter change calculated for a specimen and with an outer radius twice the inner radius and with a height 1.33 times as high as its inner radius. Data from J. B. Hawkyard and W. Johnson, *Int. J. Mech. Sci.* v. 17, (1967).

8.4 GALLING

Galling is the wear and transfer of work material onto the tooling during sliding between the tools and work material. The heat generated at the mating surfaces causes bonding of material to the tools. Galling is progressive, the rate of material transferred to the tools accelerating as forming progresses. Eventually the friction becomes so high that the process must be terminated and the tools cleaned. Galling often occurs during the forming of aluminum and causes tool breakdown. The tendency to gall can be decreased by the use of lubricants. The galling onto coated steel tools during forming depends on the work material. For stainless steel, carbon-based coatings provide the best protection, whereas aluminum and titanium alloys require nitride-type coatings, such as TiN. TiC and CrN are also useful.

8.5 ULTRASONICS

It has been suggested that ultrasonic vibrations applied during forming can reduce frictional forces.

NOTE OF INTEREST

Leonardo da Vinci (1452–1519) proposed that the frictional force on a body is proportional to the load and that it is independent of the area of contact. Guillaume Amontons (1663–1705) rediscovered the two laws that daVinci proposed and is generally given credit for them. Charles Coulomb (1736–1804) realized that, at high loads, these rules are not followed. Our present understanding of friction is due in a large part to the work of F. Philip Bowden and David Taylor.

REFERENCES

W. R. D. Wilson, *Mechanics of Sheet Metal Forming*, Plenum Press, 1978, pp. 157–177.

F. Philip Bowden and David Taylor, *Friction and Adhesion*, Springer Netherlands, 2001.

PROBLEMS

8.1. If a pencil is inclined to a piece of paper and pushed with a given force, F (as shown in Figure 8.9). The eraser on its end may either slide or lock, depending on the angle, θ. Find an expression for the critical value of θ.

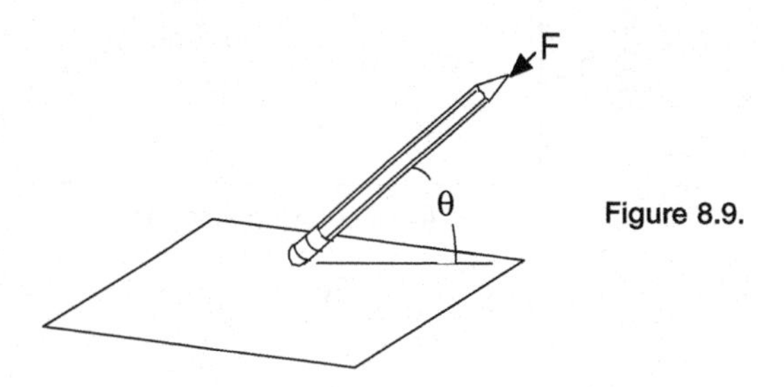

Figure 8.9.

8.2. If the force in problem 8.1 were increased, how would the critical angle change?

8.3. It has been said that if plastic deformation is occurring, the coefficient of friction can never be much greater than 0.5. Explain why this is true.

8.4. Suggest an explanation for the effect of ultrasonic vibration on friction.

9 Upper-Bound Analysis

Calculation of exact forces to cause plastic deformation in metal forming processes is often difficult. Exact solutions must be both *statically* and *kinematically* admissible. That means they must be geometrically self-consistent as well as satisfying required stress equilibrium everywhere in the deforming body. Frequently it is simpler to use limit theorems that allow one to make analyses that result in calculated forces that are known to be either correct or too high or too low than the exact solution.

Lower bounds are based on satisfying stress equilibrium, while ignoring geometric self-consistency. They give forces that are known to be either too low or correct. As such they can assure that a structure is "safe." Conditions in which $\eta = 0$ are lower bounds. Upper-bound analyses on the other hand predict stress or forces that are known to be too large. These are usually more important in metal forming. Upper bounds are based on satisfying yield criteria and geometric self-consistency. No attention is paid to satisfying equilibrium.

9.1 UPPER BOUNDS

The upper-bound theorem states that any estimate of the forces to deform a body made by equating the rate of internal energy dissipation to the external forces will equal or be greater than the correct force. The analysis involves:

1. Assuming an internal flow field that will produce the shape change.
2. Calculating the rate at which energy is consumed by this flow field.
3. Calculating the external force by equating the rate of external work with the rate of internal energy consumption.

The flow field can be checked for consistency with a velocity vector diagram or *hodograph*. In the analysis, the following simplifying assumptions are usually made:

1. The material is homogeneous and isotropic.
2. There is no strain hardening.
3. Interfaces are either frictionless or sticking friction prevails.
4. Usually only two-dimensional (plane-strain) cases are considered with deformation occurring by shear on a few discrete planes. Everywhere else the material is rigid.

9.2 ENERGY DISSIPATION ON PLANE OF SHEAR

Figure 9.1a shows an element of rigid material, ABCD, moving at a velocity V_1 at an angle θ_1 to the horizontal. AD is parallel to yy'. When it passes through yy' it is forced to change direction and adopt a new velocity V_2 at an angle θ_2 to the horizontal. It is sheared into a new shape A′B′C′D′. The corresponding hodograph is shown in Figure 9.1b. The absolute velocities V_1 and V_2 are drawn from the origin, O. Because this is a steady state process, they both have the same horizontal component, V_x. The vector, V_{12}^* is the difference between V_1 and V_2 and must be parallel to the line of shear, yy'.

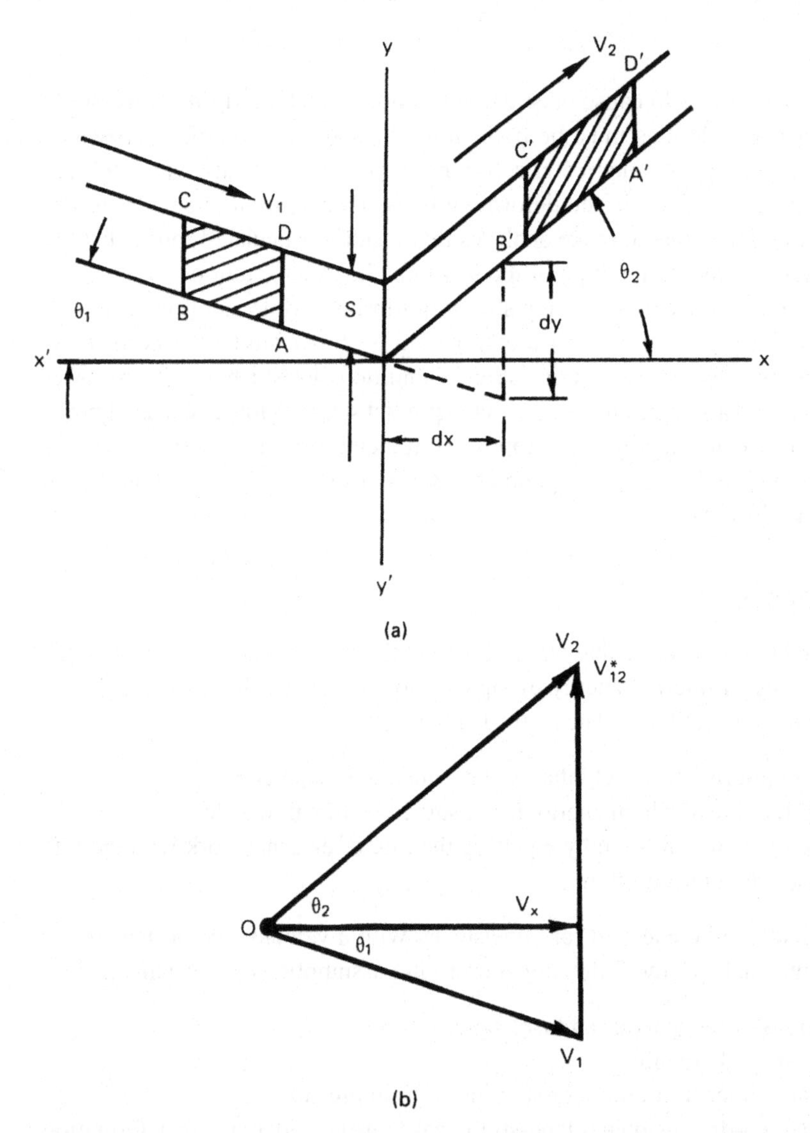

Figure 9.1. (a) Drawing for calculating energy dissipation on a velocity discontinuity and (b) the corresponding hodograph.

The rate of energy dissipation along the discontinuity equals the volume of material crossing the discontinuity per time, SV_x times the work per volume. The work per volume is $w = k\mathrm{d}y/\mathrm{d}x = kV_{12}^*/V_x$, so the rate of energy dissipation along the discontinuity is

$$\mathrm{d}W/\mathrm{d}t = (kV_{12}^*/V_x)SV_x = kSV_{12}^*. \tag{9.1}$$

For deformation fields with more than one shear discontinuity,

$$\mathrm{d}W/\mathrm{d}t = \sum_{1}^{i} kS_i V_i^*. \tag{9.2}$$

9.3 PLANE-STRAIN FRICTIONLESS EXTRUSION

Consider the plane-strain extrusion through frictionless dies as illustrated in Figure 9.2a. Only half of the field is shown. There are two planes of discontinuity, AB and BC. The corresponding hodograph, Figure 9.2b, is constructed by drawing horizontal

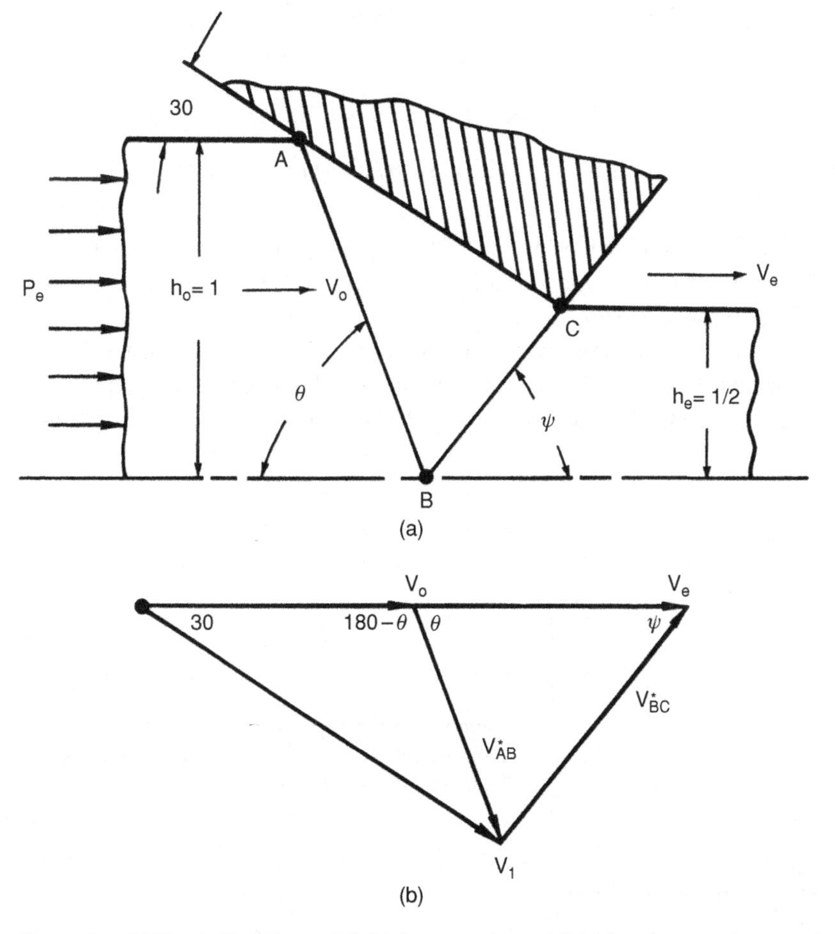

Figure 9.2. (a) Top half of Figure 9.2 (b) An upper-bound field for plane-strain extrusion and (b) the corresponding hodograph.

vectors representing the initial velocity V_0 and the exit velocity V_e. Both start at the origin. The velocity in triangle ABC, V_1, drawn parallel to AC, also starts at the origin. The velocity discontinuity, V^*_{AB} is the difference between V_0 and V_1, and V^*_{BC} is the difference between V_1, and V_e.

The rate of internal work is

$$dw/dt = k(V^*_{AB}\overline{AB} + V^*_{BC}\overline{BC}). \tag{9.3}$$

The rate of external work is $dw/dt = P_e h_0 V_0$. Equating and solving for $P_e/2k$,

$$P_e/2k = \frac{1}{2h_0 V_0}(V^*_{AB}\overline{AB} + V^*_{BC}\overline{BC}). \tag{9.4}$$

Equation 9.4 may be evaluated by physically measuring V^*_{AB} and V^*_{BC} relative to V_0 on the hodograph and measuring $\overline{AB}$ and $\overline{BC}$ relative to V_0 on the physical field. However, it is easier to evaluate $P_e/2k$ analytically. For a 50% reduction with a half-die angle of 30°, $\overline{AB}/h_0 = \csc\theta\overline{AB}/h_0 = \csc\theta, \overline{BC}/h_0 = \csc\psi/2$.

With the law of sines,

$$V^*_{AB}/\sin 30° = V_0/\sin(\theta - 30°) \quad \text{or} \quad V^*_{AB}/V_0 = \sin 30°/\sin(\theta - 30°) \quad \text{and}$$

$$V^*_{BC}/\sin\theta = V^*_{AB}/\sin\theta \quad \text{so} \quad V^*_{BC}/V_0 = (\sin\theta/\sin\psi)/V^*_{AB}/V_0.$$

The magnitude of $P_{ext}/2k$ depends on θ. If $\theta = 90°$, $V^*_{AB}/V_0 = \sin 30°/\sin(90° - 30°) = 0.577$. $V^*_{BC}/V_0 = (\sin 90°/\sin 30°)0.577 = 1.154, \overline{AB} = \overline{BC} = h_0$. Therefore $P_{ext}/2k = 0.577 + 1.154 = 0.866$.

Figure 9.3 shows the calculated variation of $P_e/2k$ with θ. The lowest value of $P_e/2k \approx 0.78$, occurs when $\theta \approx 72°$. A lower bound can be found as $P_e = \int \sigma d\varepsilon = 2k\ln(2)$ so $P_e/2k = 0.693$. The true solution of $P_e/2k = 0.762$ (see Chapter 10) lies between these.

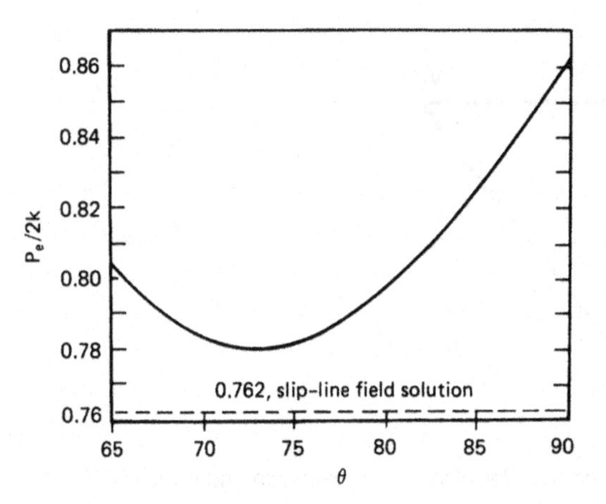

Figure 9.3. Variation of calculated extrusion pressure with the angle, θ, in the upper-bound field of Figure 9.2.

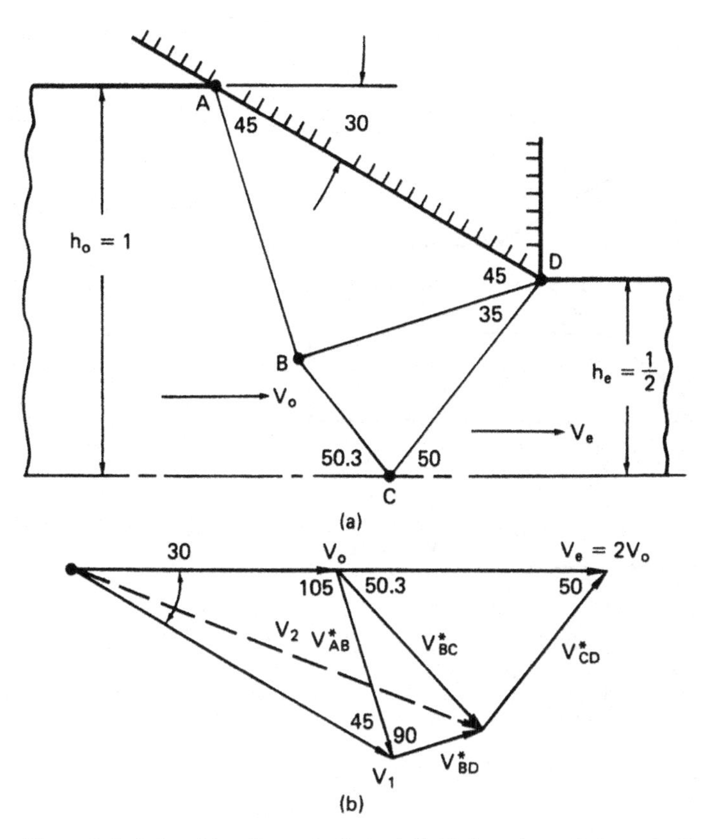

Figure 9.4. A two-triangle upper-bound field for plane-strain extrusion and the corresponding hodograph.

It is not essential to assume frictionless conditions. If sticking friction is assumed along AC, an additional term, $k(V_1/V_0)(\overline{AC}/h_0)$, must be added to the upper bound so

$$P_e/2k = (1/2)(V_1\overline{AC} + V^*_{AB}\overline{AB} + V^*_{BC}\overline{BC})/(V_0h_0) \quad \text{or}$$

$$P_e/2k = (P_e/2k)_{\text{frictionless}} + (1/2)(V_1/V_0)(\overline{AC}/h_0) \tag{9.5}$$

A plot of $P_e/2k$ vs, θ in equation 9.5 has a minimum of $P_e/2k \approx 1.43$ at $\theta \approx 83°$. If a shear stress, mk, is assumed along AC, $(1/2)(V_1/V_0)(\overline{AC}/h_0)$ in equation 9.5 is replaced by $m(V_1/V_0)(\overline{AC}/h_0)$.

More complex fields may give lower values of $P_e/2k$. Figure 9.4 shows a field composed of two triangles and the corresponding hodograph. With frictionless conditions on AD, this field gives $P_e/2k = 0.768$ which is close to the exact solution.

The distortion of the extruded slab predicted by a proposed field can be found by following several points through the field.

EXAMPLE 9.1: Construct the distortion of a vertical grid line in Figure 9.5.

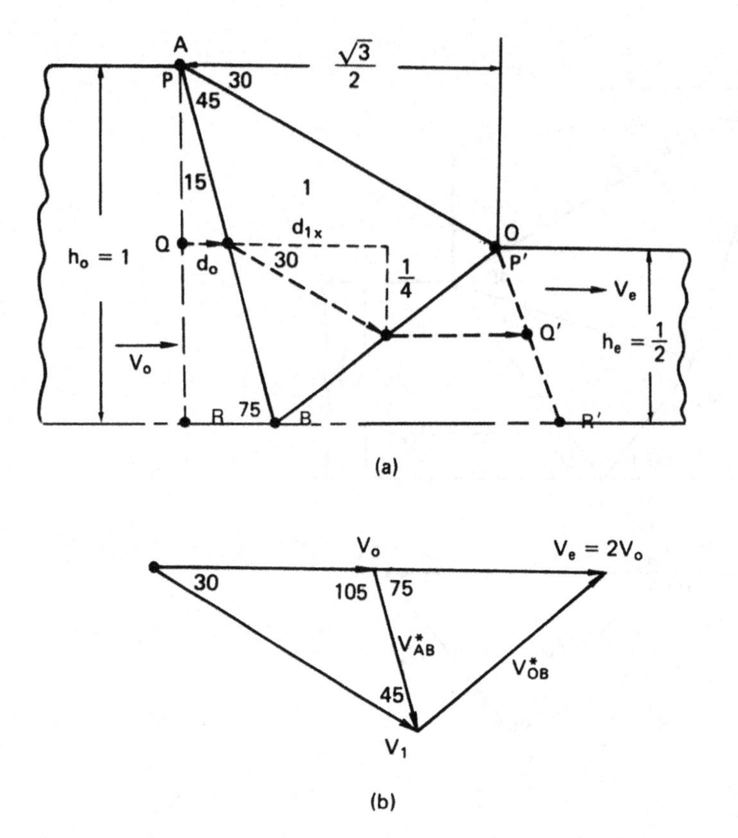

Figure 9.5. Construction of the distortion caused by extrusion through an upper-bound field.

SOLUTION: For point P, $d_0 = 0$, $d_{1x} = \sqrt{(3)}/2 = 0.866$. $V_{1x} = \cos 30° \sin 105°/\sin 45° = 1.183$. Choosing the time increment, t, as the time for point P to arrive at P′, $t = 0.866/1.183 = 0.732$.

For point Q, $d_0 = (1/2)\tan 15° = 0.134$. $t_0 = 0.134/1 = 0.134$.

$d_{1x} = (1/2)\sqrt{(3)}/2 = 0.433$, $t_{1x} = 0.433/1.183 = 0.366$

$t_e = t - t_0 - t_{1x} = 0.732 - 0.134 - 0.366 = 0.232$

$d_e = 2(0.232) = 0.464$

$d_1 = 0.134 + 0.366 + 0.464 = 0.964$

For point R, $d_0 = \tan 15° = 0.268$, $t_0 = 0.268$, $d_{1x} = 0$

$t_e = t - t_0 - t_{1x} = 0.732 - 0.268 - 0 = 0.464$

$d_e = 2(0.464) = 0.928$, $d_1 = 0.268 + 0 + .928 = 0.464$.

These points are constructed on Figure 9.5 as points P′, Q′ and R′.

Now consider Figure 9.6, which shows a two-triangle field for the same reduction. Particles crossing AB pass through two constant velocity fields, whereas particles crossing BC pass through only one.

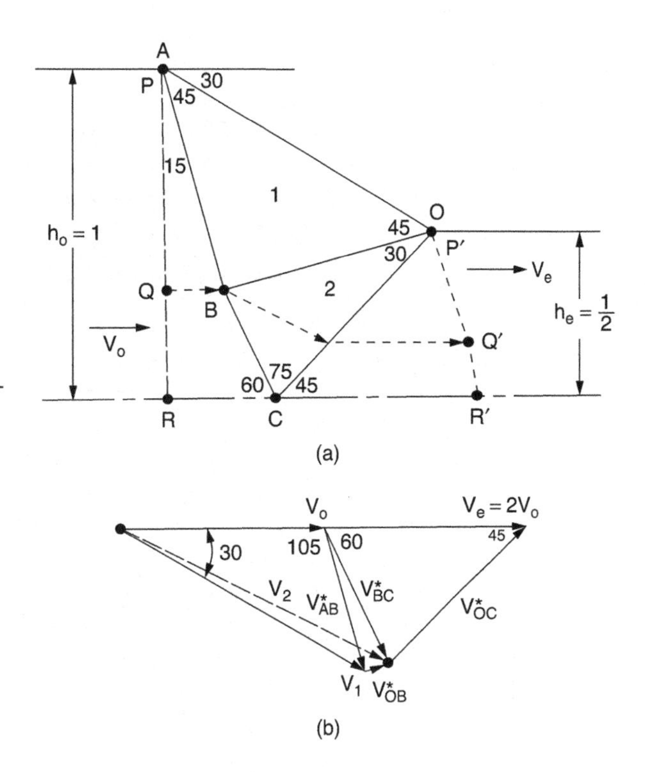

Figure 9.6. A two-triangle field and a hodograph for analyzing the distortion of a grid.

9.4 PLANE-STRAIN FRICTIONLESS INDENTATION

Figure 9.7 is a possible upper-bound field and corresponding hodograph for plane-strain frictionless indentation. Considering the right-hand side of the field, shear occurs along AB, BC and CD. The metal outside the triangles is rigid. From the hodograph, the velocity discontinuities, $V^*_{OA} = V^*_{AB} = V^*_{AC} = V^*_{BC} = V^*_{CD} = 1/\cos 30^\circ = 1.155$. The length of lines are $\overline{OA} = \overline{AB} = \overline{AC} = \overline{BC} = \overline{CD} = w/2$, so

$$\mathrm{d}W/\mathrm{d}t = k(\overline{OA}V^*_{OA} + \overline{AB}V^*_{AB} + \overline{AC}V^*_{AC} + \overline{BC}V^*_{BC} + \overline{CD}V^*_{CD}) \quad \text{or}$$
$$\mathrm{d}W/\mathrm{d}t = 5k(w/2)(2/\sqrt{3})$$

The rate of at which external energy is expended on the right-hand half of the field is

$$P_\perp/2k = 5/\sqrt{3} = 2.89 \tag{9.6}$$

If there is sticking friction along OB, $P_\perp/2k = 6/\sqrt{3} = 3.46$ $P_\perp/2k = 6/\sqrt{3} = 3.46$

9.5 PLANE-STRAIN COMPRESSION

An upper-bound analysis of plane-strain can be made with the field shown in Figure 9.8. Discrete shear occurs along OA, OB, OC and OD. The lengths of these discontinuities

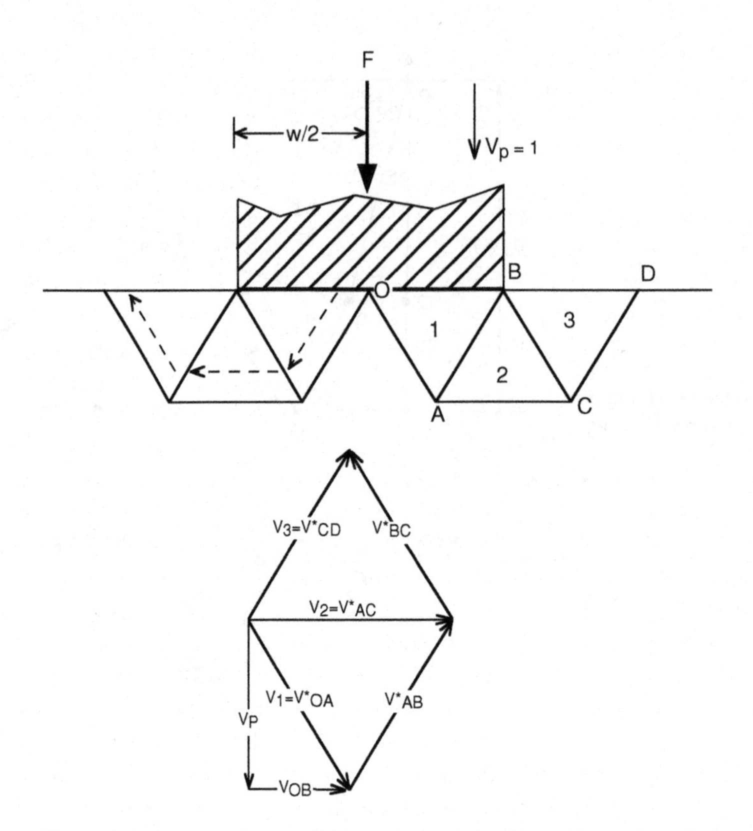

Figure 9.7. An upper-bound field for plane-strain, frictionless indentation and the hodograph for the right-hand side of the field. The triangles are equilateral so the angles are 60°.

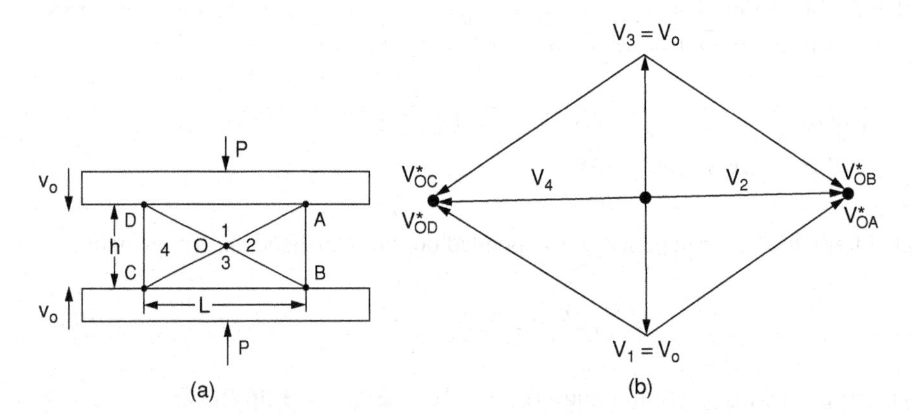

Figure 9.8. (a) A possible field for plane-strain compression and (b) the corresponding hodograph. From R. M. Caddell and W. F. Hosford, *Int. J. Mech. Eng. Educ,* 8 (1980). Reprinted by permission of the Council of the Institute for Mechanical Engineers.

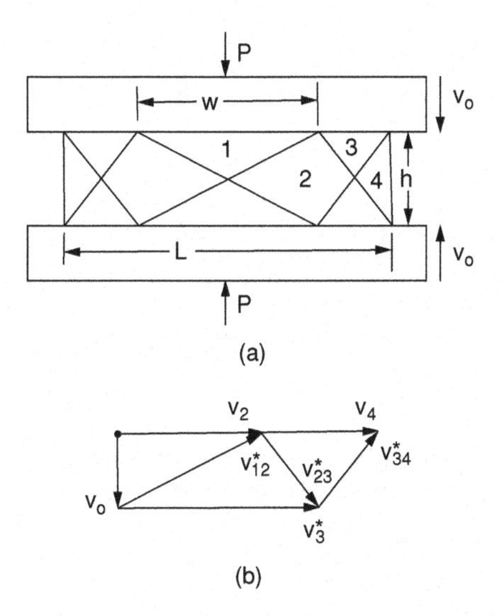

Figure 9.9. A different upper-bound field for plane-strain compression and the corresponding hodograph for the upper right-hand quarter. From R. M. Caddell and W. F. Hosford, *ibid.*

equal $\overline{AO} = (h^2 + L^2)^{1/2}$ and the velocity discontinuity along them is $V^*_{AO} = V_0(h^2 + L^2)^{1/2}/h$ so $2P_\perp L V_0 = 4k\overline{AO}V^*_{AO}$. Substituting

$$P_\perp/2k = (h/L + L/h)/2. \tag{9.7}$$

For large values of L/h, fields consisting of more than one triangle give better solutions. If there is friction, the lowest solutions have an odd number of triangles, because the middle triangle doesn't slide. The field in Figure 9.9 has three triangles. The general solution for this field is

$$P/2k = (3/2)h/L + (1/2)L/h + w^2/(2hL)w/(2h). \tag{9.8}$$

The lowest value of $P/2k$ occurs when $w = L/2$ and is

$$P/2k = (3/2)h/L + (3/8)L/h. \tag{9.9}$$

A field consisting of five triangles is shown in Figure 9.10. If $\overline{AB} = \overline{BC}$, the lowest solution occurs when $w = L/3$ and is

$$P/2k = (5/2)h/L + (1/3)L/h. \tag{9.10}$$

A general minimum solution for this class of upper bounds is

$$P/2k = \left[\frac{nh^2}{2} + \left(\frac{L}{n+1}\right)^2\right] \bigg/ (hL), \tag{9.11}$$

where $C = \frac{3n+1}{2} + \sum_1^{(n-1)/2}(2i-1)$ and the number of triangles, n, is an odd integer ≥ 3. The minimum occurs when $w = 2L/(n+1)$.

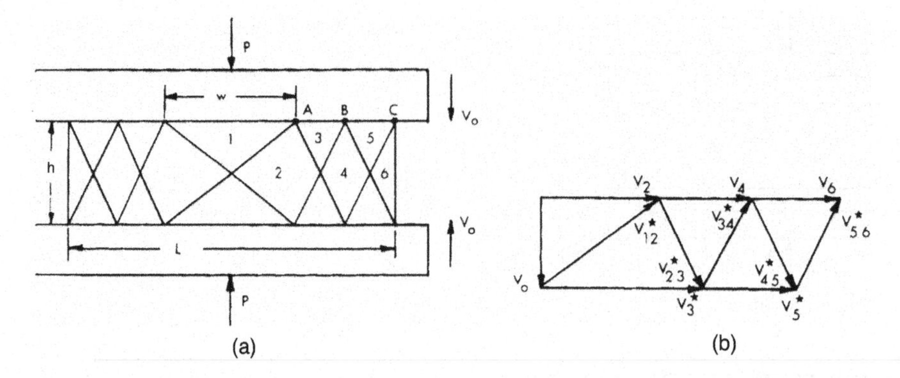

Figure 9.10. A five-triangle upper-bound field for plane-strain compression and the corresponding hodograph for the upper right-hand quarter. From R. M. Caddell and W. F. Hosford, *ibid.*

9.6 ANOTHER APPROACH TO UPPER BOUNDS

Other deformation fields can be used for upper bounds as long as they are kinematically admissible. An acceptable field may contain regions undergoing homogeneous deformation. For example consider a slab analysis of plane-strain compression as shown in Figure 9.11. Each element is assumed to deform homogeneously as it slides away from the centerline. The rate of homogeneous work on an element is $2k\dot{\varepsilon}h\,\mathrm{d}x$. Substituting $\dot{\varepsilon} = 2V_0/h$, the homogeneous work rate is $\dot{W} = 4kV_0\,\mathrm{d}x$

The velocity of the elements in admissible velocity field is

$$V_x = 2V_0x/h, \tag{9.12}$$

so the rate of energy dissipation on both tool-work piece interfaces is

$$2kV_x\,\mathrm{d}x = 4kV_0(x/h)\,\mathrm{d}x. \tag{9.13}$$

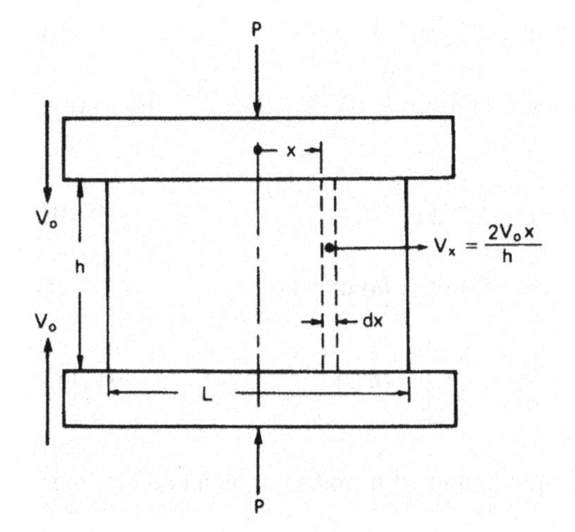

Figure 9.11. Drawing for a slab energy-balance upper bound of plane-strain compression.

Equating the external work rate, $2PLV_0$, with the interface and homogeneous work rate,

$$2PLV_0 = 2\int_0^{L/2} 4kV_0(1 + x/h)\mathrm{d}x, \quad \text{or}$$

$$P/2k = 1 + L/(4h). \tag{9.14}$$

This is identical to the solution obtained by the slab analysis, where b in equation 7.20 is the same as L here. With a constant interfacial stress, mk,

$$P/2k = 1 + mL/(4h). \tag{9.15}$$

A similar upper-bound slab analysis for axisymmetric compression with a constant interfacial stress gives

$$P/2k = \sigma_0/2k + mR/(3h), \tag{9.16}$$

which is identical to equation 7.32 if m is taken as unity.

9.7 A COMBINED UPPER-BOUND ANALYSIS

The traditional upper-bound and slab analyses can be combined to take advantage of the dead-metal cap. Figure 9.12 shows such a field along with the hodograph for the upper right-hand quarter of the field. The central region is a simple upper-bound field of width, w, while the rest of the material undergoes homogeneous deformation. The rate of energy dissipation along OA, OB, OC and OD is

$$\dot{W}_c = 2kV_0(h^2 + w^2)/h.$$

Energy is also dissipated along AD and GH. The velocity discontinuity here varies with the distance, z, from the centerline, $V_{23}^* = 2zV_0/h$, so the total rate of energy dissipation on AD and GH is

$$\dot{W}_{23} = 4\int_0^{h/2} kV_{23}^*\mathrm{d}z = khV_0. \tag{9.17}$$

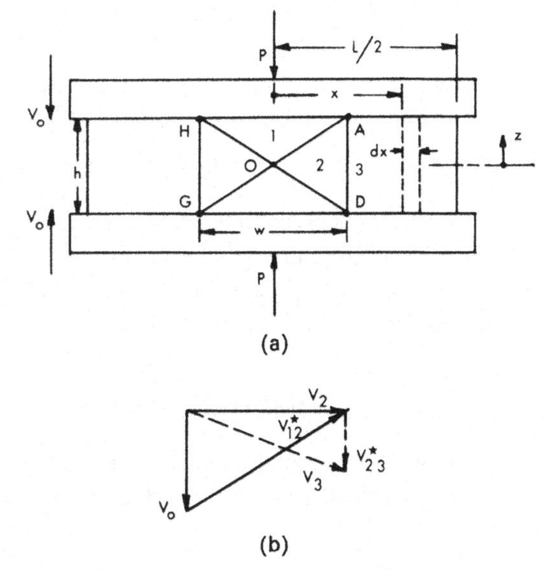

Figure 9.12. Combined conventional upper-bound and slab-energy fields for sticking friction (left) and partial hodograph (right)

In the region, $x > w/2$, the velocity, $V_x = 2V_0x/h$, so the rate of frictional work is

$$\dot{W}_f = 4\int_{w/20}^{L/2} kV_x\,\mathrm{d}x = kV_0(L^2 - w^2)/h. \tag{9.18}$$

The strain rate in the region between $w/2 \le x \le L/2$ is $\dot{\varepsilon} = 2V_0/h$, so the rate of homogeneous work is

$$\dot{W}_h = 2k\dot{\varepsilon}(L - w)h = 4kV_0(L - w). \tag{9.19}$$

Equating the internal work with the rate of external work, $2PLV_0$, $2PLV_0 = \dot{W}_c + \dot{W}_{23} + \dot{W}_f + \dot{W}_h$ or

$$P/2k = 1 + (3/4)h/L + (1/4)L/h - w/L + (1/4)w^2/(hL). \tag{9.20}$$

The minimum occurs if $w = 2h$ so

$$P/2k = 1 + (L/h - h/L)/4. \tag{9.21}$$

9.8 PLANE-STRAIN DRAWING

Figure 9.13 shows the streamlines and hodograph $\dot{W}_a = \dot{W}_h + \dot{W}_f + \dot{W}_r$ for plane-strain drawing of a sheet through a die of semi-angle α. All paths are horizontal before crossing AA′ and after crossing BB′. All particles on a vertical line such as CC′ have the same horizontal component of velocity, V_x, but V_x increases from AA′ to BB′. When a particle crosses AA′ it suffers a velocity discontinuity, V_A^* that depends on the distance, y, from the centerline. At the outer surface ($y = t/2$), $V_A^* = V_0\tan\alpha$. and at the centerline, $V_A^* = 0$. At other points, $V_A^* = V_0y/(t/2)\tan\alpha$.

The rate of work along this discontinuity is

$$\dot{W}_r = 2\int_0^{t/2} \frac{kV_0\tan\alpha}{(t_0/2)} y\,\mathrm{d}y = kV_0t_0\tan\alpha/2. \tag{9.22}$$

The rate of sliding along the die is $V_s = V_x/\cos\alpha = V_0t_0/(t\cos\alpha)$. The rate of frictional work is then

$$\dot{W}_f = \int_{t_f}^{t_0} \frac{mkV_0t_0}{\cos\alpha\sin\alpha}(\mathrm{d}t/t) = \frac{mkV_0t_0\varepsilon}{\cos\alpha\sin\alpha}, \tag{9.23}$$

where $\varepsilon = \ln(t/t_0)$. The external work rate is $\dot{W}_a = \sigma_d V_f t_f = \sigma_d V_0 t_0$. Equating external and internal work rates, $\dot{W}_a = \dot{W}_h + \dot{W}_f + \dot{W}_r$, and simplifying

$$\sigma_d/2k = (1 + m/\sin 2\alpha)\varepsilon + (1/2)\tan\alpha. \tag{9.24}$$

This can be interpreted as $\sigma_d = w_h + w_f + w_r$ where $w_h/2k = \varepsilon$, $w_f/2k = m\varepsilon/\sin 2\alpha$, and $w_r/2k = (1/2)\tan\alpha$.

The force balance produces equation 9.24 without the (1/2)tan α term if μP in equation 7.1 is replaced by *mk*.

9.9 AXISYMMETRIC DRAWING

Consider drawing a rod of diameter, D_0, to a diameter D_f, through a die of semi-angle a with a constant interface shear stress, *mk*. For a slab of radius, *R*, the horizontal

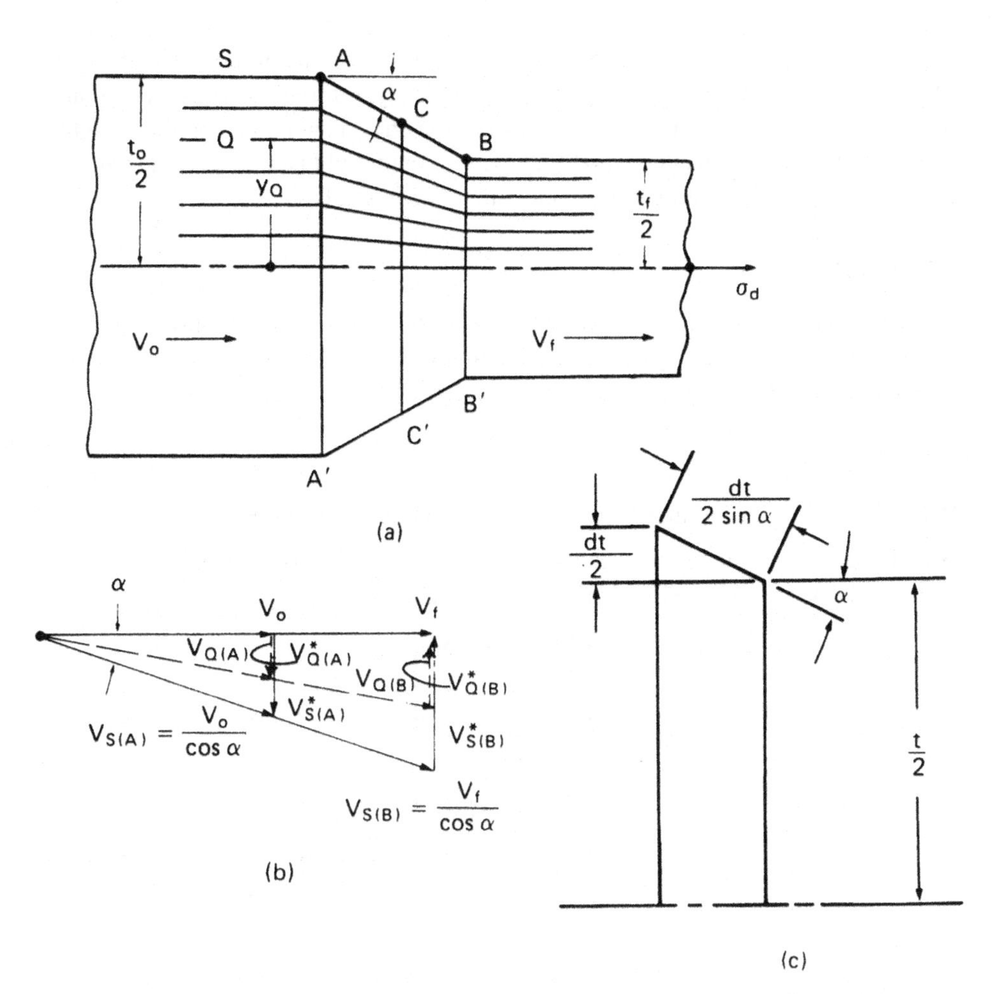

Figure 9.13. Plane-strain drawing. (a) flow lines, (b) partial hodograph and (c) a differential element.

component of velocity is $V_x = V_0(R_0/R)2$ so the sliding velocity at the interface is $V_x = V_0(R_0/R)^2/\cos\alpha$. The area of the element in contact with the die is $2\pi R\,\mathrm{d}R/\sin\alpha$ and the interface stress is mk so

$$\dot{W}_f = \int_{R_f}^{R_0} \frac{2\pi mkR_0^2V_0}{R\sin\alpha\cos\alpha}\mathrm{d}r = 2\pi mkR_0^2V_0\varepsilon/\sin 2\alpha. \tag{9.25}$$

The velocity discontinuity, $V_r^* = V_0(r/R_0)\tan\alpha$, on entering the field depends on the radial distance, r, so the rate of energy dissipation is

$$\dot{W}_r = \int_0^{R_0} 2\pi rkV_0(r/R_0)\tan\alpha\,\mathrm{d}r = (2/3)\pi kV_0R_0^2\tan\alpha. \tag{9.26}$$

The homogeneous work rate is

$$\dot{W}_h = (\sigma_d/2k)R_f^2V_f = (\sigma_d/2k)R_0^2V_0. \tag{9.27}$$

Equating the rates of external and internal work,

$$\sigma_d/2k = (\sigma_0/2k + m/\sin 2\alpha)\varepsilon + (2/3)\tan\alpha. \tag{9.28}$$

Again this is equal to the slab analysis that produced equation 9.28 without the redundant work term, $(2/3)\tan\alpha$. Other kinematically admissible fields may be analyzed. Avitzur derived an upper bound for axisymmetric drawing that predicts slightly lower drawing stresses than equation 9.28. His velocity field is more complex and the difference between his prediction and equation 9.28 is small.

REFERENCES

W. Johnson and P. B. Mellor, *Engineering Plasticity*, Van Nostrand Reinhold, 1973.
C. R. Calladine, *Engineering Plasticity*, Pergamon Press, 1969.
B. Avitzur, *Metal Forming: Processes and Analysis*, McGraw-Hill, 1968.

PROBLEMS

9.1. Find $P_e/2k$ for Figure 9.2 if θ is 80° and compare with Figure 9.3.

9.2. Calculate $P_e/2k$ for the plane-strain frictionless extrusion illustrated in Figure 9.14. Triangles ABC and CDE are equilateral.

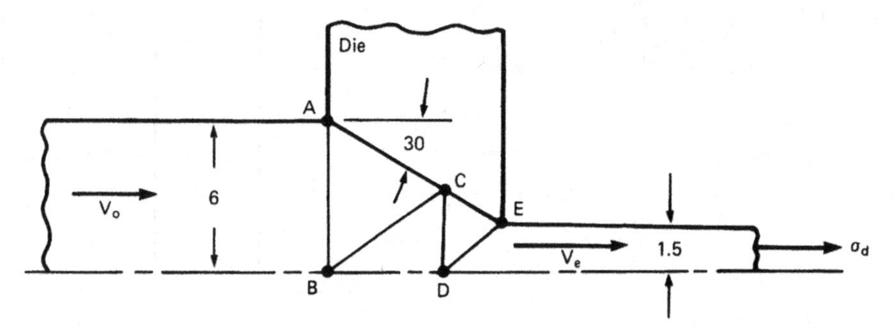

Figure 9.14. Upper-bound field for plane-strain drawing for Problem 9.2.

9.3. On which discontinuity in Figure 9.14 is the largest amount of energy expended?

9.4. Draw the hodograph corresponding to the frictionless indentation illustrated in Figure 9.15.

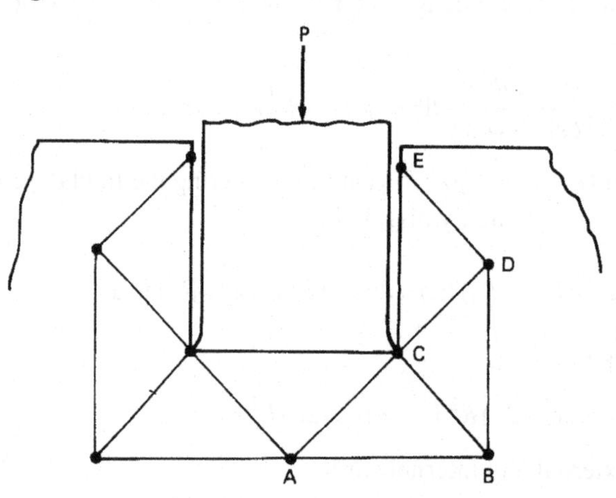

Figure 9.15. Upper-bound field for indentation for Problem 9.4.

9.5. For the plane-strain compression illustrated in Figure 9.16, calculate $P_e/2k$ for L/H values of 1, 2, 3, and 4. Assume sticking friction.

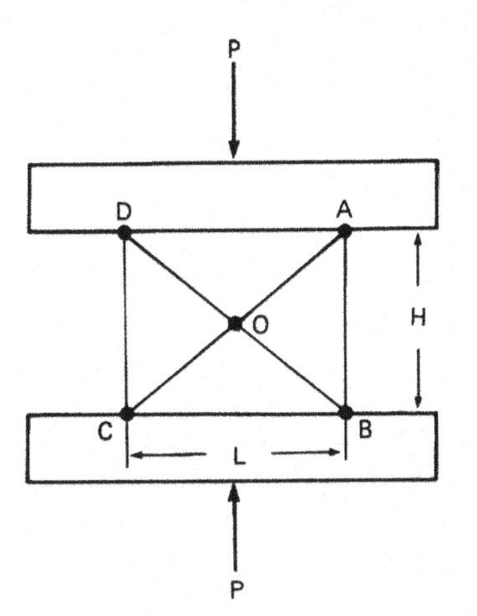

Figure 9.16. Upper-bound field for plane-strain compression for Problem 9.5.

9.6. Reanalyze Problem 9.4 if frictionless conditions prevailed.

9.7. For the indentation shown in Figure 9.7, $P_e/2k = 2.89$ if all the angles were 60°. Find $P_e/2k$ if the angles OAB, ABC, and BCD are 90° and the other angles are 45°.

9.8. Figure 9.17 shows an upper-bound field for a plane-strain extrusion. There are two dead-metal zones ADF and FEG.

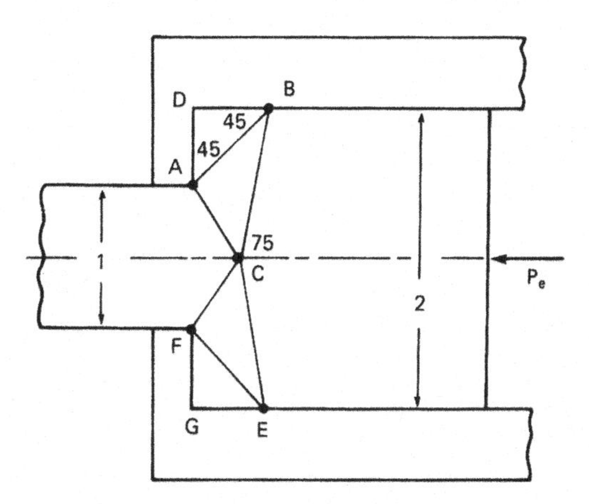

Figure 9.17. Upper-bound field for Problem 9.8.

a) Calculate $P_e/2k$ for the field.
b) Determine the velocity inside triangle ABC.
c) Determine V^*_{AC}.
d) Compute the deformation efficiency.

9.9. **a)** Use equation 9.28 to find the drawing stress, σ_d, for an axisymmetric rod drawing (Figure 9.18) with reduction of 30%, a semi-die angle of 10°, and a constant interfacial shear stress of 0.1 k. Assume the Tresca criterion.
b) Predict σ_d using the von Mises criterion.

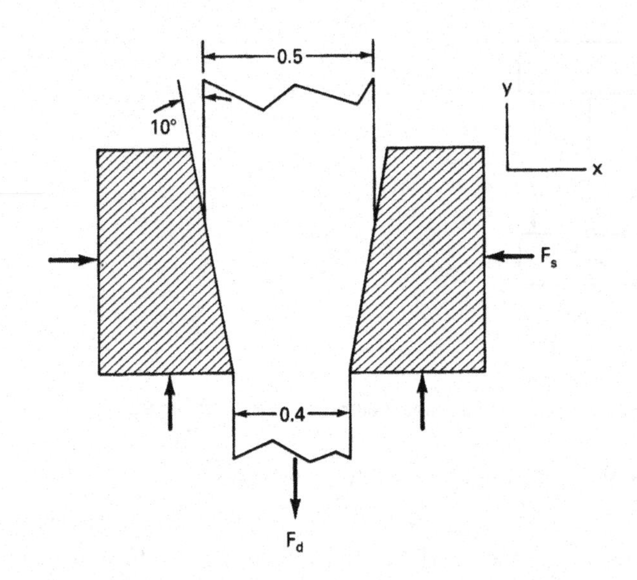

Figure 9.18. Illustration of axisymmetric drawing for Problem 9.9.

9.10. Consider the upper-bound field in Figure 9.19 for an asymmetric extrusion.
a) Draw the corresponding hodograph.
b) Determine the angle, θ.

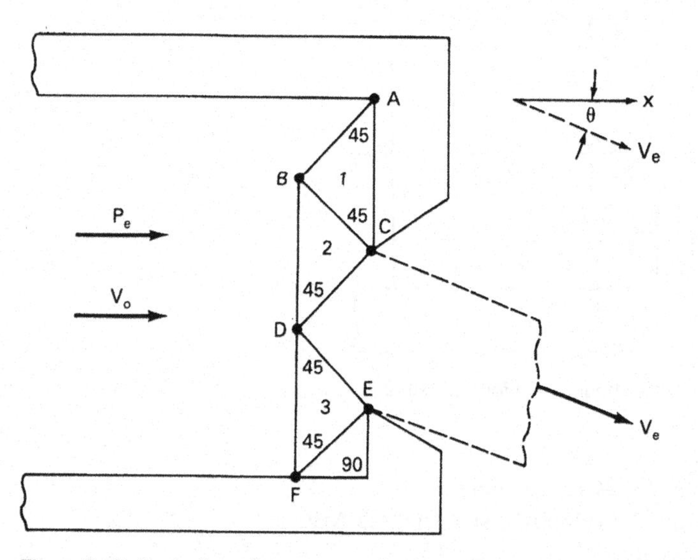

Figure 9.19. Illustration of an asymmetric plane-strain drawing for Problem 9.10.

9.11. For the plane-strain compression illustrated in Figure 9.20, calculate $P_e/2k$ for L/H values of 1, 2, 3, and 4. Assume sticking friction.

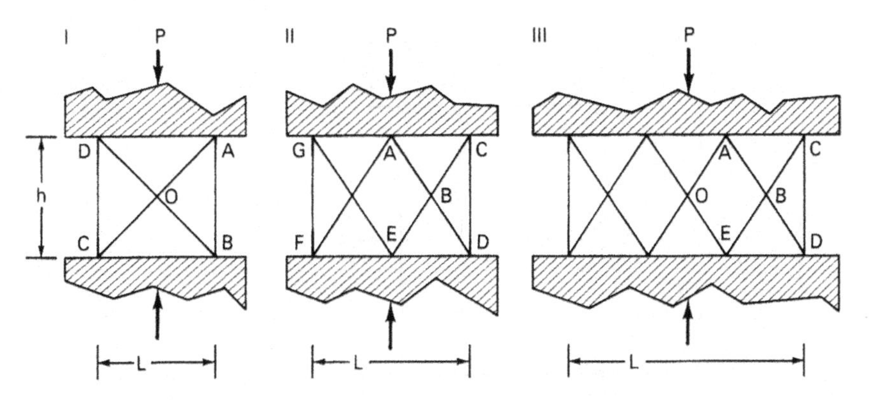

Figure 9.20. Upper-bound fields for plane-strain compression in Problem 9.11.

9.12. For the indentation shown in Figure 9.7, $P_e/2k = 2.89$ if all the angles were 60°. Find $P_e/2k$ if the angles OAB, ABC, and BCD are 90° and the other angles are 45°.

9.13. A proposed upper-bound field for extrusion is shown in Figure 9.21. Draw a hodograph to scale and determine the absolute velocity of particles in the triangle bounded by BCD.

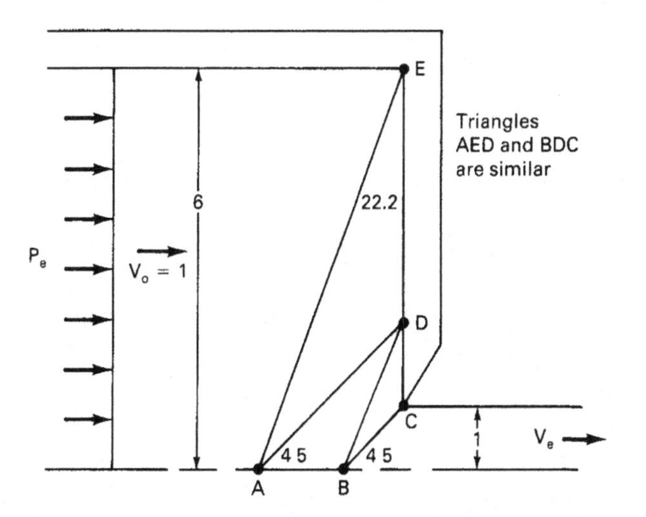

Figure 9.21. Upper-bound field for the plane-strain extrusion of Problem 9.13.

9.14. Figure 9.22 shows an upper-bound field for a plane-strain extrusion. There are two dead-metal zones: ADF and FEG.

a) Calculate $P_e/2k$ for the field.

b) Determine the velocity inside triangle ABC.

c) Determine V^*_{AC}.
d) Compute the deformation efficiency.

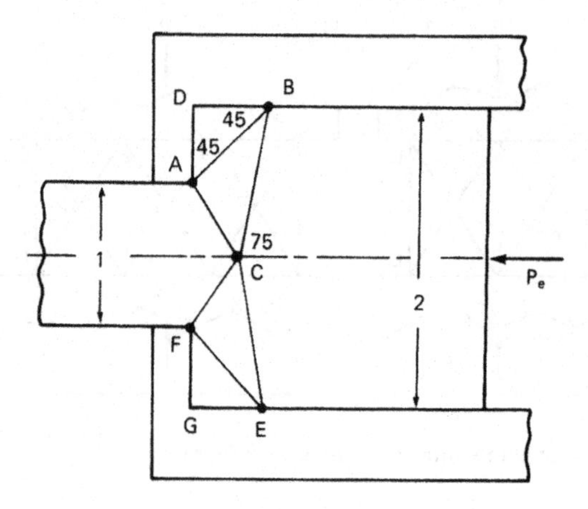

Figure 9.22. A 2:1 extrusion field for Problem 9.14.

9.15. Consider the plane-strain indentation illustrated in Figure 9.23. Assume that the deformation in region AA′B′B is homogeneous. There are discontinuities along AA′ and BB′.

a) Write an expression for $V^*_{AA'}$ and $V^*_{BB'}$ in terms of V_0, z, and t.
b) What is the ratio of the energy expended on these discontinuities to the homogeneous work?

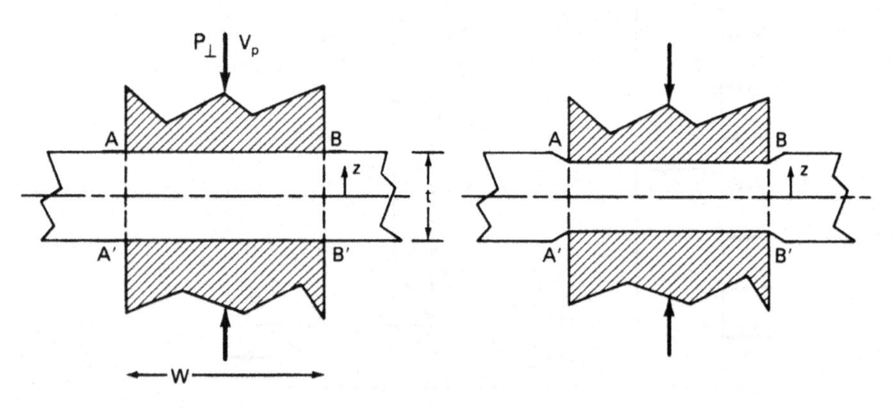

Figure 9.23. Figure for Problem 9.16.

9.16. Figure 9.24 shows two different upper-bound fields for a 2:1 reduction by extrusion. Regions ABC and EFG are dead-metal zones.

a) Calculate $P_e/2k$ for both fields.
b) Determine the deformation efficiency, η, for both cases.
c) What is the absolute velocity of a particle in triangle JGH?

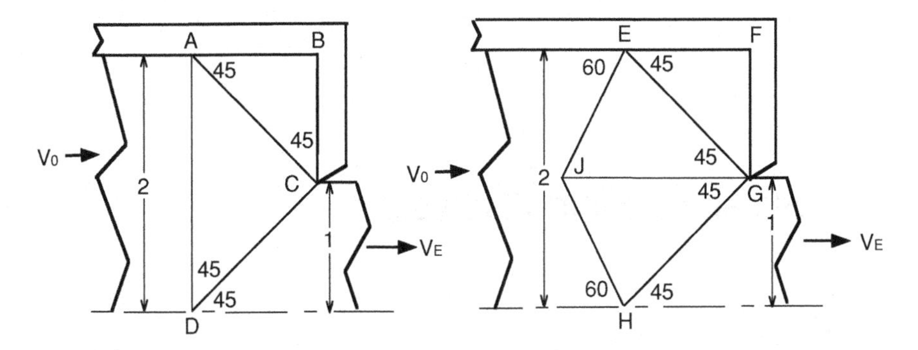

Figure 9.24. Two proposed upper-bound fields for a plane-strain extrusion with a 50% reduction for Problem 9.16.

9.17. In Problem 9.16, either the slip-line field or the upper bound gives a lower solution. However, as discussed in Section 10.12, a pipe may form at the end of an extrusion. Figure 9.25 gives an upper-bound field that leads to pipe formation. Calculate $P_{ext}/2k$ as a function of t for $0.25 \leq t \leq 5$ and compare with the solution to Problem 9.16.

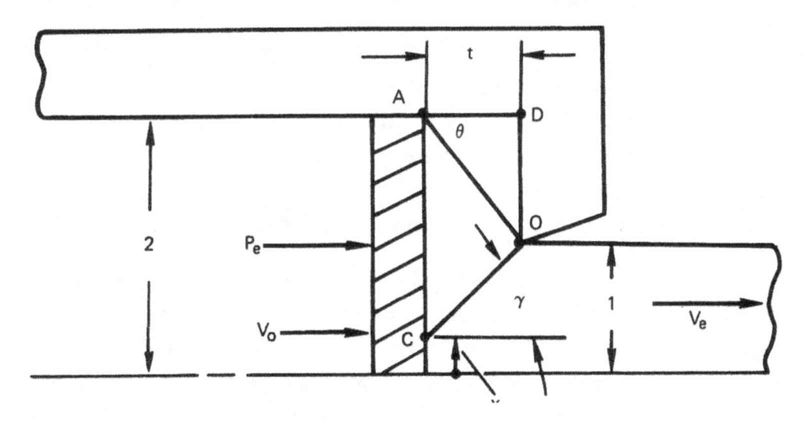

Figure 9.25. An upper-bound field for pipe formation. (Problem 9.17).

10 Slip-Line Field Analysis

10.1 INTRODUCTION

Slip-line field theory is based on analysis of a deformation field that is both geometrically self-consistent and statically admissible. Slip lines* are planes of maximum shear stress and are therefore oriented at 45° to the axes of principal stress. Basic assumptions are:

1. The material is isotropic and homogeneous.
2. The material is rigid-ideally plastic (i.e. no strain hardening).
3. Effects of temperature and strain rate are ignored.
4. Plane-strain deformation.
5. The shear stresses at interfaces are constant, usually frictionless or sticking friction.

Figure 10.1 shows the very simple slip line for indentation where the thickness, t, equals the width of the indenter, b. The maximum shear stress occurs on line DEB and CEA. The material in triangles DAE and CEB is rigid. As the indenters move closer together the field must change. However, for now, we are concerned with calculating the force when the geometry is as shown. The stress, σ_y, must be zero because there is no restrain to lateral movement. The stress, σ_z, must be intermediate between σ_x and σ_y. Figure 10.2 shows the Mohr's circle for this condition. The compressive stress necessary for this indentation, $\sigma_x = -2k$. Few slip-line fields are composed of only straight lines. More complicated fields will be considered.

10.2 GOVERNING STRESS EQUATIONS

With plane-strain, all of the flow is in the x–y plane. This means that $d\varepsilon_y = -d\varepsilon_x$ and $d\varepsilon_z = 0$ so $\sigma_z = \sigma_2 = (\sigma_x + \sigma_y)/2$. Therefore according to the von Mises criterion, σ_z is always the mean or hydrostatic stress.

$$\sigma_2 = (\sigma_1 + \sigma_2 + \sigma_3)/3 = \sigma_{mean} \tag{10.1}$$

* The term *slip lines* used here should not be confused with the microscopic slip lines found on the surface of crystals.

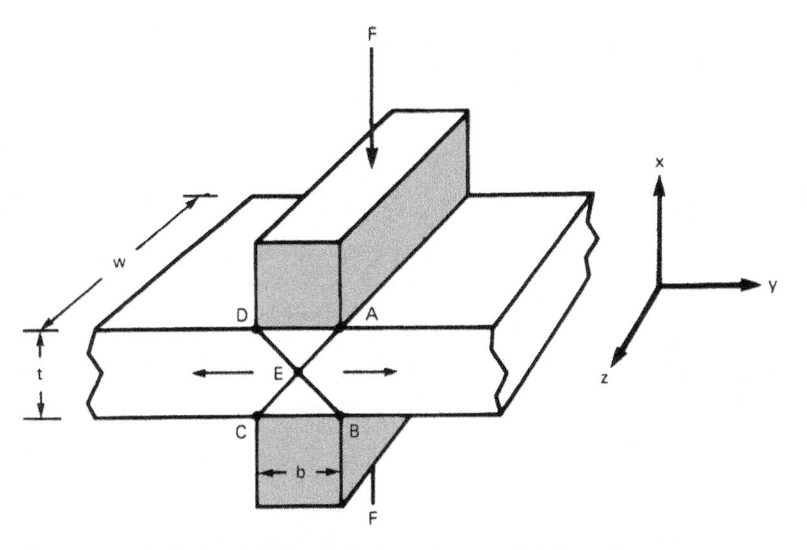

Figure 10.1. A slip-line field for frictionless plane-strain indentation.

and

$$\sigma_1 = \sigma_2 + k, \quad \sigma_3 = \sigma_2 - k. \tag{10.2}$$

Thus plane-strain deformation can be considered as pure shear with a super-imposed hydrostatic stress, σ_2.

Planes of maximum shear stress are mutually perpendicular. The projection of these planes form a series of orthogonal lines called slip lines. Figure 10.3 illustrates a section of a field of slip lines. The shear stress acting on these lines is k, while the mean stress, σ_2, acts perpendicular to the slip lines. The slip lines are rotated at some angle, ϕ, to the x and y axes.

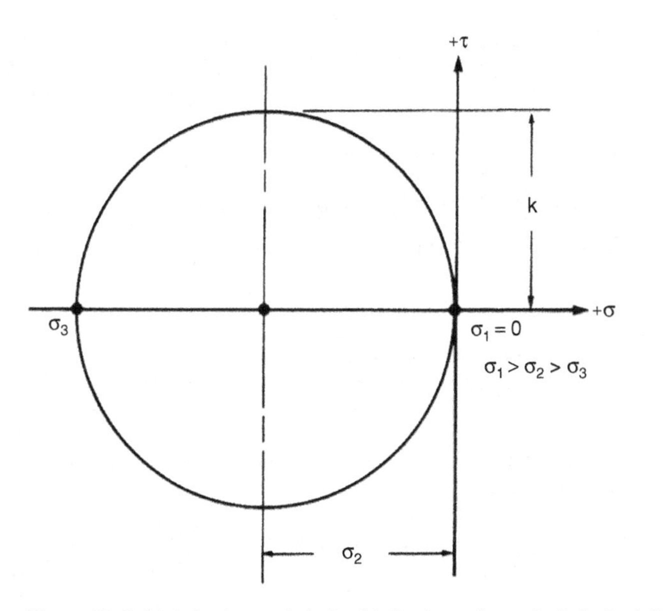

Figure 10.2. Mohr's stress circle for frictionless plane-strain indentation in Figure 10.1.

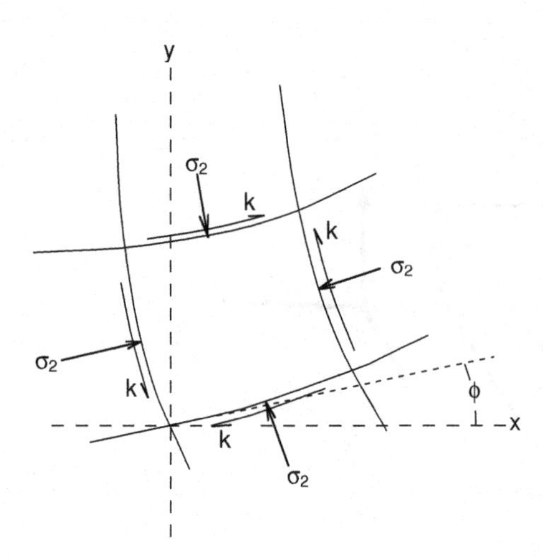

Figure 10.3. Stresses acting on a curvilinear element.

To develop the necessary equations it is necessary to adopt a convention for slip-line identification. The families of slip lines are labeled either α or β. The convention is that the largest principal stress (most tensile) lies in the first quadrant formed by α and β lines as illustrated in Figure 10.4. If all of the stresses are compressive, the least negative is σ_1.

For plane-strain, τ_{xy} and τ_{zx} are zero, so the equilibrium equations (equation 1.40) reduce to

$$\partial\sigma_x/\partial x + \partial\tau_{yx}/\partial y = 0$$

$$\text{and} \quad \partial\sigma_y/\partial y + \partial\tau_{xy}/\partial x = 0. \tag{10.3}$$

From the Mohr's stress circle diagram, Figure 10.5,

$$\sigma_x = \sigma_2 - 2k\sin\phi,$$
$$\sigma_y = \sigma_2 + 2k\sin\phi,$$
$$\tau_{xy} = k\cos\phi. \tag{10.4}$$

Differentiating equations 10.4 and substituting into equations 10.3,

$$\partial\sigma_2/\partial x - 2k\cos 2\phi\,\partial\phi/\partial x - 2k\sin 2\phi\,\partial\phi/\partial y = 0 \quad \text{and}$$
$$\partial\sigma_2/\partial y + 2k\cos 2\phi\,\partial\phi/\partial y - 2k\sin 2\phi\,\partial\phi/\partial x = 0. \tag{10.5}$$

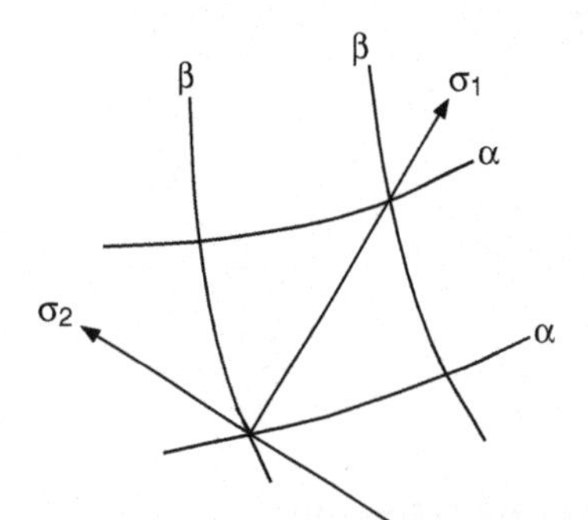

Figure 10.4. The 1-axis lies in the first quadrant formed by the α and β lines.

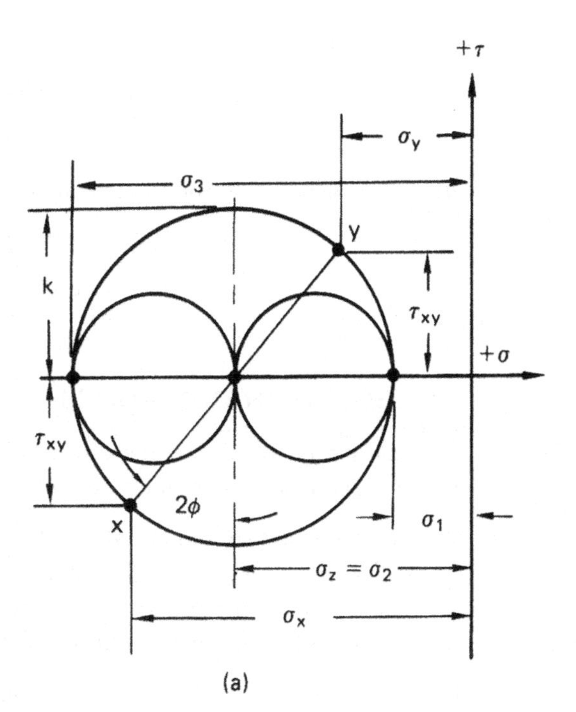

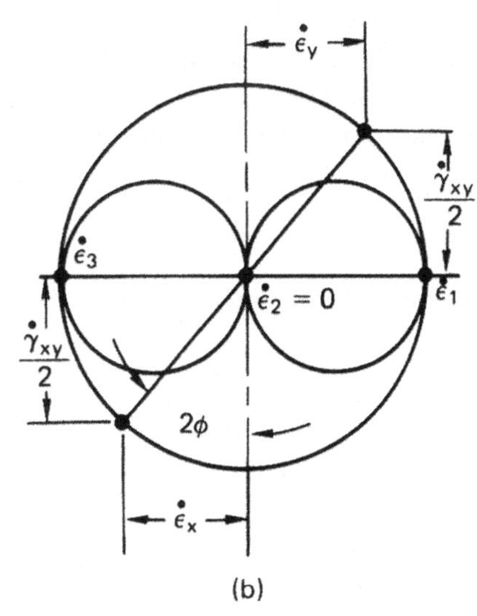

Figure 10.5. (a) Mohr's stress and (b) strain-rate circle for plane-strain.

A set of axes, x' and y' can be oriented so that they are tangent to the α and β lines at the origin. In that case, $\phi = 0$ so equations 10.5 reduce to

$$\partial\sigma_2/\partial x' - 2k\partial\phi/\partial x' = 0$$
$$\text{and} \quad \partial\sigma_2/\partial y' - 2k\partial\phi/\partial y' = 0 \tag{10.6}$$

Integrating,

$$\sigma_2 = 2k\phi + C_1 \text{ along an } \alpha\text{-line}$$
$$\text{and} \quad \sigma_2 = -2k\phi + C_2 \text{ along a } \beta\text{-line.} \tag{10.7}$$

Physically this means that moving along an α-line or β-line causes σ_2 to change by

$$\Delta\sigma_2 = 2k\Delta\phi \text{ along an } \alpha\text{-line}$$
$$\text{and} \quad \Delta\sigma_2 = -2k\Delta\phi \text{ along a } \beta\text{-line.} \tag{10.8}$$

If σ_2 is replaced by $-P$ (pressure) equations 10.8 are written as

$$\Delta P = -2k\Delta\phi \text{ along an } \alpha\text{-line}$$
$$\Delta P = +2k\Delta\phi \text{ along a } \beta\text{-line} \tag{10.9}$$

10.3 BOUNDARY CONDITIONS

One can always determine the direction of one principal stress at a boundary. The following boundary conditions are useful:

1. The force and stress normal to a free surface is a principle stress, so the α- and β-lines must meet the surface at 45°.
2. The α- and β-lines must meet a frictionless surface at 45°.
3. The α- and β-lines meet surfaces of sticking friction at 0 and 90°.

Equations 10.7 establish a restriction on the shape of statically admissible fields. Consider the field in Figure 10.6. The difference between σ_2 at A and C can be found by traversing either of two paths, ABC or ADC. On the path through B, $\sigma_{2B} = \sigma_{2A} - 2k(\phi_B - \phi_A)$ and $\sigma_{2C} = \sigma_{2B} + 2k(\phi_C - \phi_B) = \sigma_{2A} - 2k(2\phi_B - \phi_A - \phi_C)$. On the other hand traversing the path ADC, $\sigma_{2D} = \sigma_{2A} + 2k(\phi_D - \phi_A)$ and

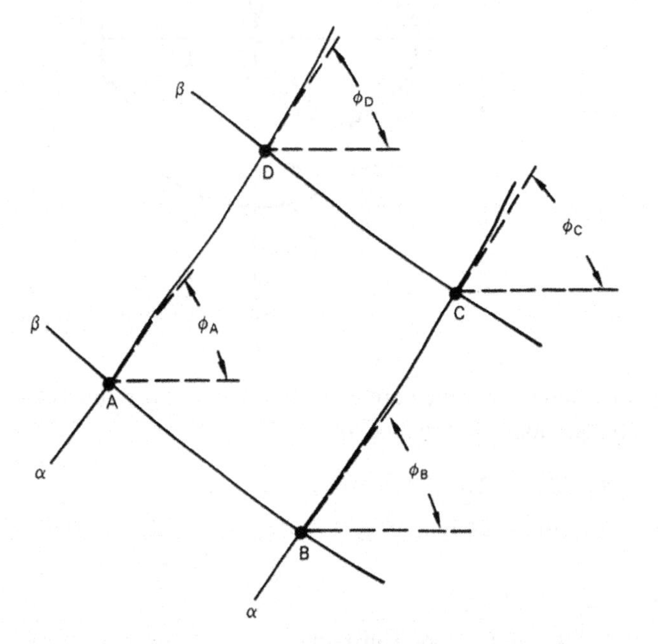

Figure 10.6. Two pairs of α- and β-lines for analyzing the change in mean normal stress by traversing two different paths.

$\sigma_{2C} = \sigma_{2D} - 2k(\phi_C - \phi_D) = \sigma_{2A} - 2k(2\phi_d - \phi_A - \phi_C)$. Comparing these two paths,

$$\phi_A - \phi_B = \phi_D - \phi_C$$
$$\text{and} \quad \phi_A - \phi_D = \phi_B - \phi_C. \tag{10.10}$$

Equation 10.10 implies that the net of α- and β-lines must be such that the change of ϕ is the same along a family of lines moving from one intersection with the opposite family to the next intersection. This together with the orthogonality requirement indicates that it is the angular change along a line rather than the length of line that is of significance.

10.4 PLANE-STRAIN INDENTATION

There are two simple fields that meet these requirements. One is a set of straight lines and the other is a centered fan (Figure 10.7). σ_2 is the same everywhere in the field of straight lines. It is a constant pressure zone. In the centered-fan field, σ_2 is the same everywhere along a given radius but varies from one radius to another.

A number of problems can be solved with these two fields. Consider plane-strain indentation. A possible field consisting of two centered fans and a constant pressure zone is shown in Figure 10.8. The α- and β-lines can be identified by realizing that

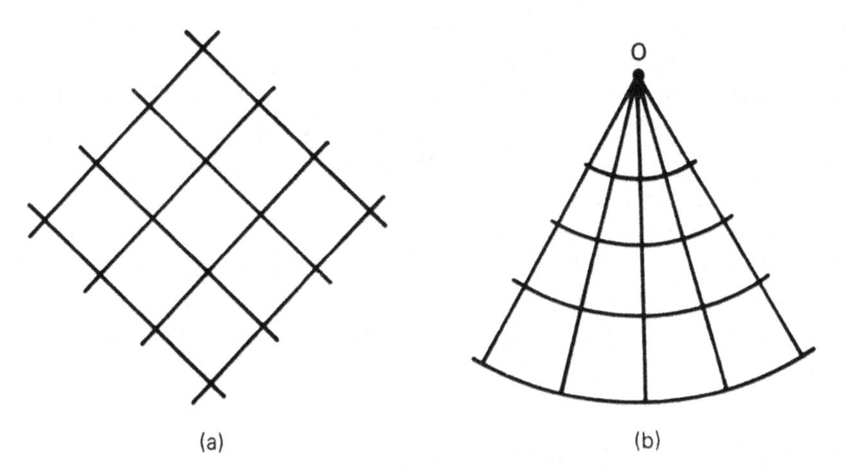

Figure 10.7. (a) Net of straight lines (b) Centered fan.

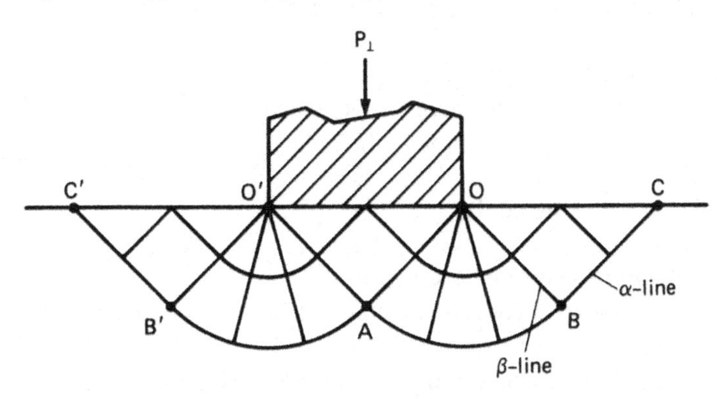

Figure 10.8. A possible slip-line field for plane-strain indentation.

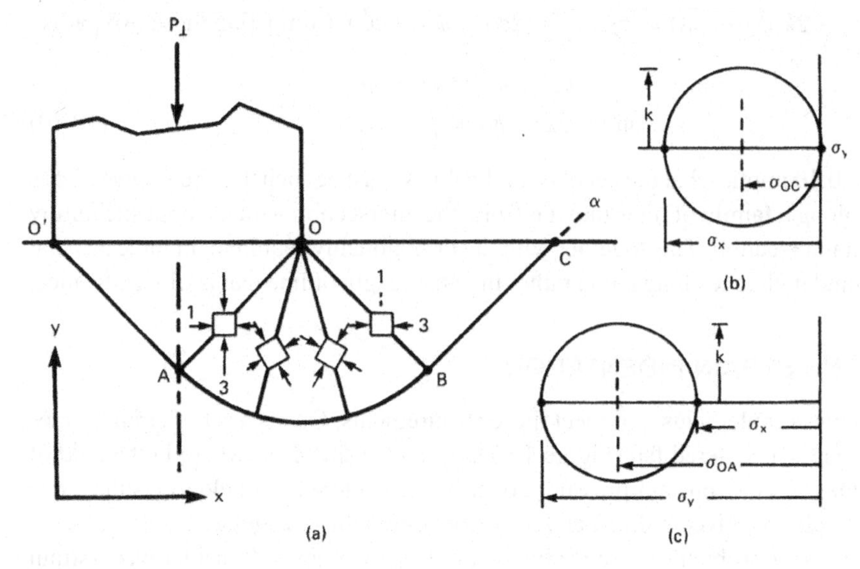

Figure 10.9. A detailed view of Figure 10.8 showing the changing stress state and the Mohr's stress circles for triangles OBC and O′OA.

parallel to OC the stress is compressive and the stress normal to it is zero. Or alternatively, that under the indenter the most compressive stress is parallel to $P_\perp$.

A more detailed illustration of the field is given in Figure 10.9. Along OC, $\sigma_y = \sigma_1 = 0, \sigma_2 = -k, \sigma_x = \sigma_3 = -2k$. Rotating clockwise along CBAO′ on the α line through $\Delta\phi = -\pi/2, \sigma_{2OO'} = \sigma_{2OC} + 2k\Delta\phi_\alpha = -k + 2k(-\pi/2)$, $P_\perp/2k = P_O/2k = -\sigma_{2OO}/2k = 1 + \pi/2 = 2.57$.

With the von Mises criterion, $2k = 1.155Y$, so $P_\perp = 2.97Y$. This plane-strain indentation is analogous to a two-dimensional hardness test so is a hardness. It is a frequently used rule of thumb that with consistent units the hardness is 3 times the yield strength. The pressure is constant but different in regions OBC and in O′OA. Although the metal is stressed to its yield stress in these regions, they do not deform.

10.5 HODOGRAPHS FOR SLIP-LINE FIELDS

Construction of hodographs for slip-line fields is necessary to:

1. Assure the field is kinematically admissible.
2. Determine where in the field most of the energy is expended.
3. Predict distortion of material as it passes through the field.

In constructing hodographs it may be noted that:

1. The velocity is constant within a constant pressure zone.
2. In leaving a field of changing σ_2 there may or may not be a sudden change of velocity.
3. The magnitude of the velocity everywhere along a given slip line is constant though the direction may change.
4. In a field of curved lines, both the magnitude and direction of the velocity change.

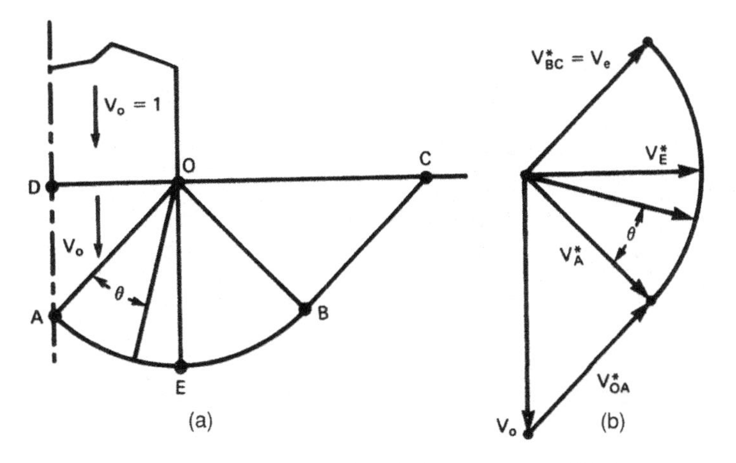

Figure 10.10. (a) A partial slip-line field for indentation and (b) the corresponding hodograph.

5. There is always shear on the boundary between the deforming material inside the field and the rigid material outside of it.
6. The vector representing a velocity discontinuity must be parallel to the discontinuity itself.

Figure 10.10 shows half of the field in Figure 10.8 and the corresponding hodograph. Region OAD moves downward with the velocity, V_0, of the punch. There is a discontinuity, V^*_{OA} along OA such that the absolute velocity is parallel to the arc AE at A, and the velocity just to the right of OA differs from that in triangle OAD by a vector parallel to OA. The discontinuity, V^*_A between the material in the field at A and outside the field is equal to the absolute velocity inside the field at A. This discontinuity between material inside and outside the field has a constant magnitude but changing direction along the arc AEB. There is no abrupt velocity discontinuity along OB.

In this field, there is intense shear along OA (V^*_{OA}), along AEB ($V^*_A = V^*_E = V^*_B$) and BC ($V^*_{BC} = V^*_A$). There is also energy dissipated by the gradual deformation in the fan OAB.

10.6 PLANE-STRAIN EXTRUSION

Consider again the frictionless plane-stain extrusion treated by upper bounds in Section 10.3 where $r = 50\%$ and $\alpha = 30°$. Figure 10.11(a) is the top half of the slip-line field and Figures 10.11(b) and (c) are Mohr's stress circle diagrams along OB and OC. The force balance on the die wall is shown in Figure 10.11(d).

At the exit the stress $\sigma_1 = \sigma_x = 0$ and the stress σ_y is compressive, so line OC is a β-line. On OC, $\sigma_{2OC} = -k$. Rotating clockwise through $\Delta\phi_\alpha = -\pi/6$ on an α line, $\sigma_{2OB} = -k + 2k(-\pi/6)$. In triangle ABO, $\sigma_{2ABO} = -k + 2k(-\pi/6)$ so $P_{ABO} = k + 2k(\pi/6)$. Acting against the die wall, $P_\perp = P_{ABO} + k = 2k(1 + \pi/6)$. $F_\perp = P_\perp(\overline{OA}) = P_\perp r/\sin\alpha = P_\perp$. $P_\perp F_x = F_\perp \sin\alpha = P_e(1/2)$ so $P_{ext}/2k = (0.5)(1 + \pi/6) = 0.762$.

The example above is a special case where the geometry is such that the slip-line field consists of a constant pressure zone and a single centered fan. Figure 10.12 shows

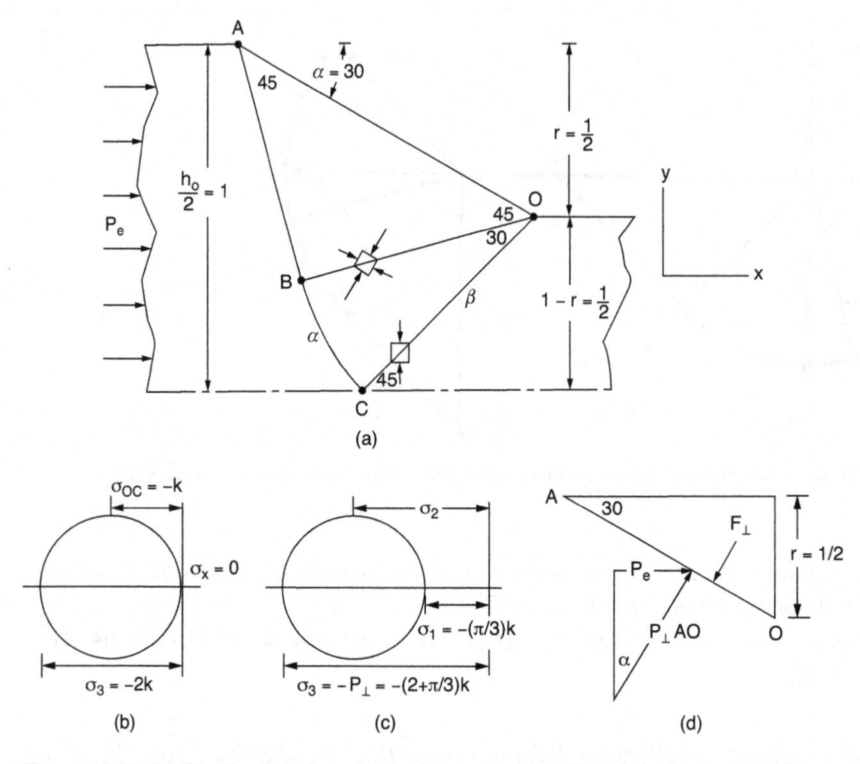

Figure 10.11. (a) Slip-line field for a frictionless extrusion, (b) Mohr's stress circle diagram along OB, (c) Mohr's stress circle diagram along OC and (d) force balance on die wall.

such a field. If the entrance thickness, $h_0 = 1$, the exit thickness is $1 - r$. In that case $\sin\alpha = r/2(1 - r)$ or

$$r = 2\sin\alpha/(1 + 2\sin\alpha). \tag{10.11}$$

Following the procedure for the 30° die,

$$P_{\text{ext}}/2k = r(1 + \alpha). \tag{10.12}$$

when $r = 2\sin\alpha/(1 + 2\sin\alpha)$.

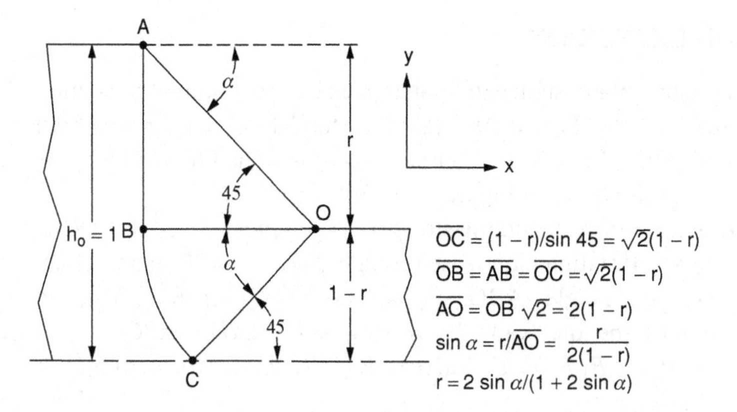

Figure 10.12. The geometry of a general field for plane-strain extrusion when $r = \sin\alpha/(1 + 2\sin\alpha)$.

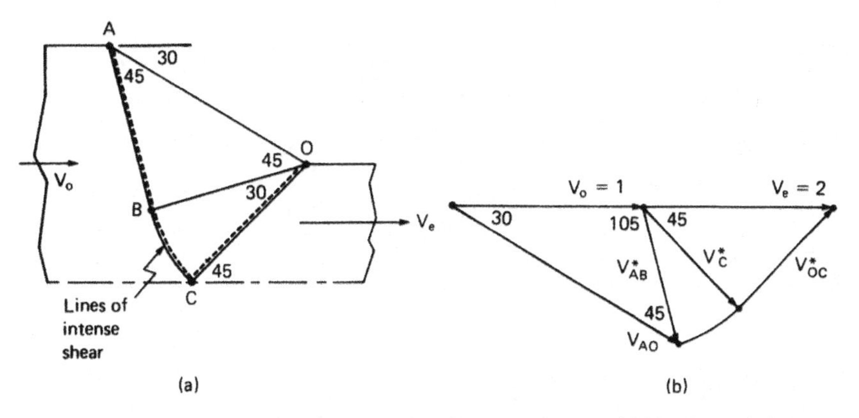

Figure 10.13. (a) Slip-line field showing lines of intense shear and (b) hodograph for an extrusion.

The homogeneous work for the reduction is $w_i = \overline{\sigma\varepsilon} = 2k\ln[1/(1-r)$. Since $w_a = P_{\text{ext}}$, the efficiency predicted by these fields

$$\eta = w_i/w_a = \ln[1/(1-r)]/[r(1+\alpha)]. \qquad (10.13)$$

If the reductions in this section had been made by drawing instead of extrusion the results would have been the same. To analyze drawing, the exit stress would be σ_{d} instead of 0, so $\sigma_{2\text{OC}} = k + \sigma_{\text{d}}$, $\sigma_{2\text{OAC}} = k(\pi/6) + k + \sigma_{\text{d}}$, $P_{\perp} == k(\pi/6) + 2k + \sigma_{\text{d}}$. $F_x = F_{\perp}\sin\alpha = rP_{\perp} = r[k(\pi/6) + 2k + \sigma_{\text{d}}]$ and $F_x = (1-r)\sigma_{\text{d}}$. Equating $\sigma_{\text{d}}/2k = r(1+\alpha)$ which is identical to equation 10.12.

10.7 ENERGY DISSIPATION IN A SLIP-LINE FIELD

Consider the plane-strain extrusion in Figure 10.12. There are velocity discontinuities along AB, BC and CO (Figure 10.13a). The hodograph (Figure 10.13b) shows that if $V_0 = 1$, $V^*_{\text{AB}} = V^*_{\text{BC}} = V^*_{\text{CO}} = 1/\sqrt{2}$. The lengths are $\overline{AB} = \overline{OC} = 1/\sqrt{2}$ and $\overline{BC} = (\pi/6)/\sqrt{2})$. So the energy dissipation along these lines is $k(\overline{AB}\,V^*_{\text{AB}} + \overline{BC}\,V^*_{\text{BC}} + \overline{OC}\,V^*_{\text{OC}}) = 1.262k$. The extrusion pressure was earlier determined as $P_{\text{ext}}/2k = 0.762$, so $P_{\text{ext}} = 1.542k$. Thus $1.262/1.542 = 81\%$ of the energy is expended along these lines of discontinuity. The other 17% is expended in the fan OBC.

10.8 METAL DISTORTION

The distortion of the metal in a steady state process can be determined from a slip-line field and its hodograph. As an example, consider the 2:1 extrusion through a 90° die illustrated in Figure 10.14a.

Figure 10.14b is the hodograph. The triangle to the right of AO is a dead-metal zone metal. A metal entering the field at A suffers a velocity discontinuity V^*_{A} parallel to the arc at A. A metal entering the field at C suffers a velocity discontinuity V^*_{C} parallel to the arc at C. All of the velocity discontinuities along the arc have the same magnitude and are parallel to the arc. There is also a velocity discontinuity, V^*_{GO} parallel to GO and of magnitude such that V_e is horizontal.

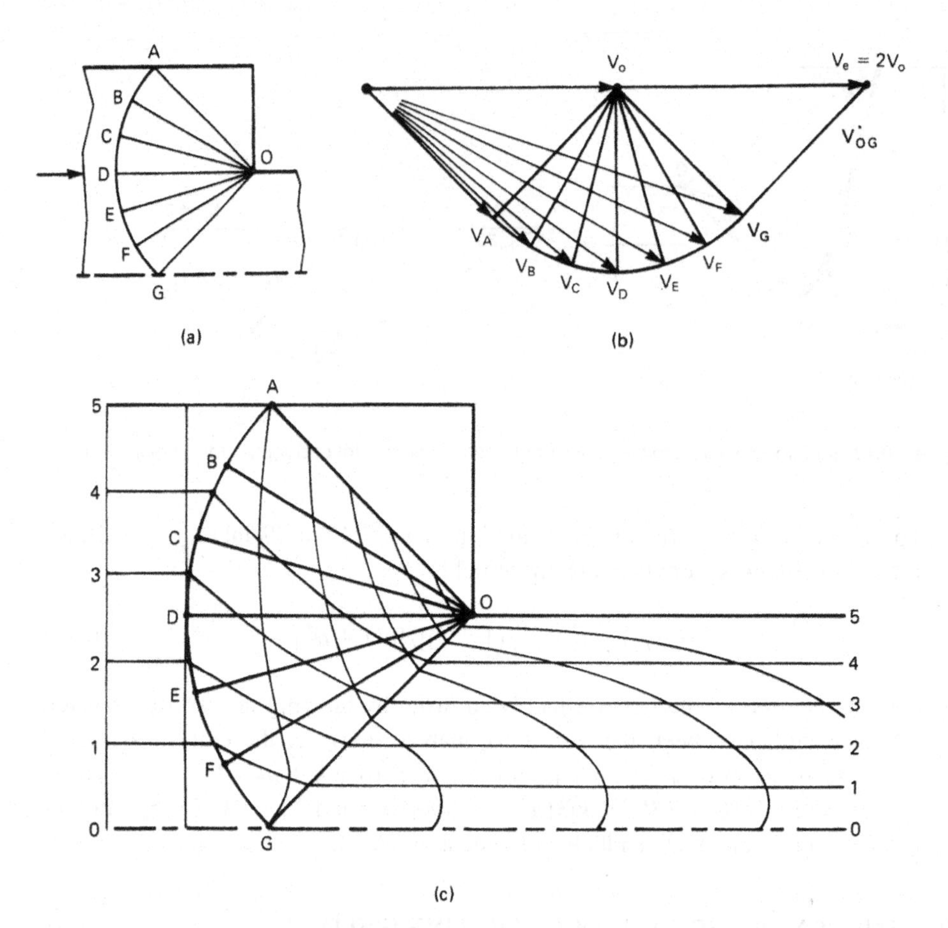

Figure 10.14. (a) Slip-line field for a 2:1 extrusion through a 90° die, (b) the hodograph for the field and (c) the predicted distortion of the metal.

Stream lines can be drawn for particles (Figure 10.14c). Consider a particle on line 3, entering the field between C and D. As it enters the field it acquires an absolute velocity midway between V_C and V_D. Its direction gradually changes as it moves through the field. Its velocity equals V_D as it crosses OD, equals V_E as it crosses OE and equals V_F as it crosses OF. When it crosses OG its velocity must become horizontal. If this construction is made correctly, it emerges on a line 3/5 of the initial thickness.

The distortion of the metal is found by considering the velocity magnitudes at each point along the path. For each time increment, Δt, the distance traveled, $\Delta s = VDt$. By following particles that are initially on a vertical grid line, the distortion of that grid line is established. The greatest distortion occurs at the surface because the velocity is least there.

10.9 INDENTATION OF THICK SLABS

The plane-strain indentation of a thick slab by two opposing indenters is shown in Figure 10.15. The simple slip-line field in Figure 10.16 is appropriate for a special

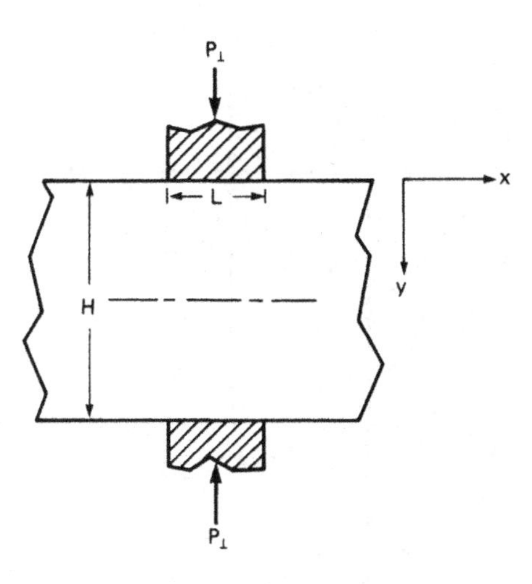

Figure 10.15. Plane-strain indentation of a thick slab by two opposing indenters.

case where the slab thickness, H, equals the indenter width, L. Along AO, $\sigma_x = 0$ so $\sigma_2 = -k$. The stress is the same everywhere in triangle O′OA so along OO′, $\sigma_2 = -k$, $P_\perp = 2k$.

$$P_\perp/2k = 1. \qquad (10.14)$$

A different field must be used for larger values of H/L. Figure 10.17 shows the field for $H/L = 5.43$. This is a field determined by two centered fans. In triangle O′OA, $\sigma_y = -P_\perp$, $\sigma_{2(\mathrm{OA})} = \sigma_y = 2 + k = -P_\perp + k$. Moving along an α-line to (0,1) $\sigma_{2(\mathrm{O},1)} = -\sigma_{2(\mathrm{OA})} + 2k\Delta\phi_\alpha$ and moving back along a β-line to (1,1), $\sigma_{2(1,1)} = -\sigma_{2(\mathrm{OA})} + 2k(\Delta\phi_\alpha - \Delta\phi_\beta)$. At 1,1, $\sigma_{x(1,1)} = -P_\perp + 2k + 2k(\Delta\phi_\alpha - \Delta\phi_\beta)$ and at every point $(n,\ n)$ along the centerline $\sigma_{x(n,n)} = -P_\perp + 2k + 2k(\Delta\phi_\alpha - \Delta\phi_\beta)_n$. Since $\Delta\phi_\alpha = -\Delta\phi_\beta$,

$$\sigma_{x(n,n)} = -P_\perp + 2k + 2k\Delta\phi_n, \qquad (10.15)$$

where $\Delta\phi_n$ is the absolute value of the rotations.

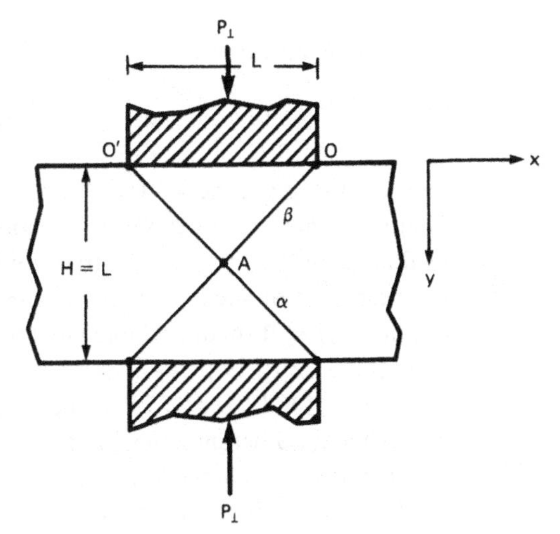

Figure 10.16. Special case of plane-strain indentation where the slab thickness H, equals the indenter width, L.

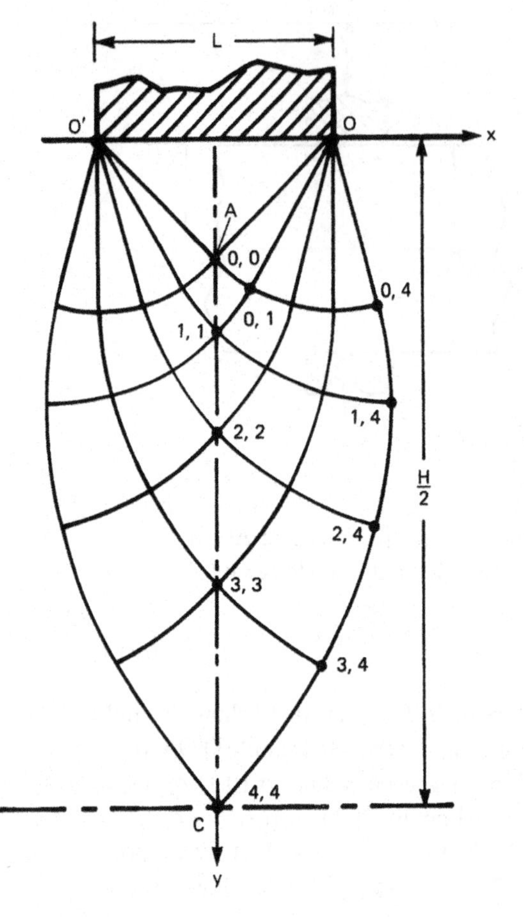

Figure 10.17. Slip-line field for mutual indentation with $H/L = 5.43$.

Because there is no x-direction constraint,

$$F_x = 0 = \int_0^{H/2} \sigma_x \, dy. \tag{10.16}$$

Substituting equation 10.15, $\int_0^{H/2} [-P_\perp = 2k(1 + 2\Delta\phi)] \, dy = 0$ or

$$P_\perp = 2k + (4k/H) \int_0^{H/2} 2\Delta\phi \, dy. \tag{10.17}$$

Figure 10.18 gives the values of $\Delta\phi$ as a function of y for nodal points on a 15° net. There is a more detailed net in the appendix. The integration in equation 10.17 can be done numerically using the trapezoidal rule or graphically by plotting $2\Delta\phi$ versus y. The results of such calculations are summarized in a plot of $P_\perp/2k$ versus H/L (Figure 10.19). It should be noted that for $H/L > 8.75$, $P_\perp/2k$ exceeds $1 + \pi/2$ so the field in Figure 10.19 gives a lower value of $P_\perp/2$. Nonpenetrating indentation should be expected for $H/L > 8.75$ and penetrating deformation for $H/L < 8.75$. A corollary is that for valid hardness testing the thickness of the material should be 4 to 5 times as thick as the diameter of the indenter. (Theoretically $H/L < 8.75/2 = 4.37$ for an indentation on a frictionless substrate).

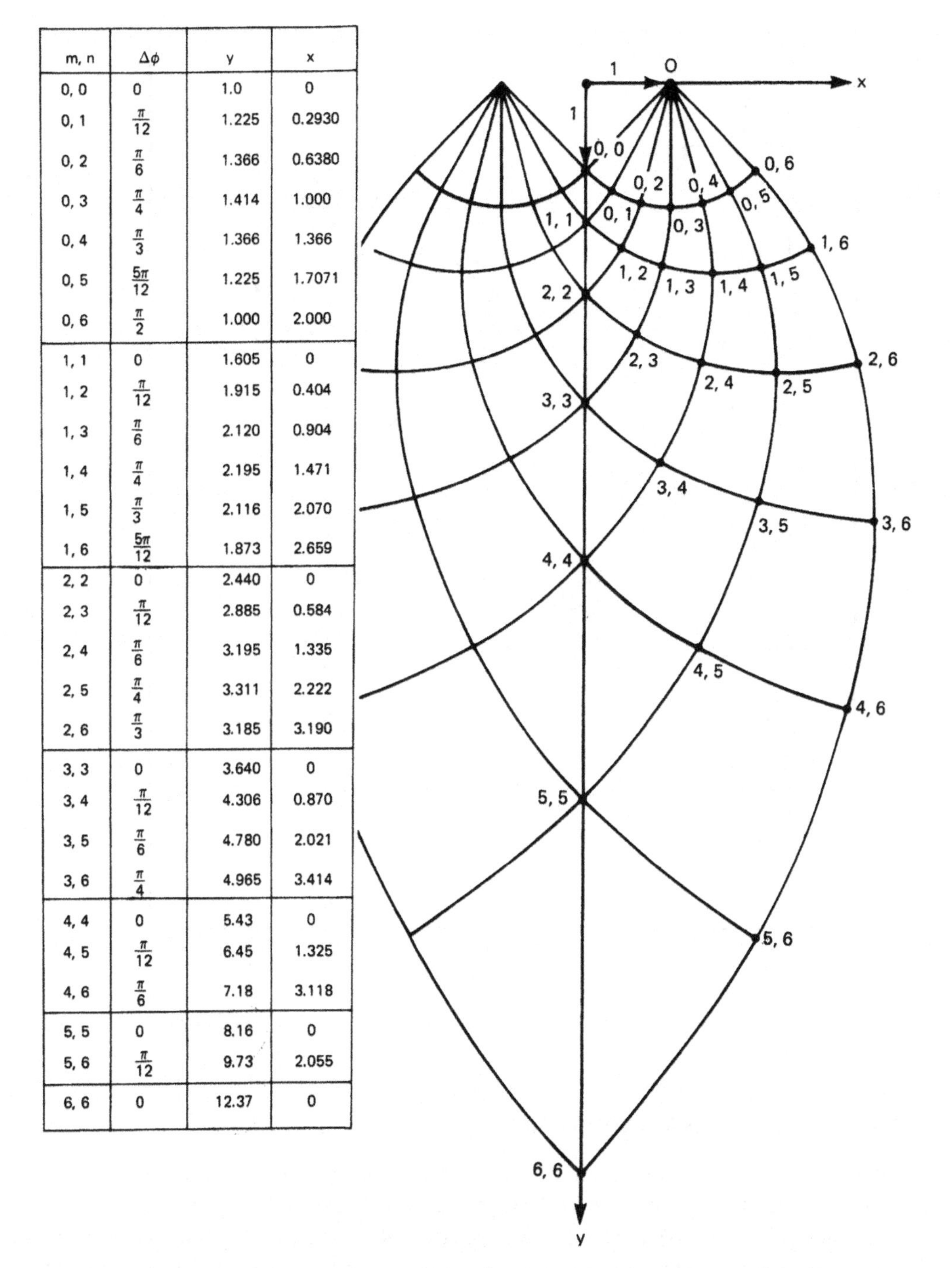

m, n	$\Delta\phi$	y	x
0, 0	0	1.0	0
0, 1	$\frac{\pi}{12}$	1.225	0.2930
0, 2	$\frac{\pi}{6}$	1.366	0.6380
0, 3	$\frac{\pi}{4}$	1.414	1.000
0, 4	$\frac{\pi}{3}$	1.366	1.366
0, 5	$\frac{5\pi}{12}$	1.225	1.7071
0, 6	$\frac{\pi}{2}$	1.000	2.000
1, 1	0	1.605	0
1, 2	$\frac{\pi}{12}$	1.915	0.404
1, 3	$\frac{\pi}{6}$	2.120	0.904
1, 4	$\frac{\pi}{4}$	2.195	1.471
1, 5	$\frac{\pi}{3}$	2.116	2.070
1, 6	$\frac{5\pi}{12}$	1.873	2.659
2, 2	0	2.440	0
2, 3	$\frac{\pi}{12}$	2.885	0.584
2, 4	$\frac{\pi}{6}$	3.195	1.335
2, 5	$\frac{\pi}{4}$	3.311	2.222
2, 6	$\frac{\pi}{3}$	3.185	3.190
3, 3	0	3.640	0
3, 4	$\frac{\pi}{12}$	4.306	0.870
3, 5	$\frac{\pi}{6}$	4.780	2.021
3, 6	$\frac{\pi}{4}$	4.965	3.414
4, 4	0	5.43	0
4, 5	$\frac{\pi}{12}$	6.45	1.325
4, 6	$\frac{\pi}{6}$	7.18	3.118
5, 5	0	8.16	0
5, 6	$\frac{\pi}{12}$	9.73	2.055
6, 6	0	12.37	0

Figure 10.18. Slip-line field for two centered fans with x and y values for each node of the 15 net.

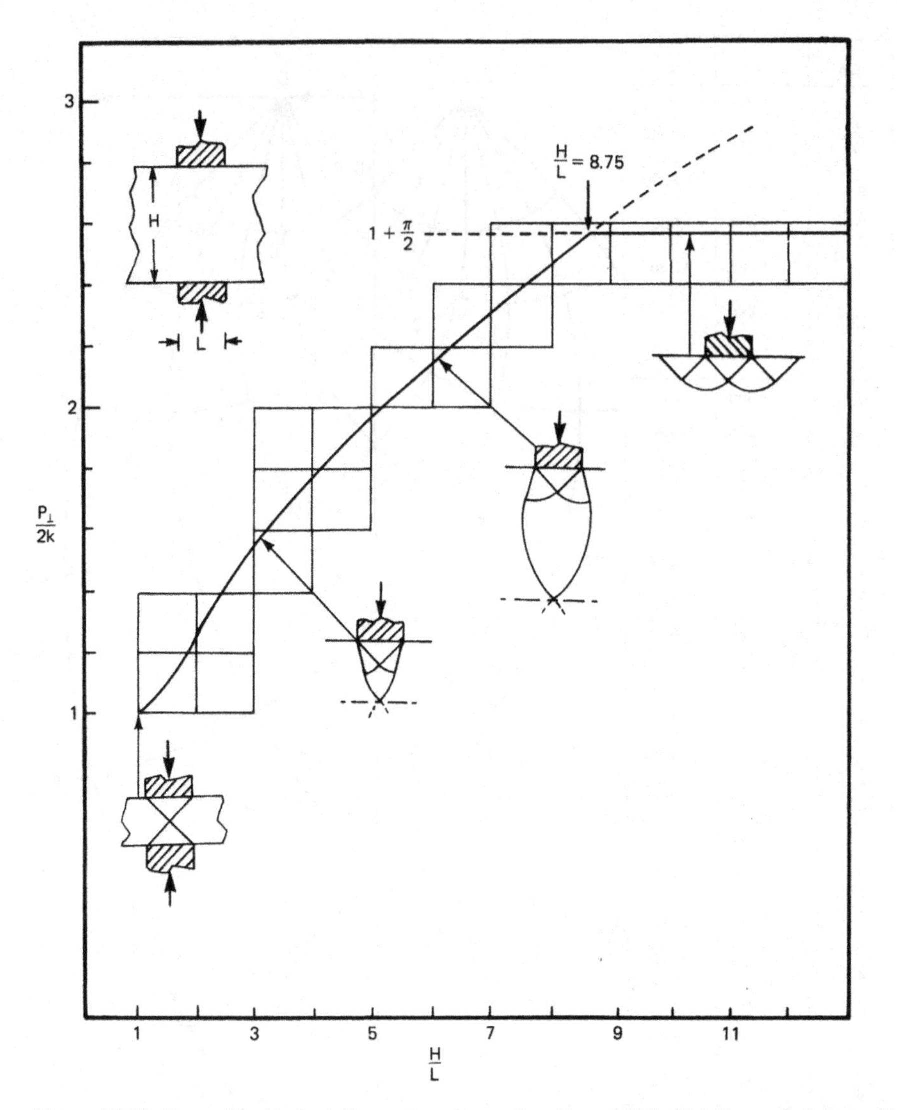

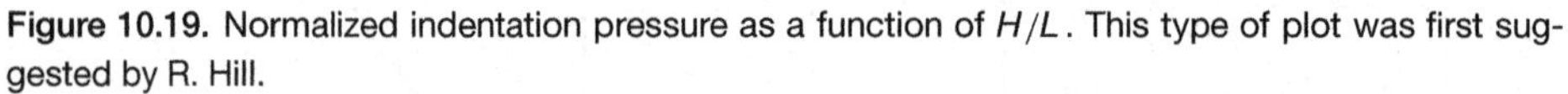

Figure 10.19. Normalized indentation pressure as a function of H/L. This type of plot was first suggested by R. Hill.

10.10 PLANE-STRAIN DRAWING

This type of field with two centered fans is useful in analyzing plane-strain drawing or extrusion operations with larger H/L values than are described by equation 10.11 (i.e., $r < 2 \sin a/(1 + 2 \sin a)$). Figure 10.20 shows the field in Figure 10.17, tilted by the die angle, α and bounded by α- and β-lines that meet the centerline at 45. The appropriate boundary condition for finding σ_{d} is $F_x = \int \sigma_x \, \mathrm{d}y + \int \tau_{xy} \, \mathrm{d}x$. Everywhere along the cut A, (0,1), $(n, n+1)$, $\sigma_x = \sigma_1$ and $\tau_{xy} = 0$, so $F_x = \int \sigma_x \, \mathrm{d}y$. The values of the nodes, A. (0,1),... $(n, n+1)$ are now designated as x' and y' and these are transformed into coordinates by

$$
\begin{aligned}
x &= (x' - 1)\cos\alpha - y'\sin\alpha \\
y &= (x' - 1)\sin\alpha + y'\cos\alpha.
\end{aligned} \tag{10.18}
$$

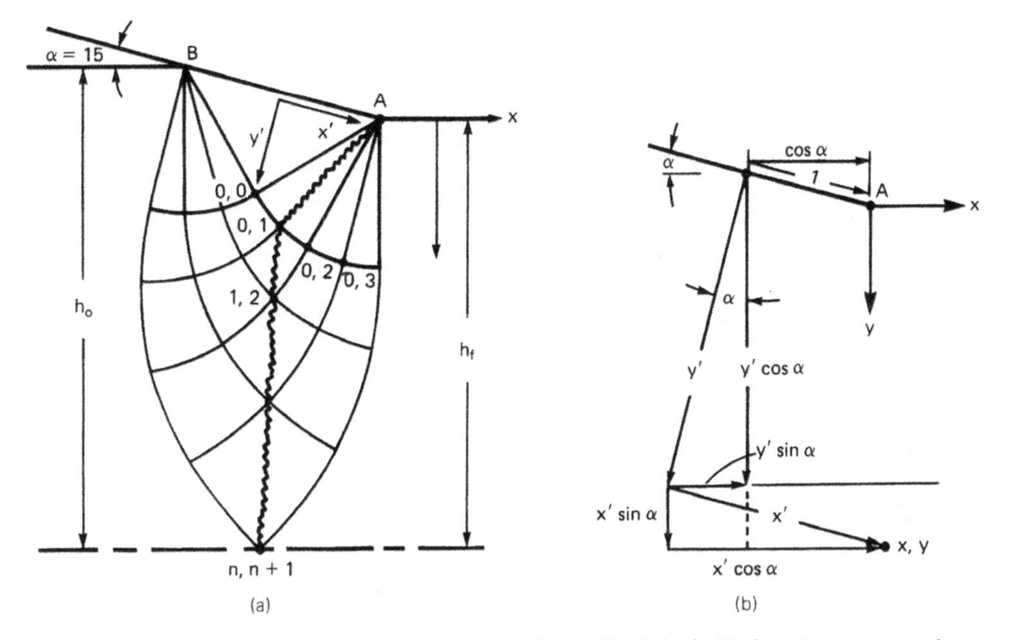

Figure 10.20. (a) Slip-line field for drawing with $r < 2\sin\alpha/(1+2\sin\alpha)$. (b) Coordinate system for analysis along the cut A, (0, 1), $(n, n+1)$.

The stress state at any node point, $(n,n+1)$, can then be found in terms of $P_\perp$. In triangle AOB, and along the line AO, $P_{AB} = P_\perp - k$. Rotating on an α-line through $\Delta\phi = +\pi/12$ to (0,1), gives $P_{(0,1)} = P_{OA} - 2k(\pi/12) = P_\perp - k - 2k(\pi/12)$. Movement to any point $(n, n+1)$ requires an additional rotation of $n(\pi/12)$ on an α-line and $-n(\pi/12)$ on a β-line so $P_{(n,n+1)} = P_\perp - k - 2k(2n+1)\pi/12$ and

$$\sigma_{x(n,n+1)} = P_\perp - k - 2k[1 + (2n+1)\pi/12]. \tag{10.19}$$

Applying the boundary condition, $F_x = \int_0^{h_f} \sigma_x \, dy = 0$, where h_f is the value of y at the centerline,

$$F_x = \int_0^{h_f} (-P_\perp + 2k[1 + (2n+1)\pi/12] \, dy = 0. \tag{10.20}$$

Since $P_\perp$ and k are independent of y, equation 10.20 can be written as

$$P_\perp/2k = \frac{\pi}{12h_f}\int_0^{h_f} (2n+1)\, dy. \tag{10.21}$$

Equation 10.21 may be evaluated either graphically or numerically for any value of h_f. Once $P_\perp$ has been evaluated, a force balance can be used to find P_{ext}.

$$P_{ext} = \frac{2P_\perp \sin\alpha}{h_f + 2\sin\alpha}. \tag{10.22}$$

The mechanical efficiency, η, for such frictionless extrusion or drawing may be found by comparing $P_{ext}/2k$ with the homogeneous work. Figure 10.21 shows how calculated values of η and $P_{ext}/2k$ vary with reduction, r, and H/L for a 15 die.

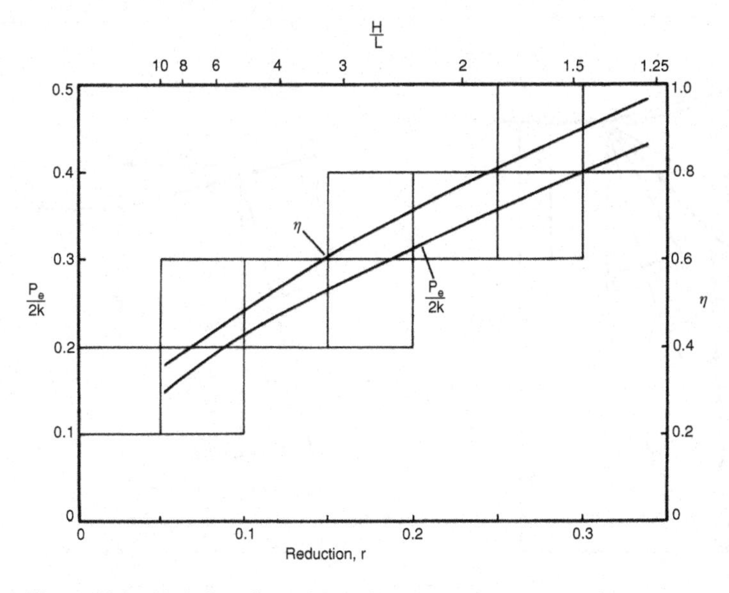

Figure 10.21. Variation of η and $P_{ext}/2k$ with r, and H/L for a 15 die.

With very low values r and high die angles deformation may not be penetrating. The alternate field shown in Figure 10.22 requires less work. This is analogous to a hardness test with a wedge-shaped indenter. This field becomes appropriate when the die pressure reaches a level

$$P_{\perp}/2k = (1 + \pi/2 - \alpha). \tag{10.23}$$

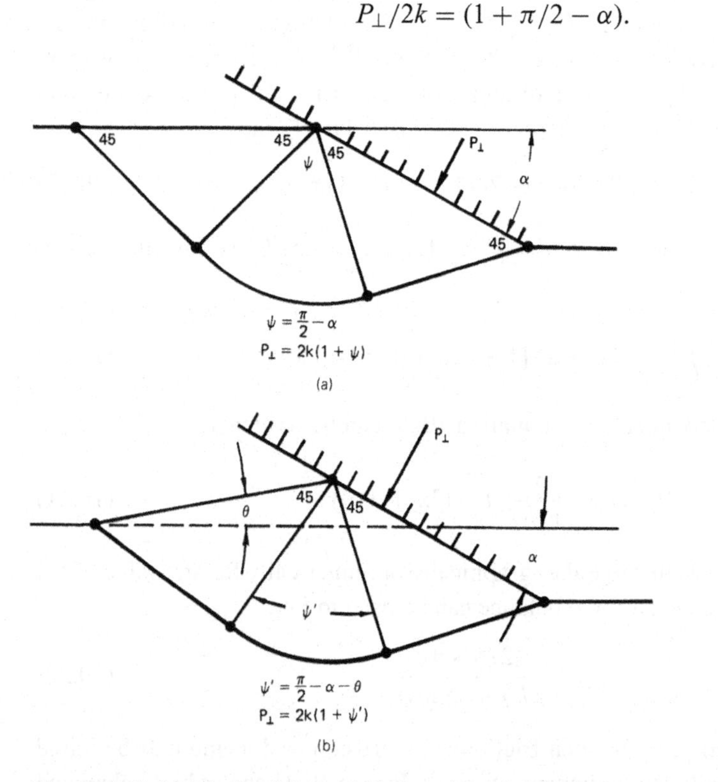

Figure 10.22. (a) Slip-line field for initiation of bulging at die entrance with large α and small r. (b) Slip-line field for continued bulging.

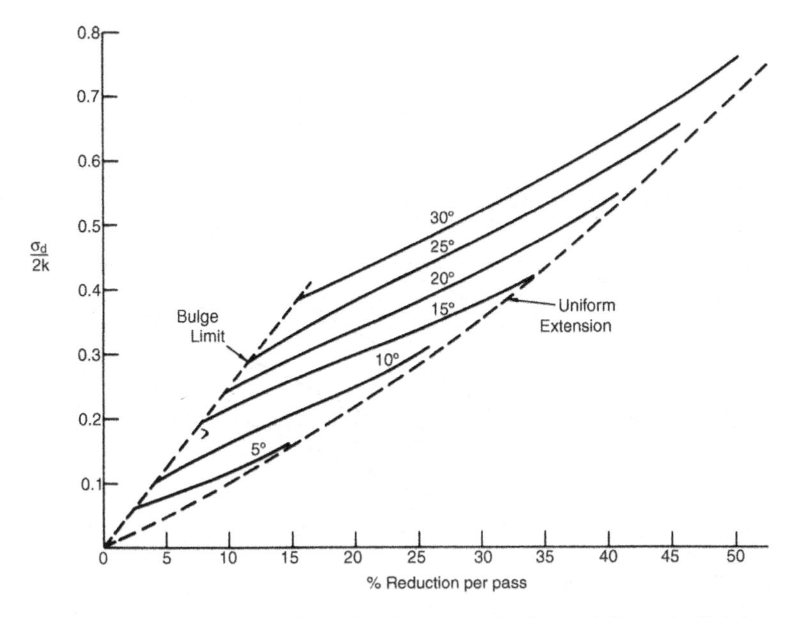

Figure 10.23. Dependence of drawing force on reduction and die angle. Bulging occurs at low reductions and high α. From H. C. Rogers and L. F. Coffin, *Trans ASM*, v. 60 (1967).

Material piles up at the die inlet causing the contact length between die and work piece to be great enough to make the penetrating field appropriate. Figure 10.23 shows the bulging limit and how the drawing stress increases with reduction.

Figure 10.24 shows the variation of $P = -\sigma_2$ in the deformation zone for plane-strain drawing with $\alpha = 15$ and $r = 0.085$. Note that σ_2 may become tensile near

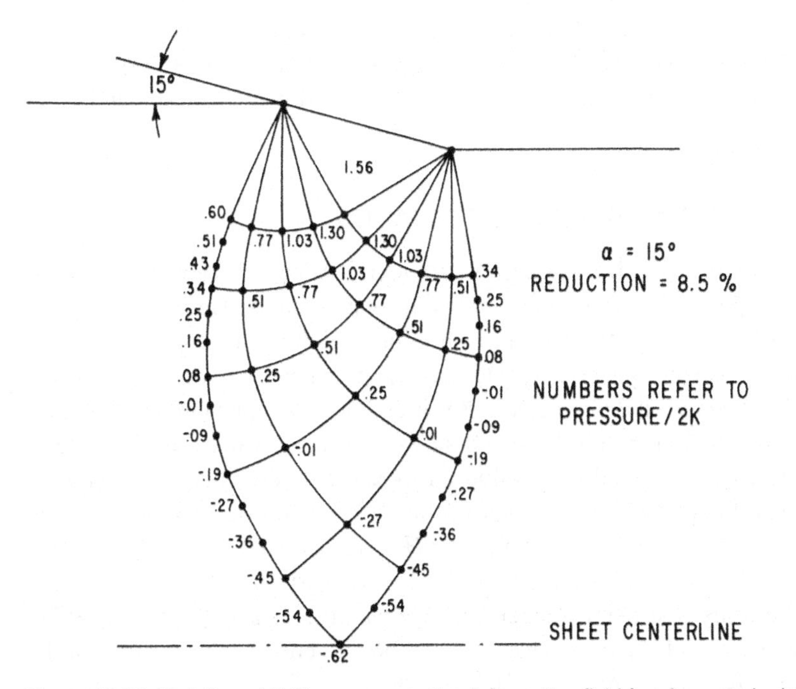

Figure 10.24. Variation of $P/2k = -\sigma_2$ over the deformation field for plane-strain drawing with $\alpha = 15$ and $r = 0.085$. From L. F. Coffin and H. C. Rogers, *ibid.*

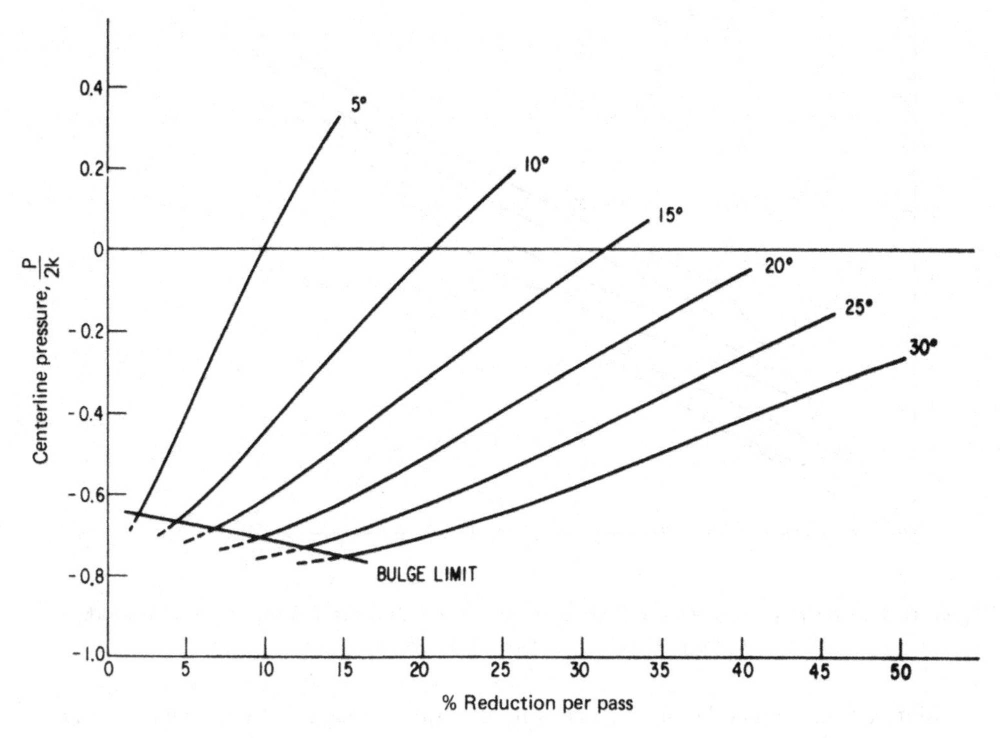

Figure 10.25. Centerline pressure as a function of α and r. Negative values of $P/2k$ indicate tension. From L. F. Coffin and H. C. Rogers, *ibid*.

the centerline. The hydrostatic tension increases as α increases and r decreases. The centerline pressure for various combinations of α and r are shown in Figure 10.25.

If $r > 2\sin\alpha/(1 + 2\sin\alpha)$, a slip-line field different than both Figures 10.11 and 10.20 is required. This is not discussed here but is discussed in W. Johnson and P. B. Mellor, *Engineering Plasticity*, 1973.

10.11 CONSTANT SHEAR STRESS INTERFACES

Slip-line fields can be used to solve problems with sticking friction or a constant friction interface. One example is the compression between rough platens. This approximates conditions during hot forging. With sticking friction the slip lines are parallel and perpendicular to the platens. There is a dead-metal zone where they do not meet the platens. Figure 10.26 shows the appropriate slip-line field. The appropriate boundary condition is $\sigma_x = \sigma_1 = 0$ along the left-hand side of the field. Values of $P_\perp = -\sigma_x$ along the centerline can be found by rotating on α- and β-lines. Then $P_{\perp\text{ave}}$ can be found by numerical integration. How much of this field should be used depends on H/L. Results of calculations for various values of H/L are shown in Figure 10.27. The slab solution of $P_\perp/2k = 1 + (1/4)H/L$ is shown for comparison.

If there is shear stress at the tool interface, $\tau = mk$, the α- and β-lines meet the interface at an angle, $\theta = (1/2)\text{arc cos } m$ and $\theta' = 90\text{-}\theta$. A general Mohr's stress circle plot for this condition is shown in Figure 10.28.

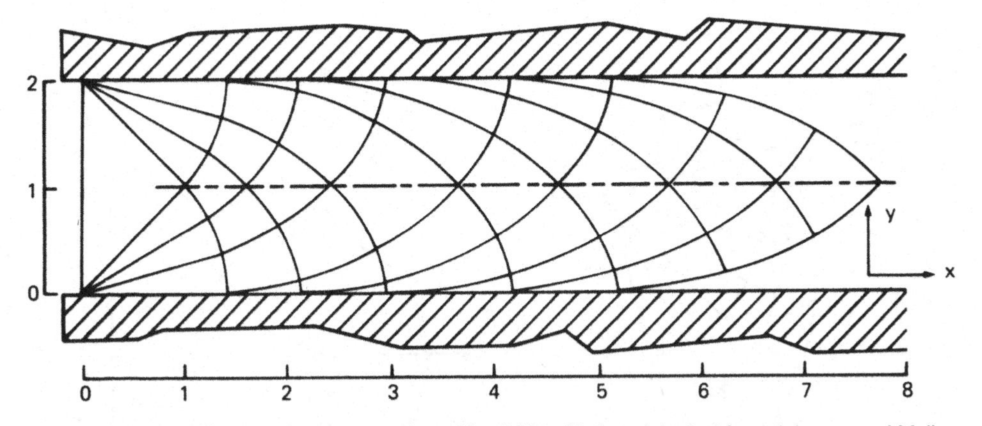

Figure 10.26. Slip-line field for compression with sticking friction. Adapted from Johnson and Mellor, *Engineering Plasticity*, Van Nostrand Reinhold, 1973.

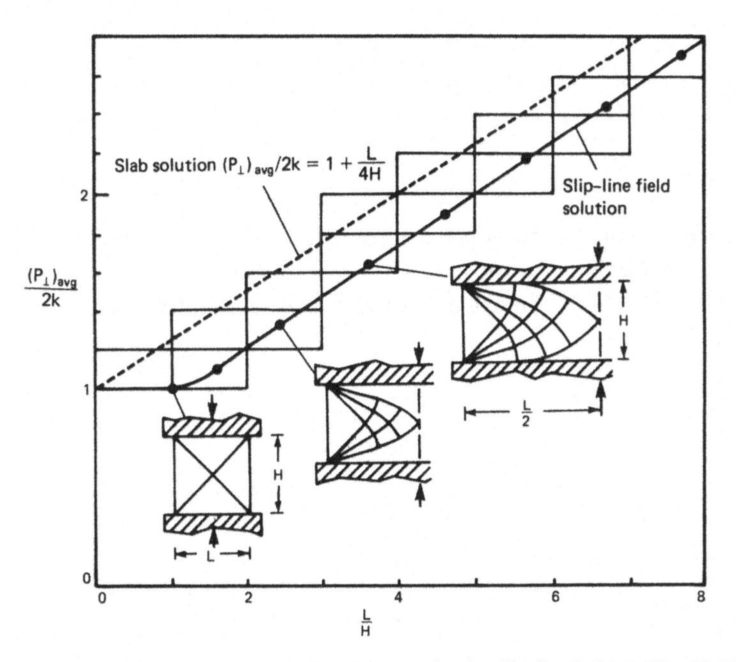

Figure 10.27. Average indentation pressure for the slip-line fields in Fig. 10.26 and slab force analysis (equation. 7.20).

10.12 PIPE FORMATION

The general variation of extrusion force with displacement is sketched in Figure 10.29. For direct or forward extrusion, after an initial rise the force drops because the friction between work material and chamber walls decreases as the amount of material in the chamber decreases. With indirect extrusion there are no chamber walls. As the ram approaches the die, there may be a drop caused by a change of direction of material flow. See Figure 10.30. As material begins to flow along the die wall, a cavity or pipe may be formed. Figure 10.31 shows such a pipe.

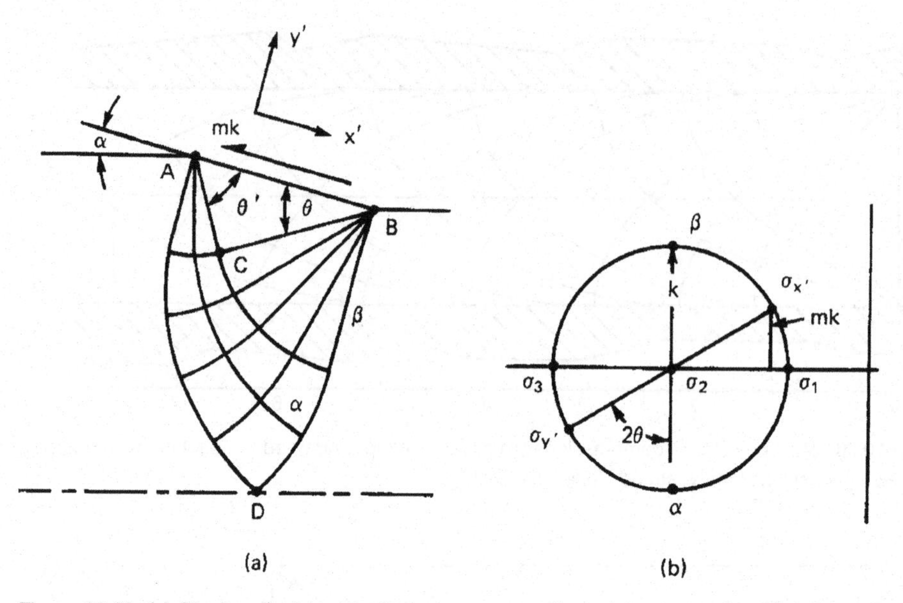

Figure 10.28. (a) Slip-line field for interface stress, $\tau = mk$, and corresponding Mohr's stress circle diagram.

NOTES OF INTEREST

W. Lüders first noted the networks of orthogonal lines that appear on soft cast steel specimens after bending and etching in nitric acid, *Dinglers Polytech. J.*, Stuttgart, 1860. These correspond to slip lines. An example of slip lines revealed by etching is

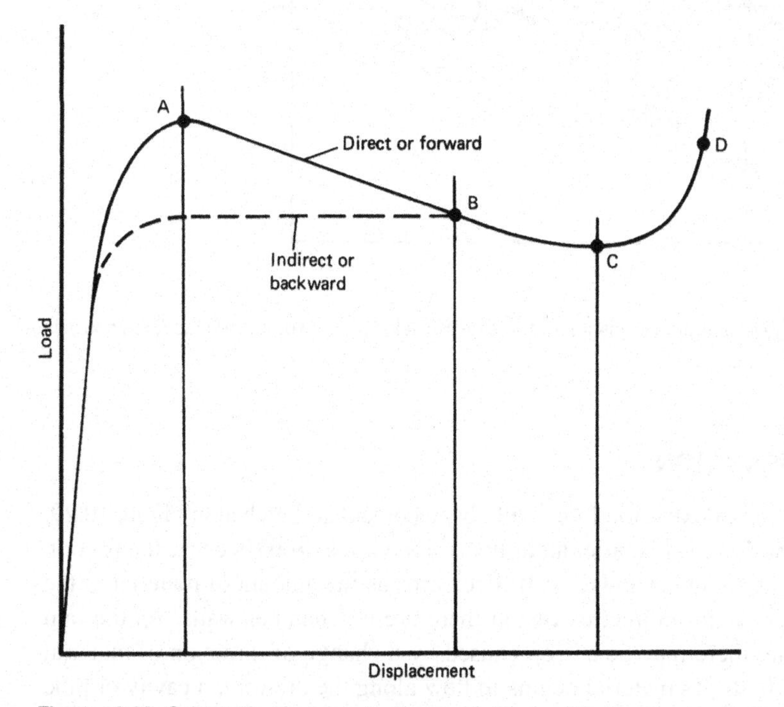

Figure 10.29. Schematic of a load-displacement diagram for direct and indirect extrusion.

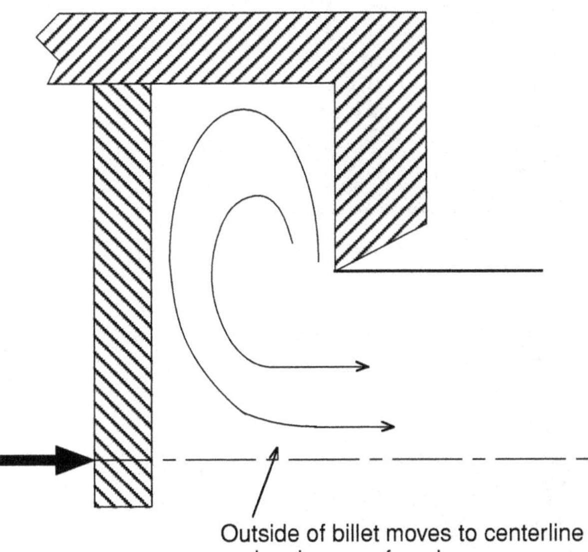

Figure 10.30. Flow pattern at the end of an extrusion. This brings material on the outside of the billet to the centerline and leads to formation of pipe.

given in Figure 10.32. Figure 10.33 shows other examples of slip lines on deformed parts.

There is a simple demonstration of the hydrostatic tension that develops at the center of a lightly compressed strip. If a cylinder of modeling clay is rolled back and forth as it is lightly compressed, a hole will start to develop at the center.

The first systematic presentations of the use of slip-line fields for solving practical problems were the books by Hill and by Prager and Hodge listed below.

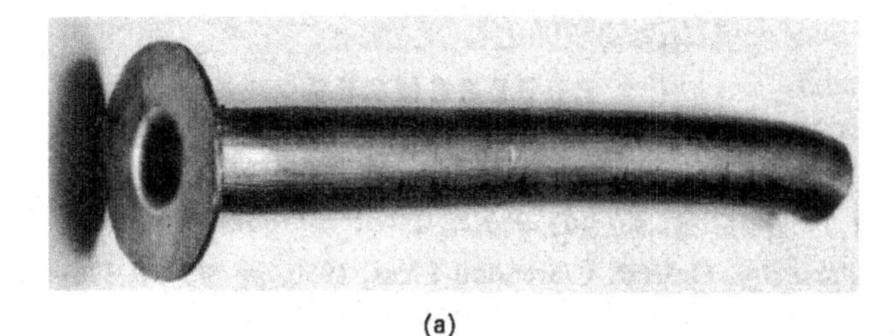

(a)

(b)

Figure 10.31. Billet made by direct extrusion (a) and the pipe at the end (b). Courtesy of W. H. Durrant.

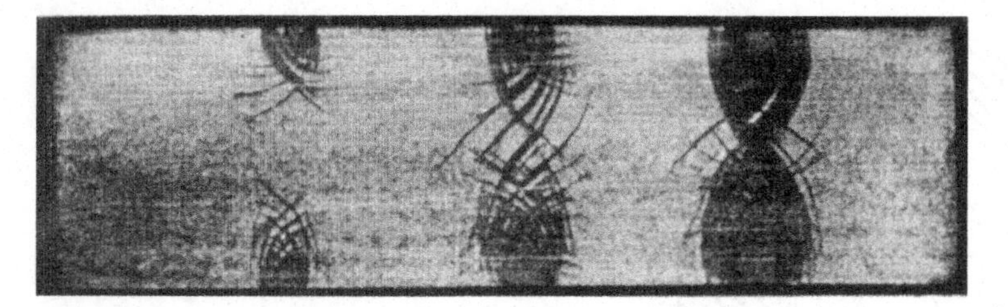

Figure 10.32. Network of lines formed by indenting a mild steel. From F. Körber, *J. Inst. Metals* v. 48 (1932), p. 317.

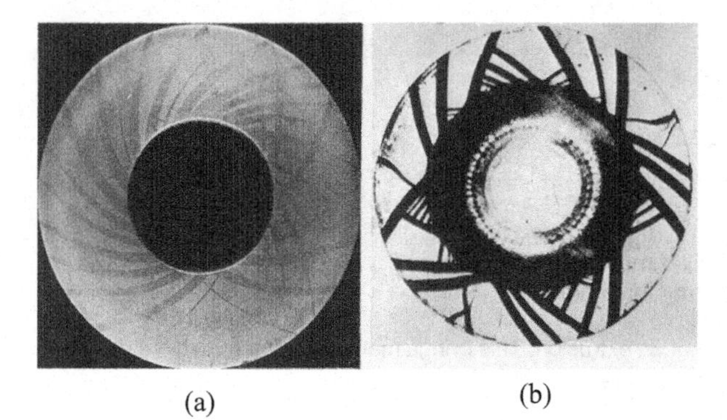

(a) (b)

Figure 10.33. (a) Thick wall cylinder deformed under internal pressure and (b) slip lines on the flange of a cup during drawing. From W. Johnson, R. Sowerby and J. B. Haddow, *Plane-Strain Slip Line Fields*, American Elsevier, 1970.

REFERENCES

B. Avitzur, *Metal Forming: Processes and Analysis*, McGraw-Hill, 1968.

L. F. Coffin and H. C. Rogers, *Trans. ASM* (1967).

H. Ford, *Advanced Mechanics of Materials*, Wiley, 1963.

H. Geringer, *Proc 3rd. Int. Congr. Appl. Mech*, 2, 1930.

A. P. Green, *Phil. Mag.* 42, (1951).

R. Hill. *Plasticity*, Oxford: Clarendon Press, 1950.

W. Johnson and P. B. Mellor, *Engineering Plasticity*, Van Nostrand Reinhold, 1973.

W. Johnson, R. Sowerby, and J. B. Haddow, *Plane-Strain Slip Line Fields*, American Elsevier, 1970.

W. Prager and P. G. Hodge Jr. *Theory of Perfectly Plastic Solids*, Wiley, 1951.

E. G. Thomsen, C. T. Yang, and S. Kobayashi, *Mechanics of Plastic Deformation in Metal Processing*, Macmillan, 1965.

APPENDIX

Table 10.1 gives the x and y coordinates of the 5° slip-line field determined by two centered fans. Figure 10.34 shows the slip-line field. These values were calculated from tables* in which a different coordinate system was used to describe the net. Here the fans have a radius of $\sqrt{2}$ and the nodes of the fans are separated by a distance of 2 and the origin is halfway between the nodes.

* E. G. Thomsen, C. T. Yang, and S. Kobayashi, *Mechanics of Plastic Deformation in Metal Processing*, Macmillan, 1965

Table 10.1. Coordinates of a 5° Net for the Slip-Line Field Determined by Two Centered Fans

$\Delta\phi_\alpha$	n	$m=n$	$m=n+1$	$m=n+2$	$m=n+3$	$m=n+4$	$m=n+5$	$m=n+6$	$m=n+7$	$m=n+8$
0°	0	y = 1.0	1.0833	1.1584	1.2247	1.2817	1.3288	1.3660	1.3926	1.4087
		x = 0	0.0910	0.1888	0.2929	0.4023	0.5163	0.6340	0.7544	0.8767
5°	1	1.1826	1.2741	1.3572	1.4312	1.4951	1.5484	1.5907	1.6214	1.6399
		0.0	0.1000	0.2083	0.3243	0.4472	0.5762	0.7101	0.8484	0.9897
10°	2	1.3831	1.4845	1.5770	1.6597	1.7320	1.7925	1.8407	1.8760	1.8975
		0.0	0.1106	0.2312	0.3613	0.4999	0.6463	0.7995	0.9583	1.1218
15°	3	1.6050	1.7177	1.8214	1.9146	1.9963	2.0653	2.1206	2.1611	2.1861
		0.0	0.1232	0.2582	0.4046	0.5617	0.7285	0.9038	1.0868	1.2760
20°	4	1.8519	1.9781	2.0946	2.2001	2.2929	2.3718	2.4351	2.4820	2.5108
		0.0	0.1377	0.2898	0.4554	0.6339	0.8243	1.0257	1.2307	1.4562
25°	5	2.1283	2.2701	2.4018	2.5215	2.6272	2.7176	2.7905	2.8446	2.8781
		0.0	0.1550	0.3267	0.5146	0.7181	0.9364	1.1680	1.4118	1.6665
30°	6	2.4390	2.5991	2.7484	2.8846	3.0056	3.1093	3.1934	3.2560	3.2948
		0.0	0.1749	0.3698	0.5839	0.8166	1.0610	1.3340	1.6162	1.9119
35°	7	2.7897	2.9713	3.1413	3.2968	3.4356	3.5549	3.6519	3.7245	3.7696
		0.0	0.1984	0.4200	0.6647	0.9314	1.2196	1.5278	1.8547	2.1985
40°	8	3.1874	3.3940	3.5879	3.7662	3.9257	4.0632	4.1755	4.2595	4.3121
		0.0	0.2257	0.4787	0.7589	1.0655	1.3977	1.7540	2.1332	2.5331
45°	9	3.6394	3.8755	4.0976	4.3023	1.1859	4.6447	4.7747	4.8732	4.9335
		0.0	0.2575	0.5472	0.8688	1.2219	1.6054	2.0182	2.4586	2.9243
50°	10	4.1561	4.4259	4.6808	4.9162	5.1281	5.3117	5.4626	5.5760	5.6472
		0.0	0.2947	0.6272	0.9973	1.4046	1.8482	2.3269	2.8389	3.3828
55°	11	4.7470	5.0565	5.3496	5.6211	5.8659	6.0786	6.2537	6.3856	
		0.0	0.3380	07205	1.1472	1.6179	2.1318	2.6873	3.2831	
60°	12	5.4248	5.7807	6.1185	6.4321	6.7154	6.9622	7.1657		
		0.0	0.3886	0.8296	1.3223	1.8670	2.4631	3.1091		
65°	13	6.2043	6.6144	7.0043	7.3671	7.6955	7.9820			
		0.0	0.4982	0.9573	1.5269	2.1584	2.8505			
70°	14	7.1023	7.5758	8.0267	8.4470	8.8281				
		0.0	0.5172	1.1055	1.7658	2.4986				
75°	15	8.1290	8.6864	9.2085	9.6961					
		0.0	0.5981	1.2794	2.0455					
80°	16	9.3375	9.9715	10.5771						
		0.0	0.6925	1.4827						
85°	17	10.726	11.460							
		0.0	0.8031							
90°	20	12.334								
		0.0								

(Continued)

Table 10.1 *(Continued)*

$\Delta\phi_\alpha$	n	$n+9$	$n+10$	$n+11$	$n+12$	$n+13$	$n=14$	$n+15$	$n+16$	$n+17$	$n+18$
0°	0	1.4141	1.4087	1.3926	1.3629	1.3288	1.2816	1.2246	1.1584	1.0833	1.000
		1.0000	1.1233	1.2456	1.3660	1.4837	1.5977	1.7071	1.8112	1.9090	2.000
5°	1	1.6463	1.6399	1.6068	1.5892	1.5449	1.4879	1.4189	1.3379	1.2455	
		1.1334	1.2718	1.4222	1.5653	1.7061	1.8434	1.9765	2.1036	2.2240	
10°	2	1.9048	1.8975	1.8751	1.8375	1.7846	1.7163	1.6330	1.5347		
		1.2891	1.4582	1.6282	1.7979	1.9658	2.1305	2.2909	2.4452		
15°	3	2.1946	2.1860	2.1595	2.1151	2.0522	1.9707	1.8707			
		1.4708	1.6686	1.8688	2.0694	2.2690	2.4658	2.6583			
20°	4	2.5207	2.5107	2.4797	2.4272	2.3527	2.2555				
		1.6828	1.9141	2.1497	2.3865	2.6233	2.8574				
25°	5	2.8895	2.8778	2.8414	2.7795	2.6913					
		1.9304	2.2014	2.4777	2.7580	3.0372					
30°	6	3.3083	3.2944	3.2519	3.1789						
		2.2196	2.366	2.8610	3.1901						
35°	7	3.7853	3.7692	3.7191							
		2.5573	2.9281	3.3089							
40°	8	4.3303	4.3114								
		2.918	3.3859								
45°	9	4.9548									
		3.4129									

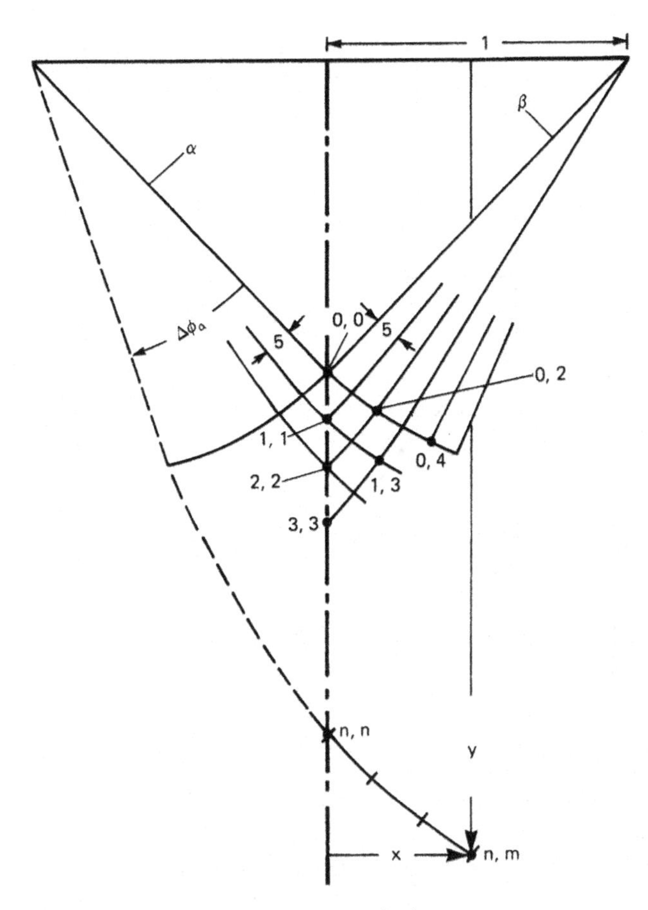

Figure 10.34. Slip-line field for Table 10.1.

PROBLEMS

10.1. Using the slip-line field in Figure 10.8 for frictionless indentation, it was found that $P_{\perp}/2k = 2.57$. Figure 10.35 shows an alternate field for the same problem proposed by Hill.

a) Find $P_{\perp}/2k$ for this field.

b) Construct the hodograph.

c) What percent of the energy is expended along lines of intense shear?

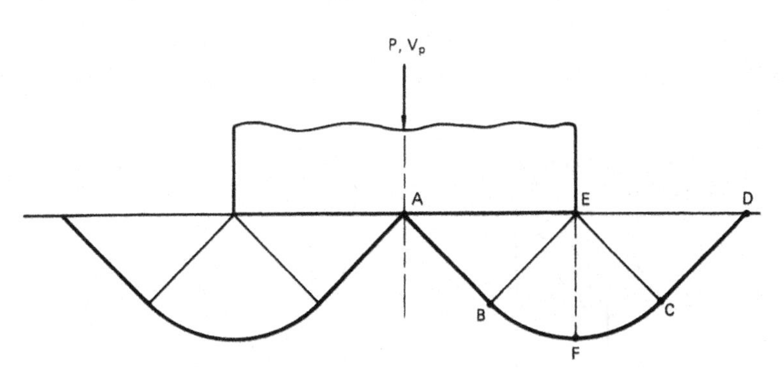

Figure 10.35. Slip-line field for plane-strain indentation.

10.2. Figure 10.36 shows a slip-line field with a frictionless punch. Construct the hodograph and find $P_{\perp}/2k$ for this field.

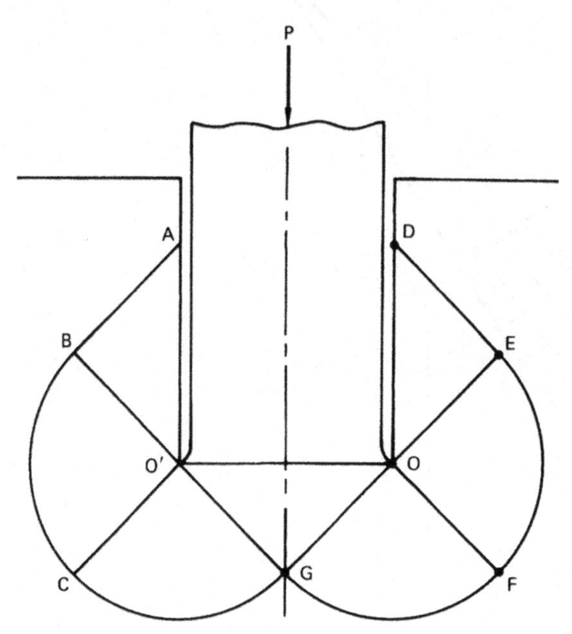

Figure 10.36. Slip-line field for Problem 10.2.

10.3. Plane-strain compression of a hexagonal rod is shown in Figure 10.37 together with a possible slip-line field.

a) Determine whether this field or penetrating deformation will occur.

b) Find $P_{\perp}/2k$.

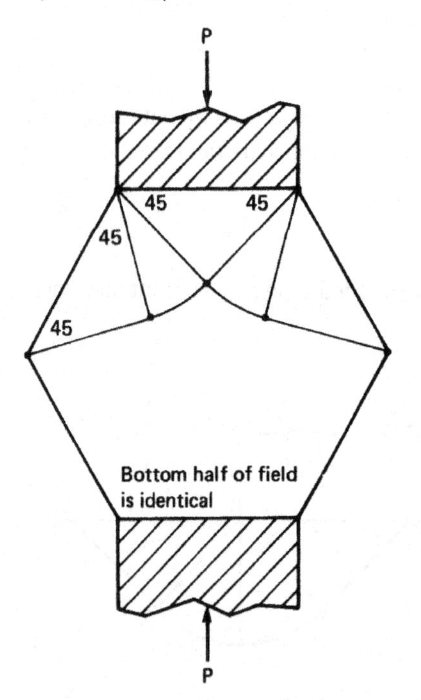

Figure 10.37. Possible slip-line field for Problem 10.3.

10.4. Figure 10.38 is the slip-line field for plane-strain wedge indentation. For the volume of the side mounds to equal the volume displaced, the angle, ψ, must be related to θ by $\cos(2\theta - \psi) = \cos\psi/(1 + \sin\psi)$. Determine F/x in terms of $2k$ for $\theta = 120°$.

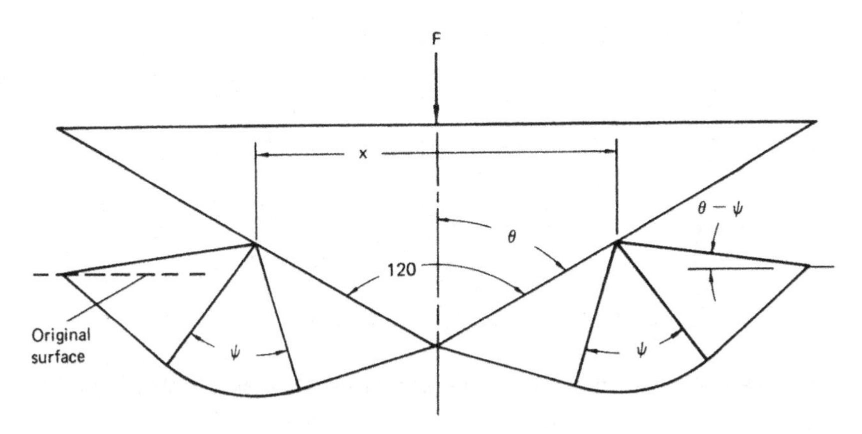

Figure 10.38. Slip-line field for plane-strain wedge indentation.

10.5. Figure 10.39 shows the slip-line field for a 2:1 reduction by indirect or backward frictionless extrusion.

a) Determine $P_{\perp}/2k$.

b) Construct a hodograph for the lower half of the field.

c) Find V^*_{AD}/V_0.

d) What percent of the energy is expended by the gradual deformation in the centered fans?

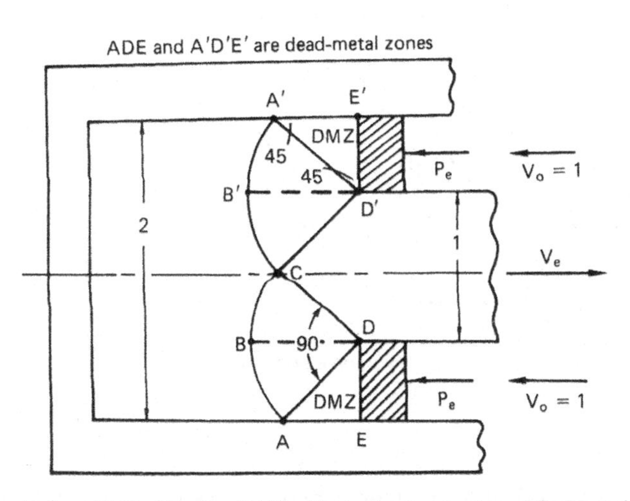

Figure 10.39. Slip-line field for the indirect extrusion of Problem 10.5.

10.6. Figure 10.40 shows the slip-line field for the 3:1 frictionless extrusion. Find $P_{\perp}/2k$ and η.

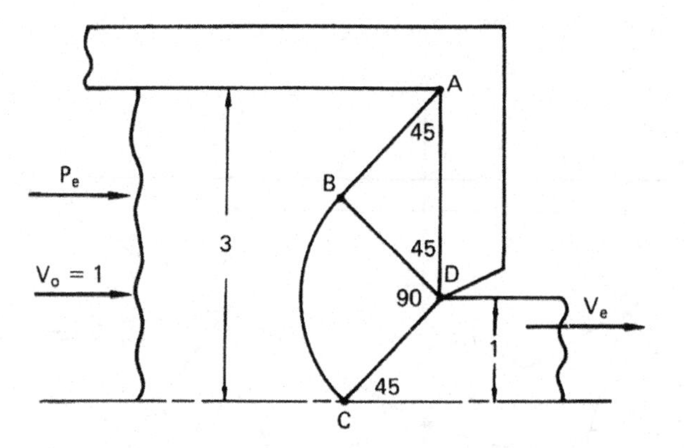

Figure 10.40. The slip-line field for the 3:1 frictionless extrusion in Problem 10.6.

10.7. Indentation of a step on a semi-infinite hill is shown in Figure 10.41. Show that this field is not possible and draw a correct field.

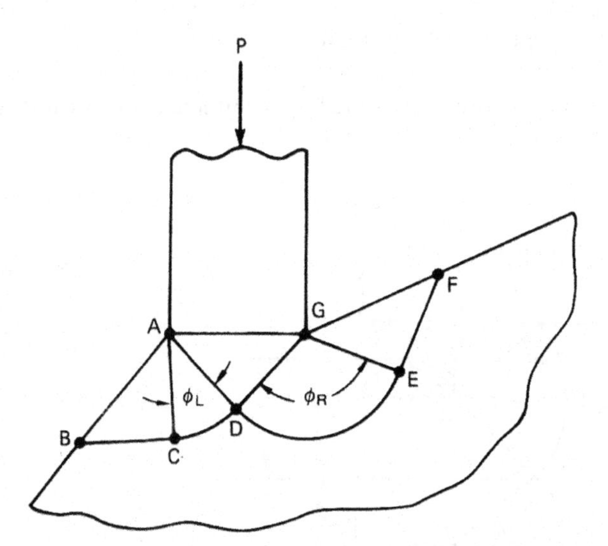

Figure 10.41. An incorrect field for indentation on a semi-infinite hill.

10.8. What is the highest level of hydrostatic tension, expressed as $\sigma_2/2k$, that can be induced by two opposing flat indenters as shown in Figure 10.42?

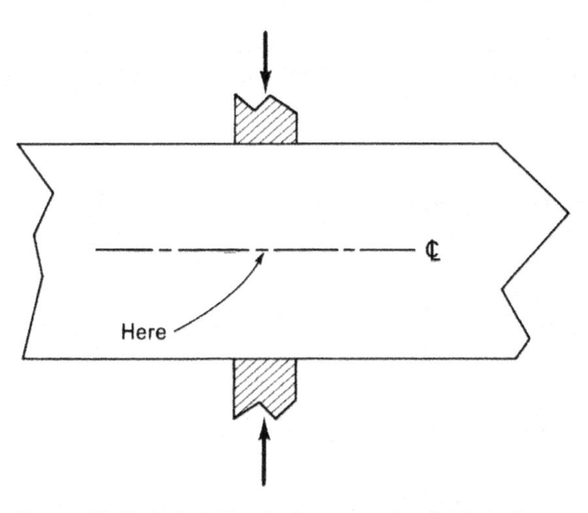

Figure 10.42. Indentation by two opposing flat indenters.

10.9. Consider the back extrusion in Figure 10.43. Assume frictionless conditions.

a) Find $P_{\perp}/2k$.

b) Construct the hodograph.

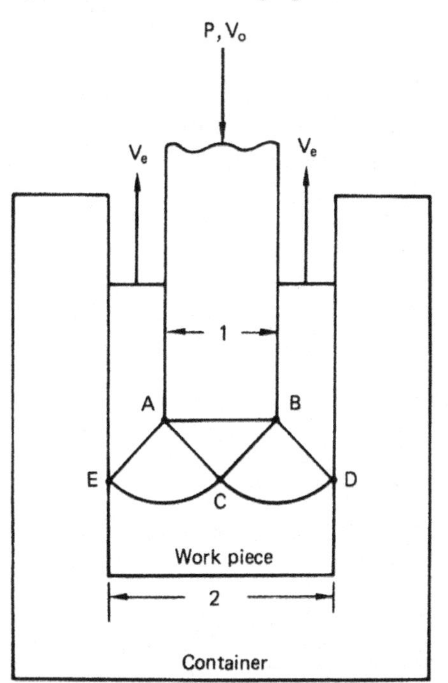

Figure 10.43. Back extrusion in Problem 10.9.

10.10. Consider punching a long, thin slot as shown in Figure 10.44. For punching, shear must occur along AB and CD.

(a) Find $P_{\perp}/2k$ as a function of t.

(b) If the ratio of t/w is too great, an attempt to punch will result in a plane-strain hardness indentation. What is the largest ratio of t/w for which a slot can be punched?

(c) Consider punching a circular hole of diameter, d. Assume a hardiness of $p/2k = 3$. What is the lowest ratio of hole diameter to sheet thickness that can be punched?

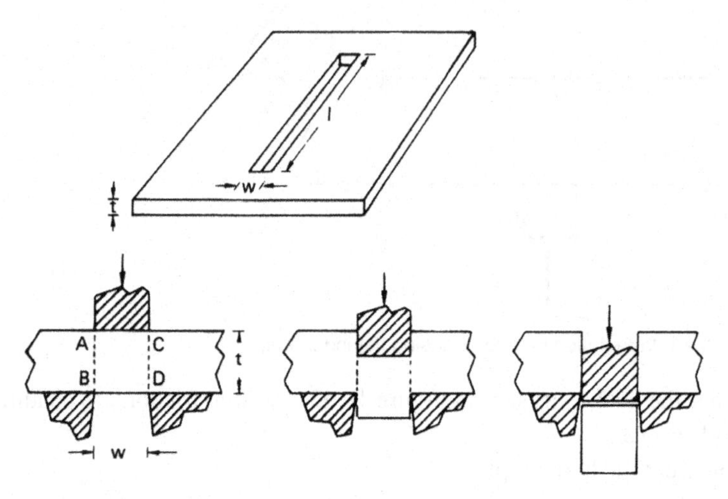

Figure 10.44. Slot punching.

10.11. Deeply notched tensile specimen much longer than its width and very deep in the direction normal to the drawing is shown in Figure 10.45. Calculate $\sigma_x/2k$ for the field where $\sigma_x = F_x/t_n$.

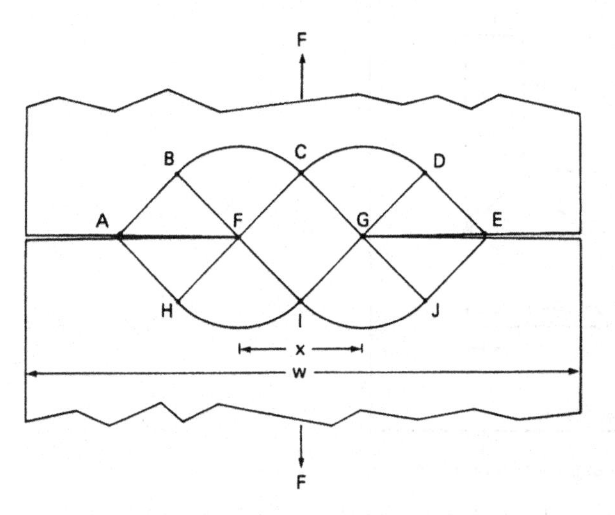

Figure 10.45. Deeply-notched tensile specimen for Problem 10.11.

10.12. Consider a plane-strain tension test on the notched specimen in Figure 10.46(a). The notch angle is 90° and $w \gg y_o$. Using the slip-line field in Figure 10.46(b), calculate $\sigma_x/2k$ where $\sigma_x = F_x/t_n$.

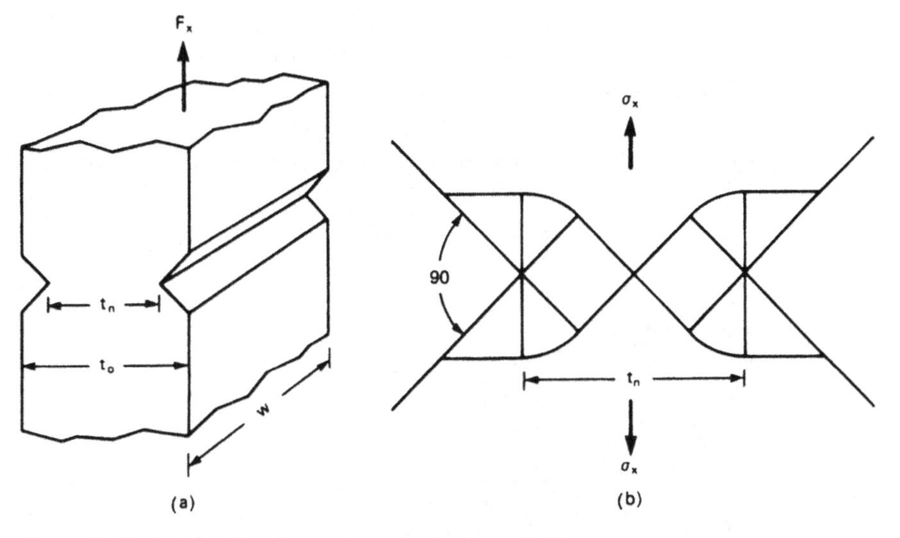

Figure 10.46. Notched tensile specimen for Problem 10.12.

10.13. If the notches in Problem 10.12 are too shallow (i.e., t_0/t_n is too small) the specimen may deform by shear between the base of one notch and the opposite side as shown in Figure 10.47. What ratio of t_0/t_n is needed to prevent this?

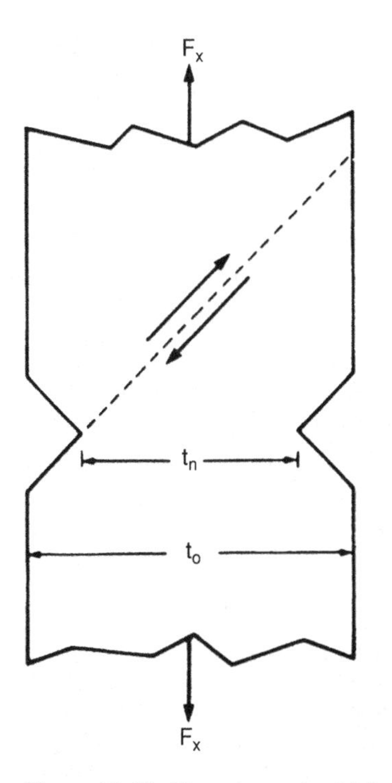

Figure 10.47. Alternate mode of failure for a notched tensile bar if t_0/t_n isn't large enough. For Problem 10.13.

10.14. Figure 10.48 shows the appropriate slip-line field for either frictionless plane-strain drawing or extrusion where $r = 0.0760$ and $\alpha = 15°$.

(a) Find the level of σ_2 at point 4,5 for extrusion.

(b) Find the level of σ_2 at point 4,5 for drawing.

(c) How might the product depend on whether this is an extrusion or a drawing?

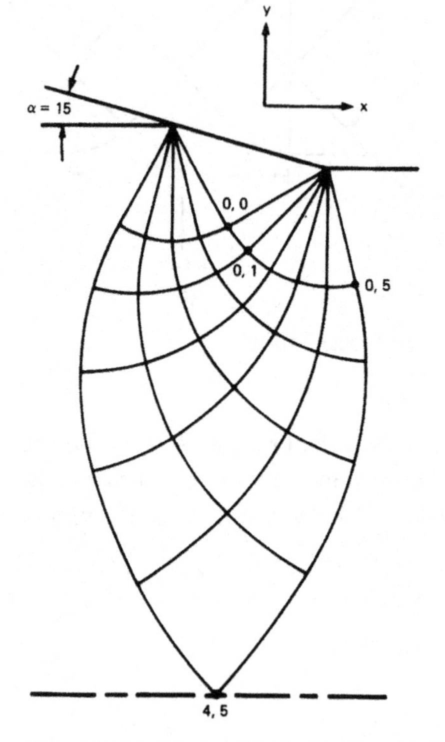

Figure 10.48. Slip-line field for Problem 10.14.

10.15. Figure 10.49 shows two slip-line fields for the compression of a long bar with an octagonal cross section. Which field is appropriate? Justify your answer.

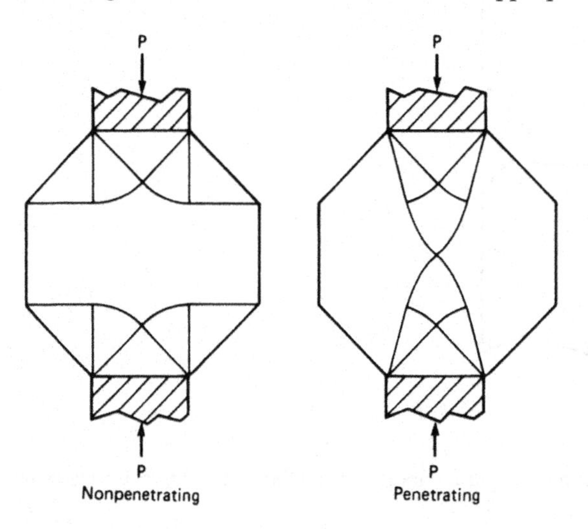

Figure 10.49. Two possible slip-line fields for plane-strain compression of an octagonal rod. For Problem 10.15.

10.16. Consider an extrusion within a frictionless die with a = 30° and such that point (2,4) in Figure 10.18 is on the centerline.

(a) What is the reduction?

(b) Calculate $P_{ext}/2k$.

(c) Calculate η.

(d) Find the hydrostatic stress, σ_2, at the centerline.

10.17. Consider the slip-line field for an extrusion with a constant shear stress along the die wall as shown in Figure 10.50.

(a) Label the α- and β-lines.

(b) Draw the Mohr's circle diagram for the state of stress in triangle ACD showing $P_\perp$, $\tau_f = mk$, α and β.

(c) Calculate the value of m.

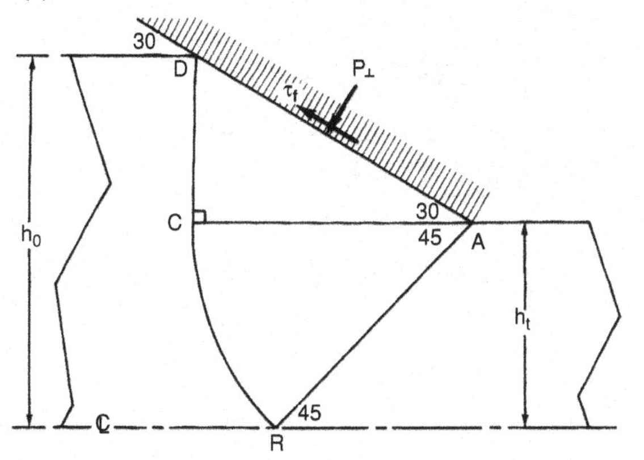

Figure 10.50. Slip-line field for extrusion with a constant shear stress die interface. See Problem 10.17.

10.18. At the end of an extrusion in a 90° die, a non-steady-state condition develops. The field in Figure 10.51(a) is no longer appropriate. Figure 10.51(b) is an

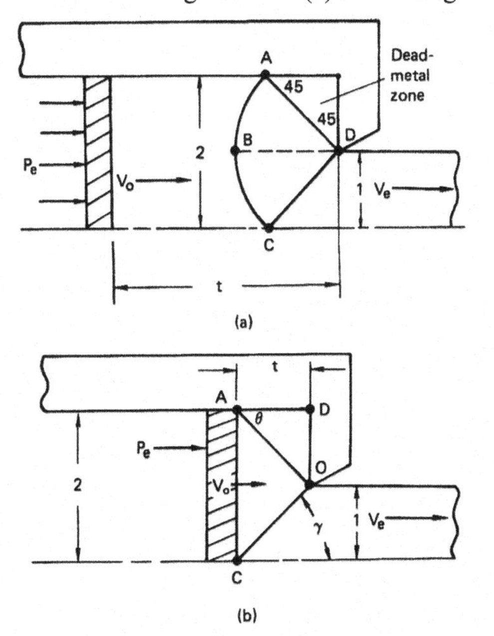

Figure 10.51. (a) Slip-line field for a 2:1 extrusion. (b) Upper-bound field for the end of a 2:1 extrusion. For Problem 10.18.

upper-bound field for $t < 1$. Calculate and plot $P_{ext}/2k$ as a function of t for $0.25 \leq t \leq 5$ for the upper-bound field. Include on your plot the value of $P_{ext}/2k$ for the slip-line field.

10.19. In Problem 10.18, either the slip-line field or the upper bound gives a lower solution. However as discussed in Section 10.12, a pipe may form at the end of an extrusion. Figure 10.52 gives an upper-bound field that leads to pipe formation. Calculate $P_{ext}/2k$ as a function of t for $0.25 \leq t \leq 5$ and compare with the solution to Problem 10.17.

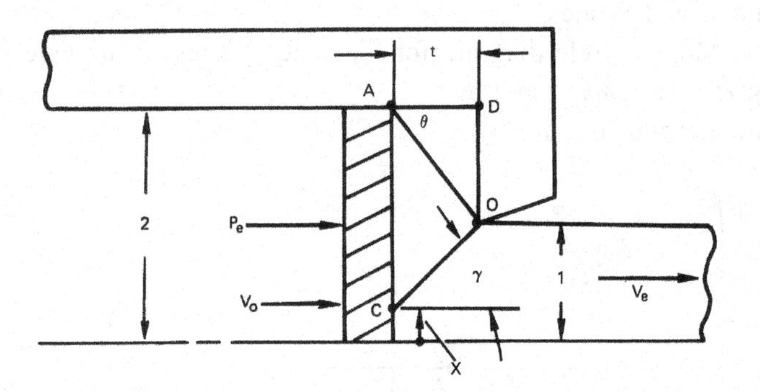

Figure 10.52. An upper-bound field for pipe formation. (Problem 10.19).

11 Deformation-Zone Geometry

11.1 THE Δ PARAMETER

The shape of the deformation zone has a strong influence on the redundant work, the frictional work and the forming forces. It also influences the properties of the product material. The homogeneity, the tendency to crack, the pattern of residual stresses and the porosity are all affected by the deformation-zone geometry. The parameter, Δ, defined as the ratio of the thickness or diameter, H, to the contact length between work piece and die, L, has a large effect on these properties.

$$\Delta = H/L. \tag{11.1}$$

For plane-strain extrusion and drawing, the contact length $L = (H_0 - H_1)/(2\sin\alpha)$ and the mean thickness, $H = (H_0 + H_1)/2$ so

$$\Delta = \frac{(H_0 + H_1)}{(H_0 - H_1)}\sin\alpha. \tag{11.2}$$

Substituting $r = (H_0 - H_1)/H_0$,

$$\Delta = \frac{2 - r}{r}\sin a. \tag{11.3}$$

Equation 11.2 holds for axisymmetric extrusion and drawing, where H is the diameter, so $r = (H_0^2 - H_1^2)/H_0^2$, so

$$\Delta = \frac{1 + \sqrt{1 - r}}{1 - \sqrt{1 - r}}\sin\alpha = \left(1 + \sqrt{1 - r}\right)^2 \sin\alpha/r. \tag{11.4}$$

Note that for the same α and r, equation 11.3 gives a Δvalue for axisymmetric deformation about twice that for plane-strain (equation 11.2.)

For flat rolling, Δ is simply the mean height $(H_0 + H_1)/2 = H_0(2 - r)/2$, divided by the contact length, $L = \sqrt{R\Delta H} = \sqrt{rRH_0}$,

$$\Delta = \frac{2 - r}{2}\sqrt{\frac{H_0}{rR}}. \tag{11.5}$$

Slightly different definitions of Δ are used elsewhere.* In all of the equations, Δ increases with die angle and decreases with reduction.

11.2 FRICTION

The frictional contribution to the total work, w_f/w_a, increases with decreasing Δ because the contact area between the die and work piece increases as Δ decreases. In both plane-strain and axisymmetric drawing, the ratio of contact area to mean cross-sectional area is

$$\text{Area ratio} = \frac{2r}{(2-r)\sin\alpha}. \tag{11.6}$$

The upper-bound analysis for constant friction interfaces in plane-strain drawing (equation 9.22) predicts that

$$w_f/w_h = m/\sin(2\alpha). \tag{11.7}$$

For small values of α, $\sin 2\alpha \approx 2\sin\alpha$, so at constant friction, w_f/w_a is proportional to Δ.

With a constant coefficient of friction, equation 7.11 for axisymmetry and equation 7.7 for plane-strain both give

$$\frac{\sigma_\mathrm{d}}{\sigma_0} = \frac{1+B}{B}[1-\exp(-B\varepsilon_h)], \tag{11.8}$$

where $B = \mu \cot\alpha$. Substituting these in the series expansion, $\exp(-x) = 1 - x + x^2/2 + \cdots$, $\sigma_\mathrm{d}/\sigma_0 = \varepsilon(1 + B - B^2\varepsilon/2 + \cdots)$. Now substituting $w_h = \sigma_0\varepsilon$ and $w_f = \sigma_\mathrm{d} - w_h$,

$$w_f/w_h = B(1 - \varepsilon/2 + \cdots) \approx \mu\cot\alpha(1-\varepsilon/2) \tag{11.9}$$

11.3 REDUNDANT DEFORMATION

It is convenient to describe the redundant strain, ε_r, by a parameter, $\phi = (\varepsilon_r - \varepsilon_h)/\varepsilon_h)$ where ε_h is the homogeneous strain. In the absence of strain hardening, $(w_r + w_h)/w_h = (\varepsilon_r - \varepsilon_h)/\varepsilon_h = \phi$. In this case

$$\phi = (\sigma_\mathrm{d} - w_f)/(\sigma_0\varepsilon_h). \tag{11.10}$$

Equations 9.21 and 9.27 from the upper-bound analysis can be used to predict ϕ for plane-strain and axisymmetric drawing. Comparing with equation 11.10,

$$\phi = 1 + (1/2)\tan\alpha/\varepsilon_h \tag{11.11}$$

for plane-strain and

$$\phi = 1 + (2/3)\tan\alpha/\varepsilon_h \tag{11.12}$$

* Backofen defined the mean height, H, as the arc length drawn through the middle of the deformation zone so that his expressions are identical to equations 11.2, 11.3 and 11.4 except that $\sin < \alpha$ is replaced by α.

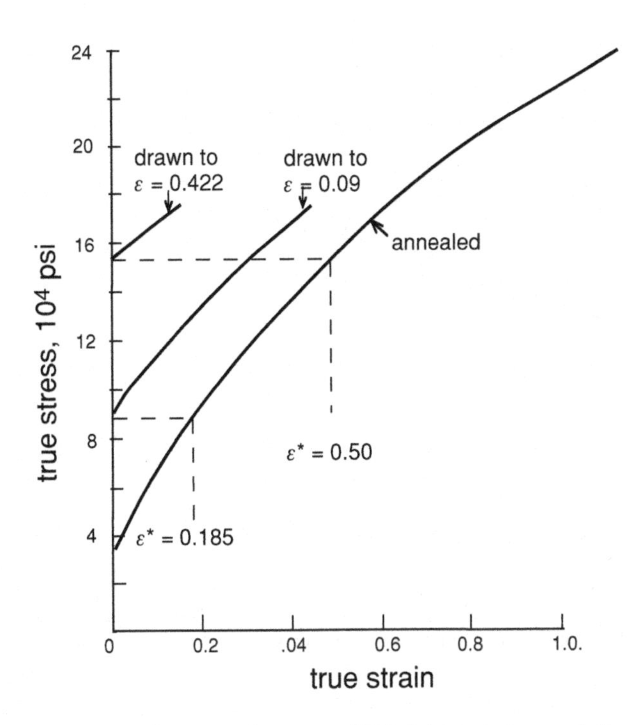

Figure 11.1. Stress–strain curves of 303 stainless before and after wire drawing. Adapted from Caddell and Atkins, *ibid*.

for axisymmetry. Numerical evaluation of equations 11.11 and 11.12 for various combinations of $\alpha < 30^\circ$ and $r < 0.5$ result in approximations

$$\phi \approx 1 + \Delta/4 \tag{11.13}$$

for plane-strain and for axisymmetric flow.

$$\phi \approx 1 + \Delta/6. \tag{11.14}$$

The slip-line solutions for plane-strain indentation (Figure 10.19) can be approximated by

$$\phi \approx 1 + 0.23(\Delta - 1) \quad \text{for} \quad \leq \Delta \leq 8.8. \tag{11.15}$$

There is no corresponding solution for axisymmetric flow.

Figure 11.1 from Caddell and Atkins* shows the true tensile stress-strain curves for a 303 stainless steel in the annealed condition and after two wire-drawing reductions. Comparing the yield strengths after drawing with the annealed stress-strain curve, the value of ϕ could be deduced. The results with other die angles and reductions (Figure 11.2) could be approximated by

$$f = C_1 + C_2\Delta. \tag{11.16}$$

where C_1 and C_2 are constants.

* R. M. Caddell and A. G. Atkins. *J. Eng. Ind. Trans ASME*, Series B (1968).

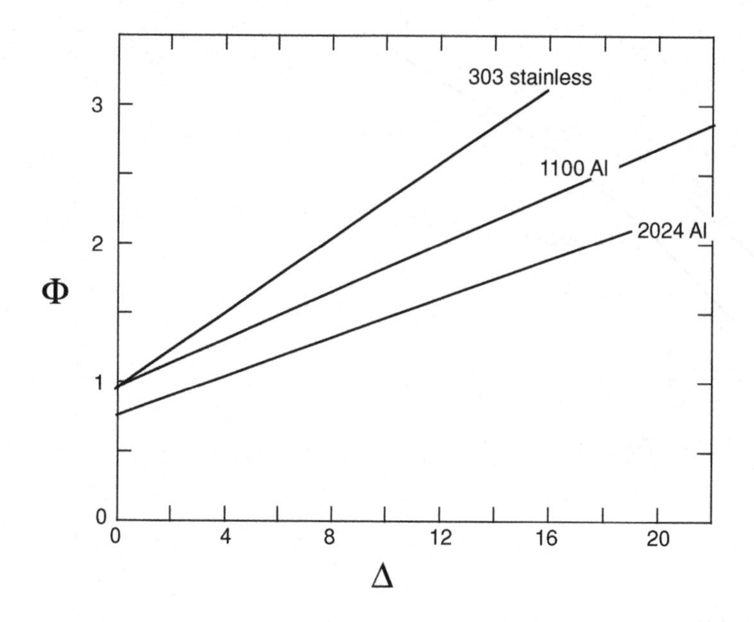

Figure 11.2. The influence of Δ on the redundant strain factor, ϕ. Adapted from R. M. Caddell and A. G. Atkins, *ibid*.

Similar results were obtained by Backofen* using hardness measurements. By comparing hardness measurements on homogeneously strained material with those after various cold-forming operations, he found results very similar to those of Caddell and Atkins (Figure 11.3). For physical reasons ϕ can never be less than unity so Backofen suggested

$$\phi = 1 + C(\Delta - 1) \quad \text{for } \Delta \geq 1 \quad \text{and} \quad \phi = 1 \quad \text{for } \Delta \leq 1, \tag{11.17}$$

where $C = 0.21$ for plane-strain and $C = 0.12$ for axisymmetric drawing. Note that the value of $C = 0.21$ is similar to the value 0.23 from slip-line theory. The results for axisymmetric and plane-strain drawing would be much closer if ϕ had been plotted against the ratio of mean cross-sectional area to contact area between work piece and die. This ratio is $\Delta/4$ for axisymmetry and $\Delta/2$ for plane-strain.

11.4 INHOMOGENEITY

The redundant strain is not evenly distributed throughout the cross section. Figure 11.4 shows the distortion of grid lines predicted by slip-line field theory for three levels of Δ and Figure 11.5 shows the actual distortion of extruded billets. The effect of Δ on inhomogeneity is also demonstrated by the hardness profiles after flat rolling (Figure 11.6). Backofen characterized the hardness gradients with an inhomogeneity factor, *IF*,

$$IF = (H_s - H_c)/H_c, \tag{11.18}$$

where H_s and H_c are the Vickers hardnesses at the surface and the center. Plots of *IF* vs. Δ indicate that *IF* increases with Δ but are also dependent on r. If these data

* W. A. Backofen, *ibid* and J. J. Burke, *Doctoral Thesis*, MIT (1968).

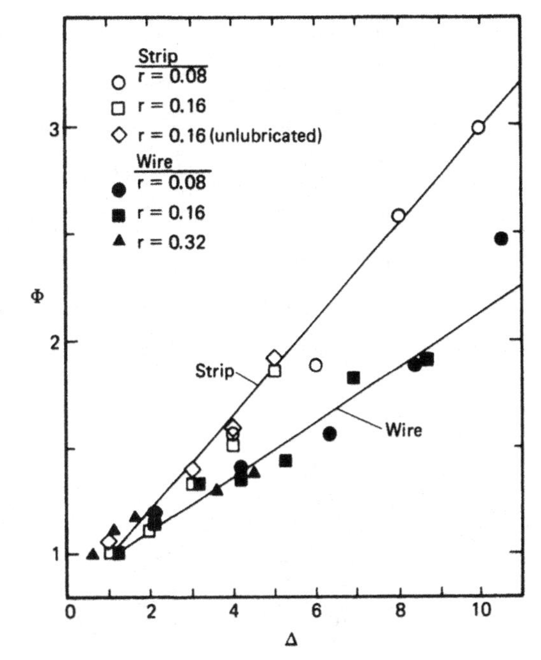

Figure 11.3. The influence of Δ on the inhomogeneity factor, ϕ, for axisymmetric and plane-strain drawing. Adapted form Backofen and Burke, *ibid.*

are plotted as *IF* versus α (Figure 11.7) there is much less dependence on r and little difference between axisymmetry and plane-strain.

Hardness variations after much larger reductions were studied in hydrostatic extrusion experiments by Miura et al.* Figure 11.8 shows that the difference between surface and centerline hardness increases with both increasing die angle and decreasing reduction. At these higher reductions the results correlate roughly with Δ.

Inhomogeneity of flow also affects how crystallographic texture changes between surface and centerline.† If Δ is high, the usual rolling texture will be found only near the centerline. The grain size after recrystallization depends on the amount of deformation prior to annealing. With high Δ the recrystallized grain size at the surface is finer than at the center. Furthermore if the deformation is rapid enough to be adiabatic, heating will depend on the amount of deformation. This can affect the residual stresses.

EXAMPLE 11.1:

(a) Using the upper-bound field from Section 9.9, derive an expression for the redundant strain, realizing that it arises from the shear discontinuities at the entrance and exit of the deformation zone.
(b) Find the ratio of the total strain at the surface to that at the center.
(c) The Vickers hardness can be expressed as $H = C\varepsilon^n$, where n is much lower than the strain-hardening exponent in tension. Derive an expression for the inhomogeneity factor in terms of α and r.

* S. Miura, Y. Saeki and T. Matushita, *Metals and Materials*, 7 (1963).
† P. S. Mathur and W. A. Backofen, *Met Trans.*, 4 (1973).

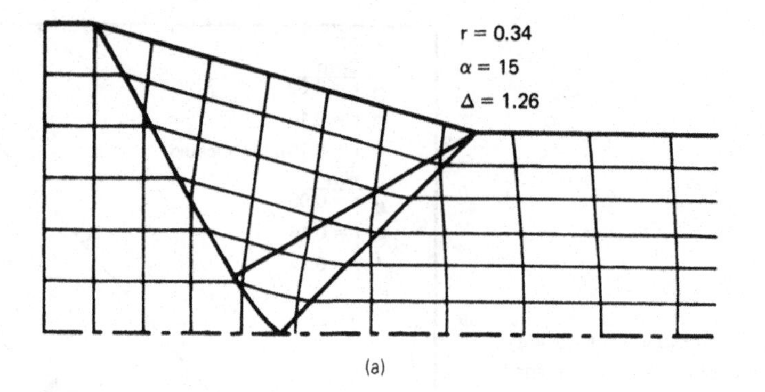

(a)

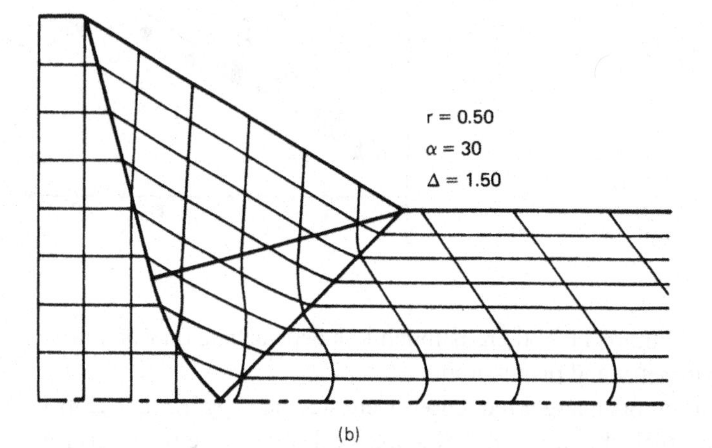

(b)

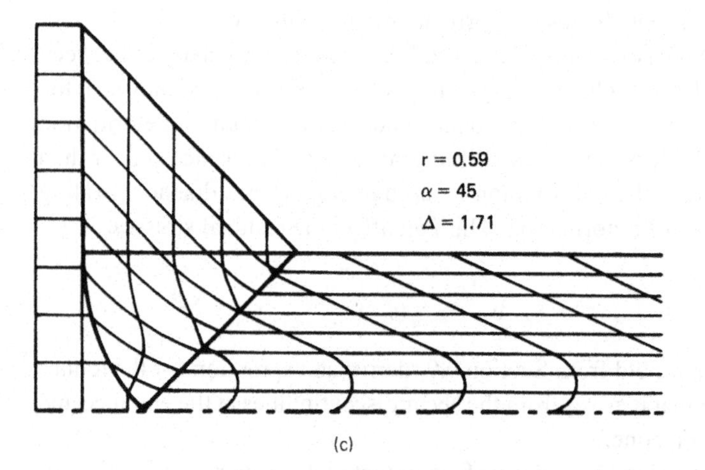

(c)

Figure 11.4. Grid distortions predicted by slip-line field theory for strip drawing. Note that the distortion increases with increasing Δ.

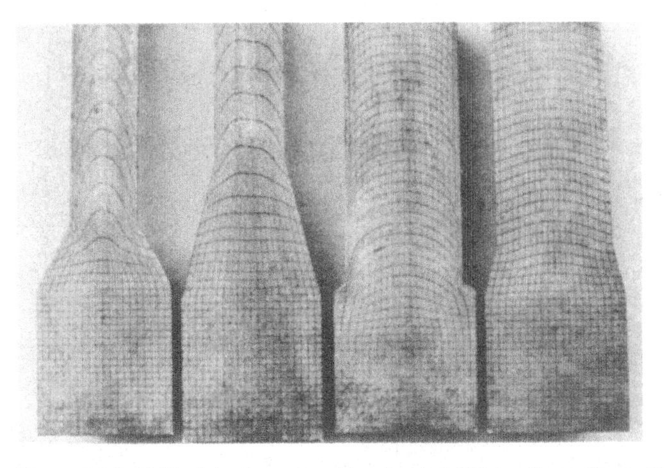

Figure 11.5. Grid distortion of cold extruded billets. From D. J. Blickwede, *Metals Progress*, 97 (May 1970).

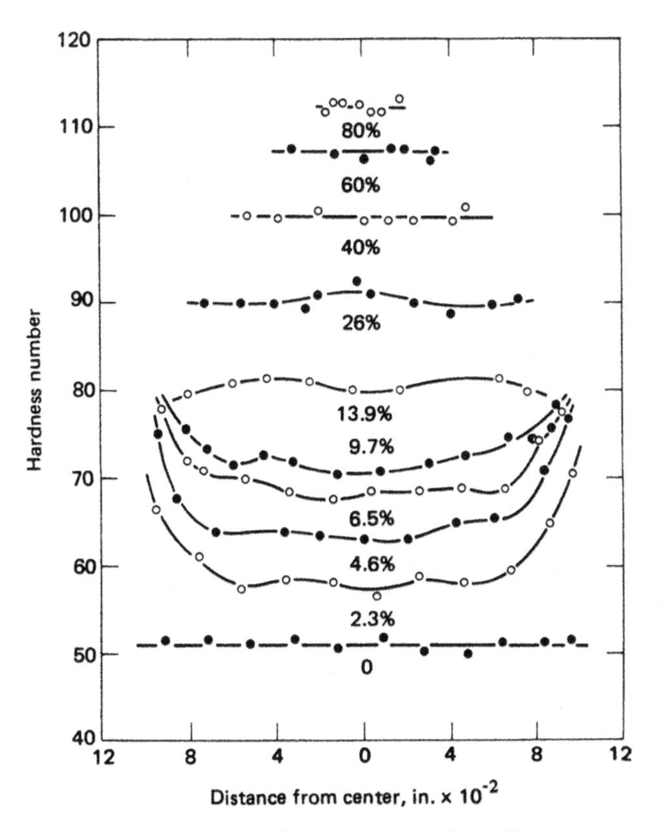

Figure 11.6. Hardness gradients in copper after rolling in a single pass to the indicated reduction. Initial thickness was 0.2 in and the roll diameter was 10 in. Note that the inhomogeneity is largest for small reductions. From B. B. Hundy and A. R. E. Singer, *J. Inst. Metals* 83 (1954–5).

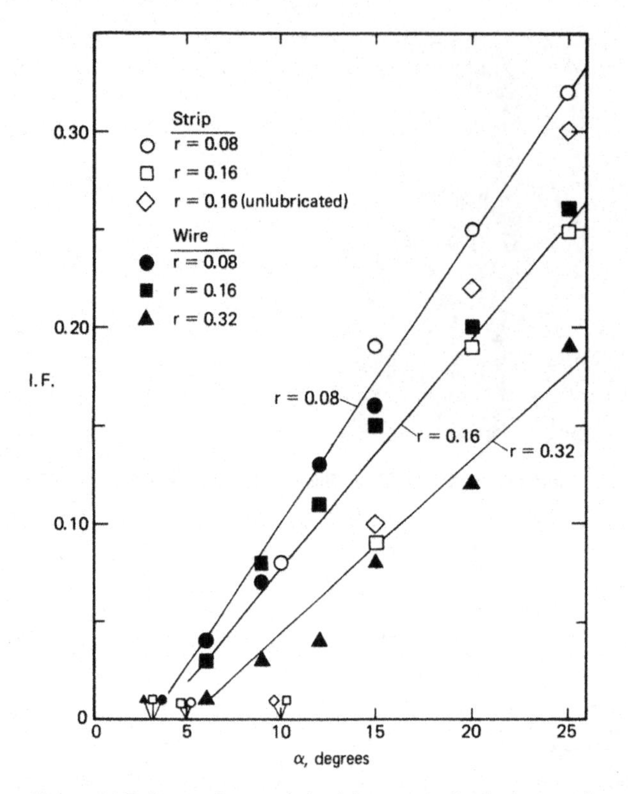

Figure 11.7. Dependence of the inhomogeneity factor on die angle. Data from W.A. Backofen and J. J. Burke, *ibid.*

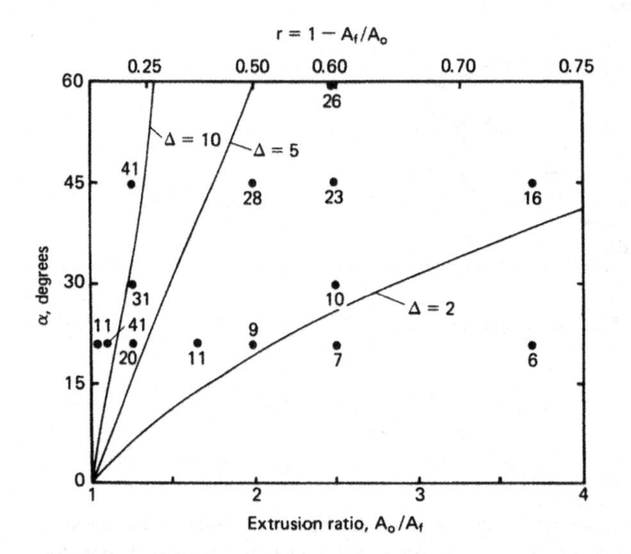

Figure 11.8. Effect of die angle and reduction on the inhomogeneity of steel extrusions. Numbers are the difference between the surface and centerline hardnesses. Lines of constant Δ were calculated from equation 11.4. Data from Miura *et al.*, *ibid.*

(d) The reference on which Figure 11.7 is based gives $H = 134\varepsilon^{0.15}$. Using this make a plot of I.F. versus α for $r = 0.08$, 0.16, and 0.32 and compare with Figure 11.7.

SOLUTION:

(a) The hodograph (Figure 9.13b) shows that the velocity discontinuity at the surface at the entrance and exit is $V_0 \tan\alpha$. This corresponds to a work rate of $kV_0 \tan\alpha dy$ in an element of thickness dy. For both the entrance and exit, the redundant work rate is $2kV_0 \tan\alpha dy$ and the work per volume is $2k_0 \tan\alpha$. Therefore the redundant strain is $\varepsilon_r = (2k/\sigma_0)\tan\alpha$. Taking $2k = \sigma_0$, $\varepsilon_r = \tan\alpha$ at the surface.
(b) The strain at the centerline is the homogeneous strain, ε_h so the total strain at the surface is $\tan\alpha + \varepsilon_h$. Therefore, $\varepsilon_s/\varepsilon_c = 1 + \tan(\alpha/\varepsilon_h)$
(c) $I.F. = H_s/H_c - 1 = (\varepsilon_s/\varepsilon_c)^n - 1 = [1 + \tan\alpha/\varepsilon_h)]^n - 1$
(d) Taking $\varepsilon_h = -\ln(1 - r)$ and evaluating, Plotting:

		(I.F.)			
r	ε	$\alpha = 5°$	$\alpha = 10°$	$\alpha = 20°$	$\alpha = 30°$
0.08	0.0834	0.114	0.186	0.287	0.364
0.16	0.1743	0.063	0.111	0.184	0.245
0.32	0.3857	0.031	0.058	0.105	0.147

This shows that the simple theory is in good agreement with the experimental data for $r = 0.16$ and 0.32 but overestimates *IF* for $r = 0.08$.

All of the previous analyses have assumed dies that are conical or wedge-shaped. However Devenpeck and Richmond* designed plane-strain dies with slip-line theory and axisymmetric dies with numerical analysis that, in principle, produce no redundant strain and therefore no surface-to-center property gradients. Figure 11.10 shows two such profiles and Figure 11.11 compares experimental and theoretical grid distortion through such dies. Although the agreement is good, it should be noted that low Δ characterizes the dies.

11.5 INTERNAL DAMAGE

The high hydrostatic tension at the centerline which occurs under high Δ conditions can lead to internal porosity and under extreme conditions to internal cracks. This is evident in Figure 10.24, where $\Delta = 5.83$ and there is mid-plane hydrostatic tension of $\sigma_2 = 0.62(2k)$. Mid-plane hydrostatic tension can exist even for extrusion. Coffin and Rogers† found significant losses of density after slab drawing as indicated in Figure 11.12. Note that the density loss increases with die angle and is cumulative, increasing from pass to pass.

The loss in density is caused by microscopic porosity being formed at hard inclusions. There is much less density loss in clean single-phase materials than in dirtier

* M. L. Devenpeck and O. Richmond, *J. Engr. Ind. Trans. ASME*, 87 (1965).

† L. F. Coffin Jr. and H. C. Rogers, *Trans. ASME*, 86 (1967).

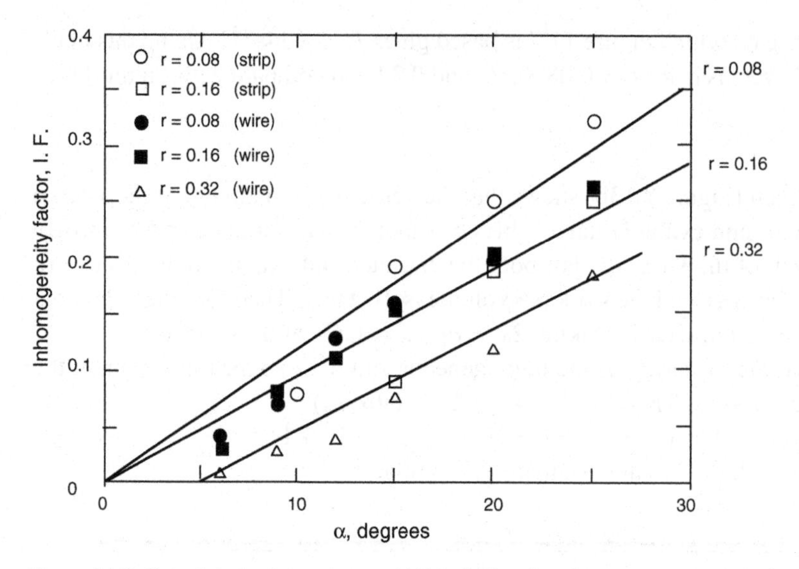

Figure 11.9. Calculated α dependence of *I.F.* (solid lines) and experimental values (symbols) for various reductions.

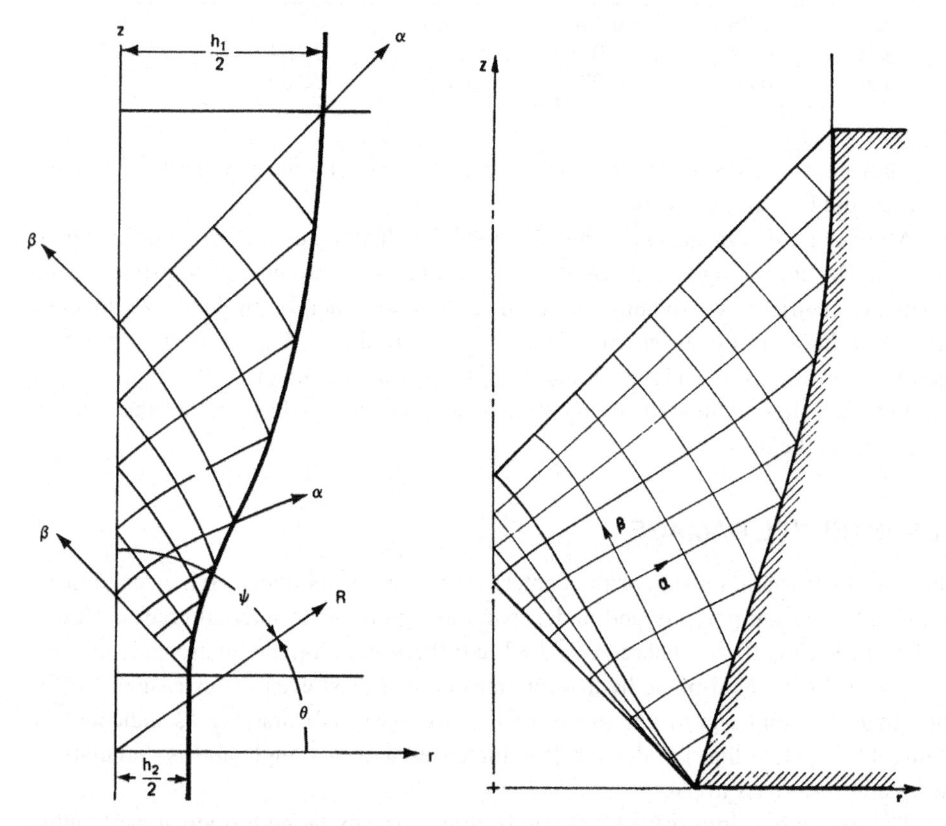

Figure 11.10. Two die profiles designed to give zero redundant strain for frictionless conditions. From (left) O. Richmond in *Mechanics of the Solid State*, F. P. J. Rimrott and J. Schwaightofer eds. (University of Toronto Press, 1968) and O. Richmond and M. L. Devenpeck. *Proc. 4th U. S. Nat. Cong. of Appl. Mech.* (1962) and (right) O. Richmond and H. L. Morrison, *J. Mech. Phys.* Solids, 15 (1967).

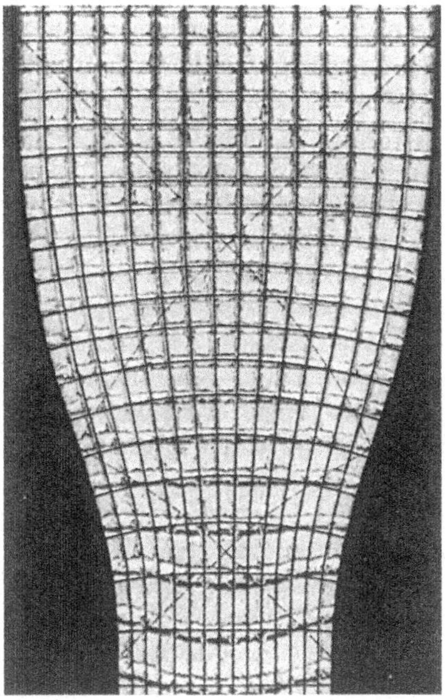

Figure 11.11. Overlay of theoretical grid distortions on actual distortions from a die like that in Figure 11.10. From O. Richmond and M. L. Devenpeck, *J. Engr. Ind. Trans. ASME*, 87 (1965).

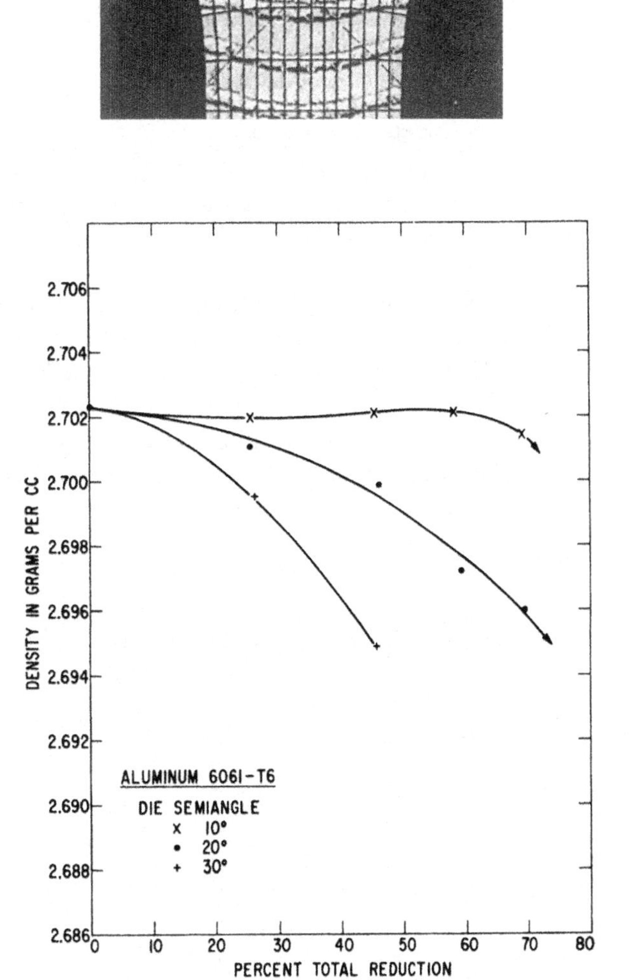

Figure 11.12. Density changes after plane-strain drawing in 6061-T6 aluminum. From H. C. Rogers, General Electric Co. Report No. 69-C-260 (July 1969).

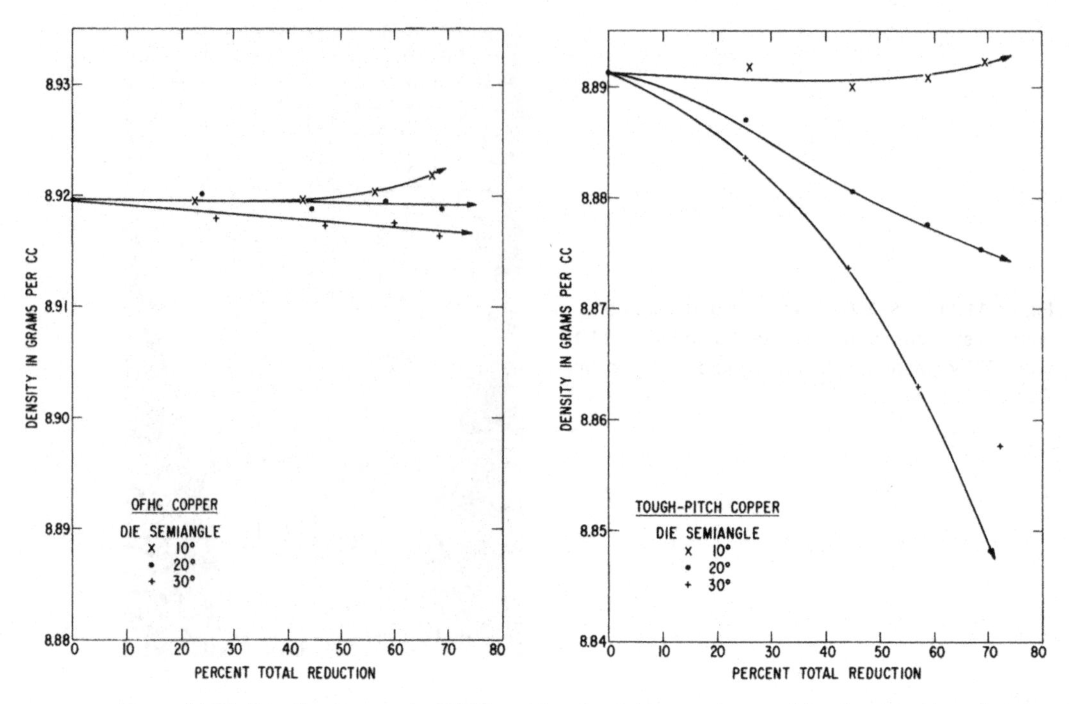

Figure 11.13. Density changes in OFHC and tough-pitch copper caused by drawing. Note that the density loss is much greater in the tough-pitch copper which contains many hard Cu_2O particles. From H. C. Rogers, *ibid.*

material with two phases. Figure 11.13 compares the density loss in tough-pitch copper with a high content of Cu_2O particles with that in oxygen-free high-conductivity (OFHC) copper. They also found that the loss of density could be decreased or eliminated by drawing under external hydrostatic pressure.

Under severe conditions pores may link up forming internal cracks. Figure 11.14 shows an early stage of such linking up and Figure 11.15 shows chevron cracks formed at very high Δ conditions in extrusion. The terms "arrowhead cracks," "centerline

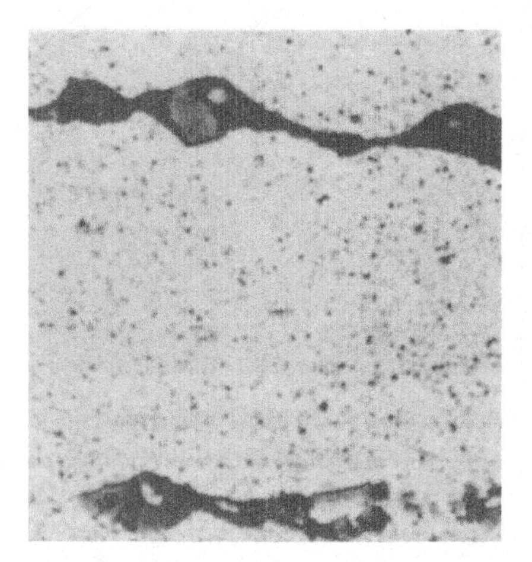

Figure 11.14. Voids formed in 6061-T6 aluminum after plane-strain drawing through a 30° die to a total reduction of 75%. From H. C. Rogers, R. C. Leech and L.F. Coffin Jr. *Final Report, Contract NOw-65-0097-f* Bureau of Naval Weapons.

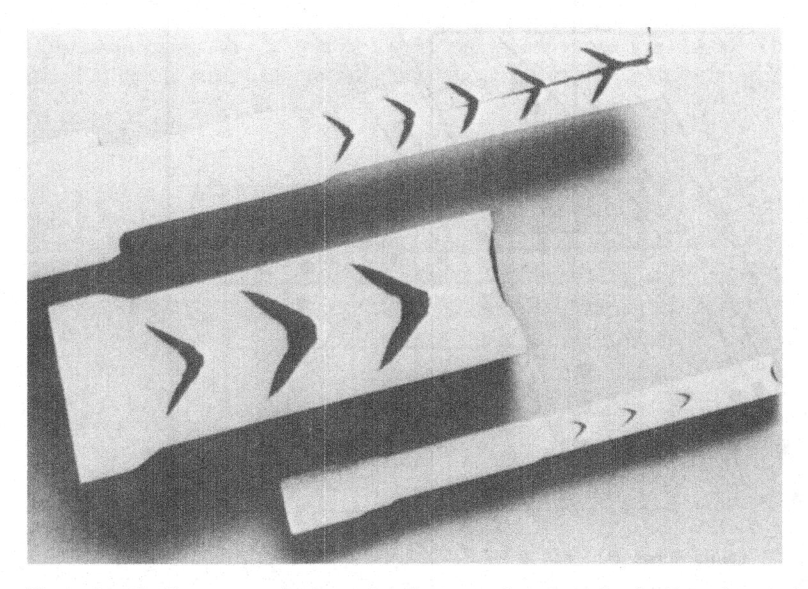

Figure 11.15. Chevron cracks formed during extrusion of steel rods. Note the very low reduction and high die angle. From D. J. Blickwede, *ibid*.

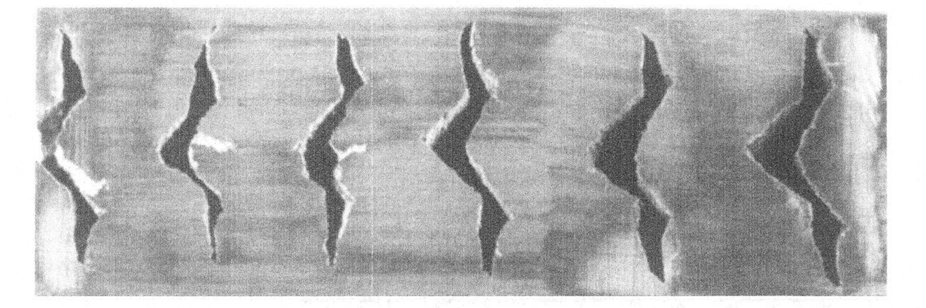

Figure 11.16. Cracks formed in a molybdenum bar during rolling under high Δ conditions.

burst," and "cuppy core" have been applied to this phenomenon. Figure 11.16 shows similar cracks formed during rod-rolling.

Avitzur* has studied the problem of centerline bursts with an upper-bound analysis and found conditions under which formation of central holes lowers the deformation energy. Figure 11.17 shows the results of his analysis and experimental results for several work-hardened steels. It is interesting to note that the conditions which produce centerline bursts can be described by $\Delta > 2$.

11.6 RESIDUAL STRESSES

The magnitude and nature of residual stresses caused by mechanical processing depend on the shape of the deformation zone. With Δ-values of one or less (high reductions and low die angles) the deformation is relatively uniform and there are only minor residual stresses. With high Δ-values the surface is left under residual tension with residual compression at the centerline. The magnitude of the stresses increases with Δ as shown in Figures 11.18 for rolling, 11.19 for drawing and 11.20 for extrusion.

* B. Avitzur, *Metal Forming: Processes and Analysis*, McGraw-Hill 1969.

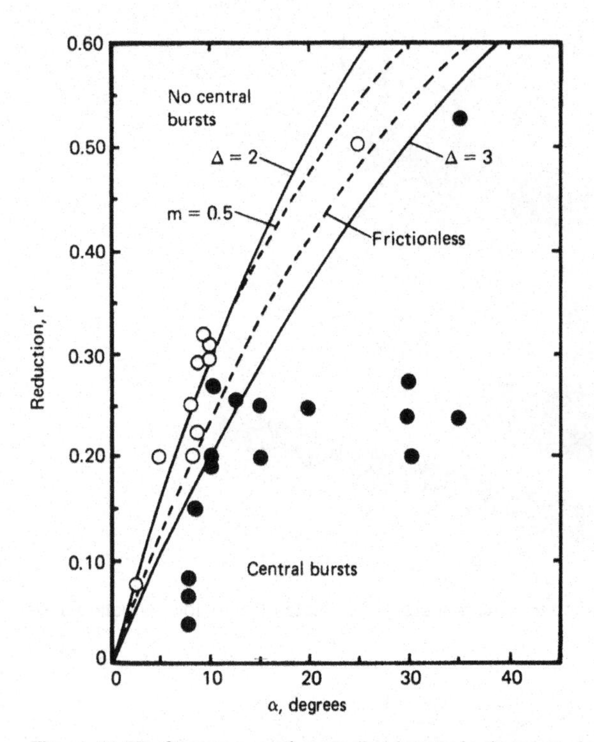

Figure 11.17. Occurrence of centerline bursts in three steels during extrusion. Filled circles indicate bursts and open circles indicate no bursts. The dashed lines are Avitzur's analysis for no friction and for interface shear stress, $mk = 0.5k$. The solid lines are for constant Δ. Data from Z. Zimmerman and B. Avitzur, *J. Eng. Ind, Trans. ASME* 92 (1970).

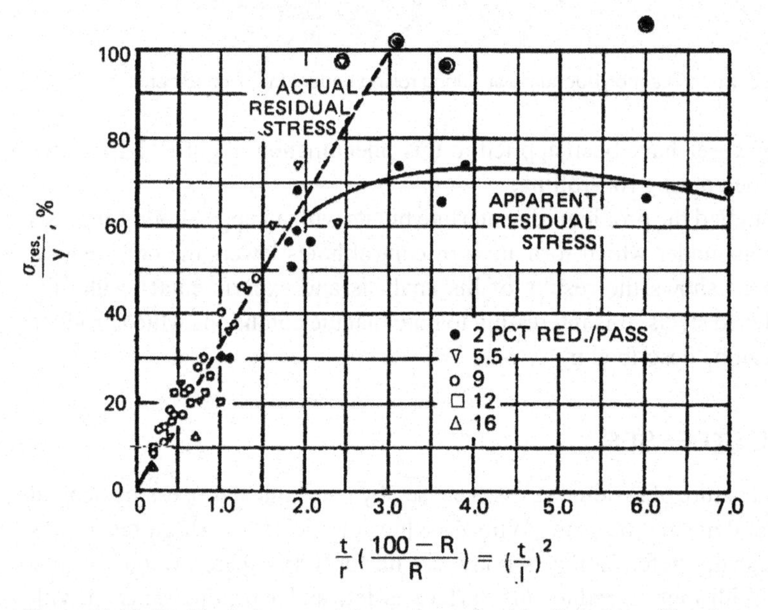

Figure 11.18. Residual stresses at the surface of cold-rolled strips. The stresses are normalized by the yield strength. Note that $(t/l)^2 = \Delta^2$. From W. M. Baldwin. *Proc. ASTM,* 49 (1949).

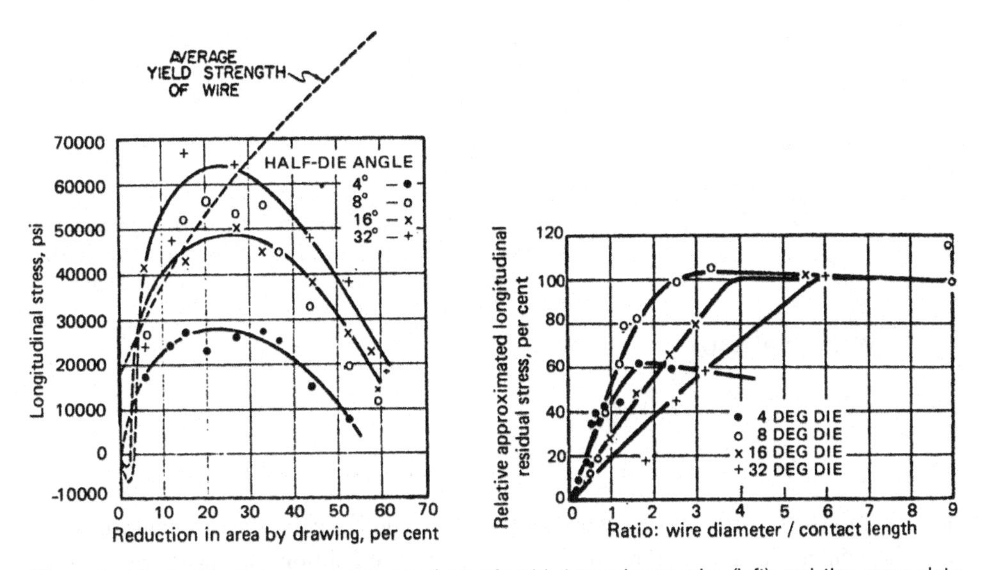

Figure 11.19. Residual stresses at the surface of cold-drawn brass wire (left) and the same data replotted as the ratio of residual stress to yield strength versus Δ (right). From W. M. Baldwin, *ibid*. Original data was from W. Linius and G. Sachs, *Mitt. Dtsch. Materialprüfunsanst*, 16 (1932).

High residual stresses, coupled with centerline damage, may cause spontaneous splitting or "alligatoring" of the work piece as it leaves the deformation zone. Figure 11.21 shows an example. In rolling this type of failure is most likely in early breakdown of ingots because the section thickness is usually large relative to the roll

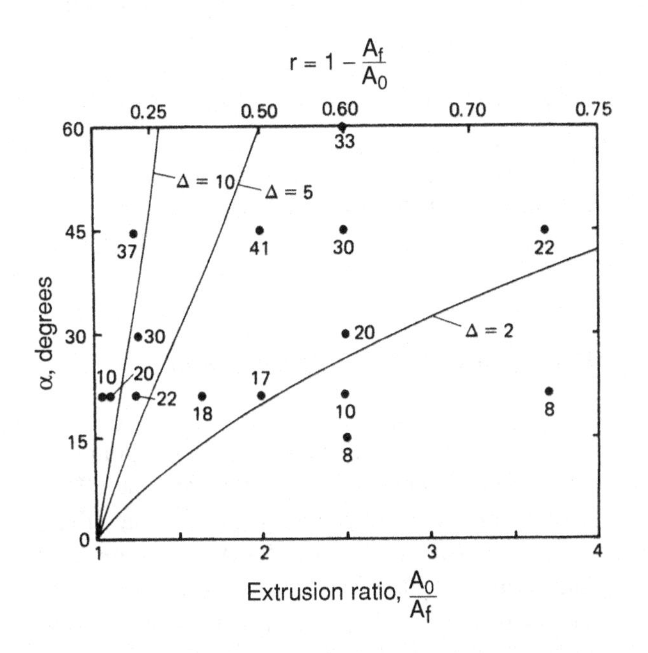

Figure 11.20. Effect of die angle and reduction on the residual stresses in extruded steel rods. The numbers are the residual stress on the surface in kg/mm^2. The lines of constant Δ fit the data well. Data from S. Miura et al, *ibid*.

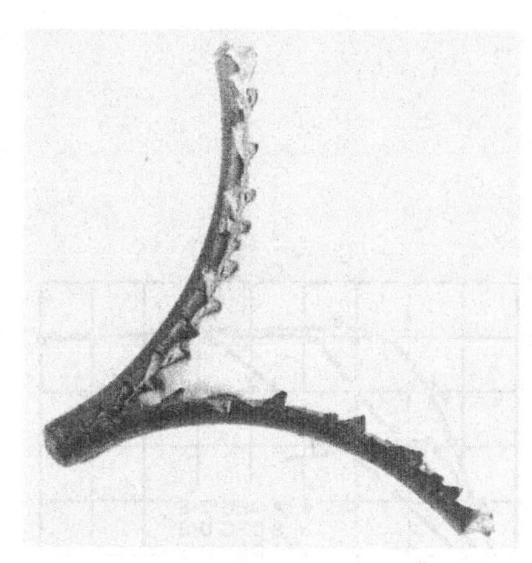

Figure 11.21. Molybdenum rod that alligatored during rolling under high Δ conditions. The "teeth" formed by edge cracking are not common in alligatored bars.

diameter. Furthermore ingots often have porosity and second phases concentrated near the centerline. Hot rolling under favorable Δ conditions tends to close such porosity. Residual tension at the surface is particularly undesirable because it leads to an increased susceptibility to fatigue and stress-corrosion cracking.

Despite the general usefulness of slip-line field theory in metal forming, it does not adequately explain the observed patterns of residual stress. For deformation under high Δ conditions, the theory predicts that during deformation, the level of hydrostatic stress is highest at the centerline and most compressive at the surface. The theory allows for no plastic deformation once the material leaves the deformation zone, so unloading must be perfectly elastic. The normal stress must go to zero and the horizontal force must go to zero. It is difficult to rationalize how there can be complete reversal of stresses on unloading.

If the deformation is rapid enough to be nearly adiabatic, the surface will be hotter than the interior. The surface and interior must undergo the same contraction on cooling so the surface will be left under residual tension.

Very low reductions lead to an opposite pattern of residual stress with the surface being left in residual compression. Figure 11.22 illustrates this. Note, however, that as soon as the reduction reaches 0.8% the normal pattern of residual tension on the surface is observed.

11.7 COMPARISON OF PLANE-STRAIN AND AXISYMMETRIC DEFORMATION

Friction, redundant strain and inhomogeneity effects on plane-strain and axisymmetric forming are summarized in Table 11.1. A general conclusion is that frictional effects depend primarily on die angle. In plane-strain and axisymmetric forming the effects are the same for the same values of α and r. These are the same for Δ in axisymmetric forming half as large as the Δ in plane-strain.

Similarly, the inhomogeneity at low α and r is the same in both types of operations. At larger reductions the data of Miura et al. indicate a Δ-dependence for axisymmetry. Redundant strain correlates best with Δ. For equal values of Δ, plane-strain operations are characterized by much higher ϕ than for axisymmetry. However

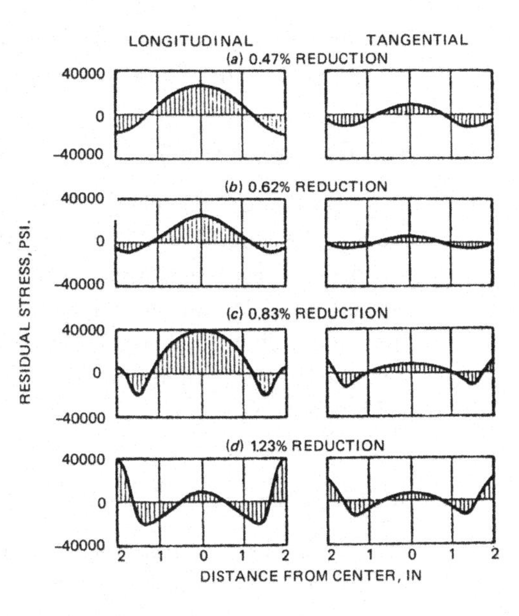

Figure 11.22. Residual stresses in lightly drawn steel rods. From W. M. Baldwin, *ibid,* after H. Bühler and H. Buchholtz, *Archiv. für des Eisenhüttenwesen*, 7, (1933–4), pp. 427–30.

Table 11.1. Comparison of Axisymmetric and Plane-Strain Drawing

	Plane-strain	Axisymmetry	Comment
Ratio of mean cross-sectional area to tool–work piece contact area	$\Delta/2$	$\Delta/4$	Equal for same *r* and α.
Frictional contribution to $\sigma_d/2k$ according to slab analysis with shear stress	$m/\sin 2\alpha$	$m/\sin 2\alpha$	Equal for same *r* and α, but with same Δ, axisymmetry constant has twice the effect.
Frictional contribution to $\sigma_d/2k$ according to slab analysis with friction coefficient	$\approx\mu\cot\alpha(1-\varepsilon/2)$	$\approx\mu\cot\alpha(1-\varepsilon/2)$	Equal for same *r* and α, but with same Δ, axisymmetry constant has twice the effect.
Redundant strain contribution to $\sigma_d/2k$ according to upper-bound slab analysis	$\approx\Delta/4$	$\approx\Delta/6$	For same *r* and α, less redundant strain in plane-strain but for same Δ less for axisymmetry.
Redundant strain according to slip-line field theory	$\approx 0.23(\Delta-1)$	no solution for axisymmetry	
Redundant strain from Burke and Backofen experiments[†]	$\approx 0.21(\Delta-1)$	$\approx 0.12(\Delta-1)$	For same Δ, nearly twice as much for plane-strain but for same α and *r*, roughly equal.
Inhomogeneity factor	$[1+\tan(\alpha/\varepsilon_h)]^n-1$		Approximately equal for axisymmetry and plane-strain at the same α.

[†]Source: W. A. Backofen, *op. cit.*

if comparisons are made at equal ratios of mean cross-sectional area to contact area, the values of ϕ are slightly higher for axisymmetry.

NOTE OF INTEREST

W. A. Backofen (December 8, 1928–November 8, 2006) was born in Rockville, Connecticut. He studied at MIT from 1943 to 1950, when he received his doctorate and was appointed as an assistant professor. He did much innovative research on metal forming in his 25 years at MIT. He introduced the basic concepts of superplasticity and of texture strengthening and made clear the importance of deformation-zone geometry. He retired in 1975 to a farm in New Hampshire where he raised Christmas trees, apples, and blueberries.

REFERENCES

W. A. Backofen, *Deformation Processing*, Addison-Wesley, 1972.
B. Avitzur, *Metal Forming: Processes and Analysis*, McGraw-Hill, 1968.
E. M. Meilnik, *Metalworking Science and Engineering*, McGraw-Hill, 1991.

PROBLEMS

11.1. The XYZ Special Alloy Fabrication Company has an order to roll a $4 \times 4 \times 15$ in billet of a nickel-base super alloy to ½ in thickness. They tried rolling it with a 5% reduction per pass, but it split on the fourth pass. At a meeting, the process engineer suggested using a lower reduction per pass, the consultant suggested applying forward and backward tension, and the shop foreman was in favor of heavier reductions per pass. With whom, if anyone, do you agree? Defend your position.

11.2. The Mannesmann process for making tubes from cylindrical billets is illustrated in Figure 11.23. It involves passing the billet between two nonparallel rolls adjusted for a very small reduction onto a mandrel positioned in the middle. Explain why the axial force on the mandrel is low and why the mandrel, which is long and elastically flexible, follows the center of the billet.

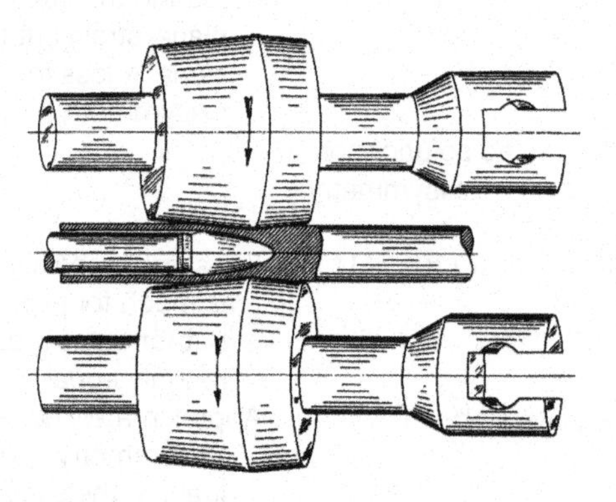

Figure 11.23. The Mannesmann process for making tubes.

11.3. **a)** Show that, as r and α approach zero, equations 11.11 and 11.12 reduce to $\phi = 1 + \Delta/4$ for plane-strain and $\phi = 1 + \Delta/6$ for axisymmetry.
b) What percent error does this simplification introduce at $r = 0.5$ and $\alpha = 30°$?

11.4. Consider wire drawing with a reduction of $r = 0.25$ and $\alpha = 6°$.

a) Calculate the ratio of average cross-sectional area to the contact area between wire and die.

b) What is the value of Δ?

c) For strip drawing with $\alpha = 6°$, what reduction would give the same value of Δ?

d) For the reduction in c), calculate the ratio of average cross-sectional area to the contact area and compare with your answer to a).

e) Explain why, for the same Δ, the frictional drag is greater for axisymmetric drawing.

11.5. Some authorities define Δ as the ratio of the length of an arc through the middle of the deformation zone and centered at the apex of the cone or wedge formed by extrapolating the die to the contact length.

a) Show that, for wire drawing with this definition, $\Delta = \alpha(1 + \sqrt{1-r})^2/r$.

b) Calculate the ratio of Δ from this definition to the value of Δ from equation 11.4 for i) $\alpha = 10°$, $r = 0.25$; ii) $\alpha = 10°$, $r = 0.50$; iii) $\alpha = 45°$, $r = 0.50$; and iv) $\alpha = 90°$, $r = 0.50$.

11.6. Backofen developed the following equation for the mechanical efficiency, η, during wire drawing: $\eta = [1 + C(\Delta - 1) + \mu \tan\alpha]^{-1}$ for $\Delta > 1$, where C is an empirical constant equal to about 0.12 and μ is the friction coefficient. Note that the term $C(\Delta - 1)$ represents w_r/w_i, and the term $\mu/\tan\alpha$ represents w_f/w_i, so $\eta = (1 + w_r/w_i + w_f/w_i)^{-1} = (w_a/w_i)^{-1}$.

a) Evaluate this expression for $\mu = 0.05$ and $\alpha = 0.3$ for $2 \leq \alpha \leq 20°$. Note that for small angles $\Delta < 1$ so the redundant strain term, $C(\Delta - 1)$, is zero.

b) Plot η vs. α and find the optimum die angle, α^*.

c) Derive an expression for α^* as a function of μ and r. Does α^* increase or decrease with μ? With r?

11.7. A slab of annealed copper with dimensions 2.5 cm × 1.5 cm × 8 cm is to be compressed between rough platens until the 2.5 cm is reduced to 2.2 cm. Describe, with a sketch, and as fully as possible, the resulting inhomogeneity of hardness.

11.8. A high-strength steel bar must be cold reduced from a diameter of 2 cm to 1.2 cm by drawing. A number of schedules have been proposed. Which schedule would you choose to avoid drawing failure and minimize the likelihood of centerline bursts? Assume that $\eta = 0.5$, and give your reasoning.

a) A single pass through a die of semi-angle 8°

b) Two passes (2.0 to 1.6 cm and 1.6 to 1.2 cm) through a die of semi-angle 8°

c) Three passes (2.0 to 1.72 cm, 1.72 to 1.44 cm, and 1.44 to 1.2 cm) through a die of semi-angle 8°

d) Four passes (2.0 to 1.8 cm, 1.8 to 1.6 cm, 1.6 to 1.4 cm, and 1.4 to 1.2 cm) through a die of semi-angle 8°

e, f, g, h) Same reductions as in schedules a, b, c, and d except through a die of semi-angle 16°

11.9. Determine Δ from the data of Hundy and Singer (Figure 11.6) for reductions of 2.3%, 6.5%, 13.9%, and 26%. Plot *I.F.* versus Δ.

12 Formability

An important concern in forming is whether a desired process can be accomplished without failure of the work material. Forming limits vary with material for any given process and deformation-zone shape. As indicated in Chapter 11, central bursts may occur at a given level of Δ in some materials and not in others. Failure strains for a given process depend on the material.

12.1 DUCTILITY

In most bulk forming operations, formability is limited by ductile fracture.* Forming limits correlate quite well with the reduction of area as measured in a tension test. Figure 12.1 shows the strains at which edge cracking occurs in rolling as a function of the tensile reduction in area. The fact that the limiting strains for strips with square edges strip higher than those with rounded edges indicates that process variables are also important. Similar results are reported for other processes.

12.2 METALLURGY

The ductility of a metal is strongly influenced both by the properties of the matrix and by the presence of inclusions. Factors that increase the strength generally decrease ductility. Solid solution strengthening, precipitation, cold work and decreased temperatures all lower fracture strains. The reason is that with higher strengths, the stresses necessary for fracture will be encountered sooner.

Inclusions play a dominant role in ductile fracture. The volume fraction, nature, shape, and distribution of inclusions are important. In Figure 12.2, the tensile ductility is seen to decrease with increased amounts of artificial inclusions. Despite the scatter, it appears that some second-phase particles are more deleterious than others. Figure 12.3 shows similar trends for oxides, sulfides, and carbides in steel.

Mechanical working tends to elongate and align inclusions in the direction of extension. This mechanical fibering reduces the fracture strength and ductility normal

* Wire drawing is an exception; the maximum reduction per pass is limited by the ability of the drawn section to carry the required drawing force without yielding and necking. Once the drawn wire necks, it can no longer support the required load, so the subsequent fracture strain is of little concern. It is also possible, though not common, that brittle fracture limits formability.

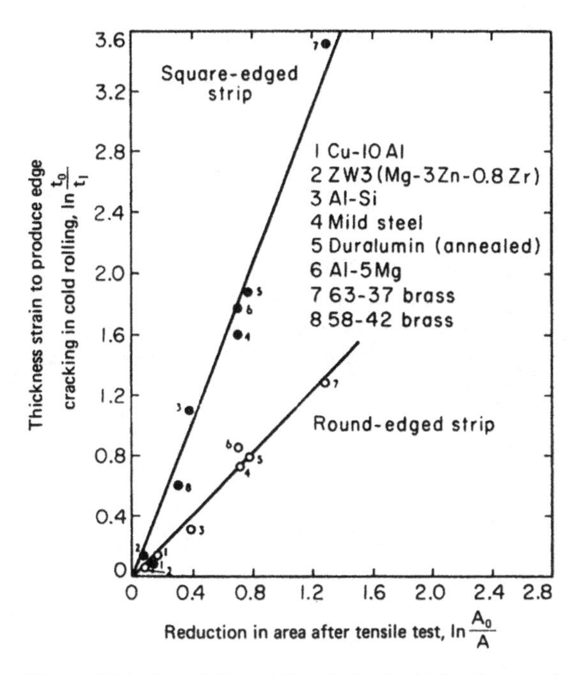

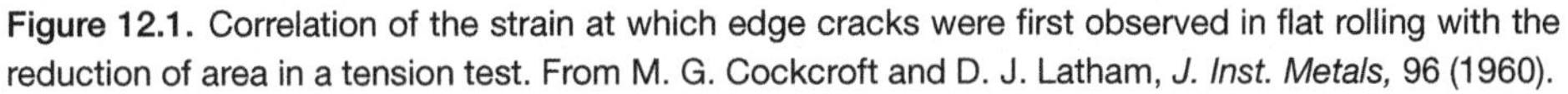

Figure 12.1. Correlation of the strain at which edge cracks were first observed in flat rolling with the reduction of area in a tension test. From M. G. Cockcroft and D. J. Latham, *J. Inst. Metals,* 96 (1960).

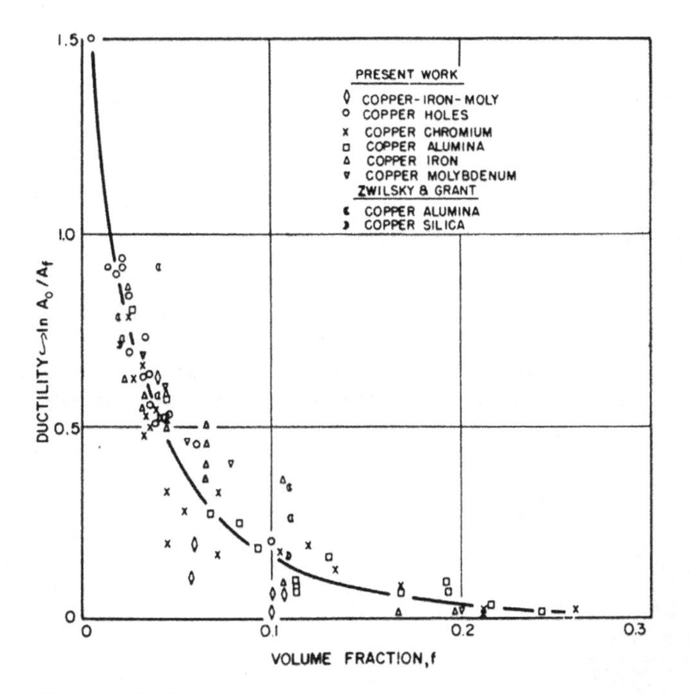

Figure 12.2. Effect of volume fraction second-phase particles in copper on tensile ductility. From B. I. Edelson and W. M. Baldwin, *Trans. Q. ASM,* 55 (1962), pp. 20–50.

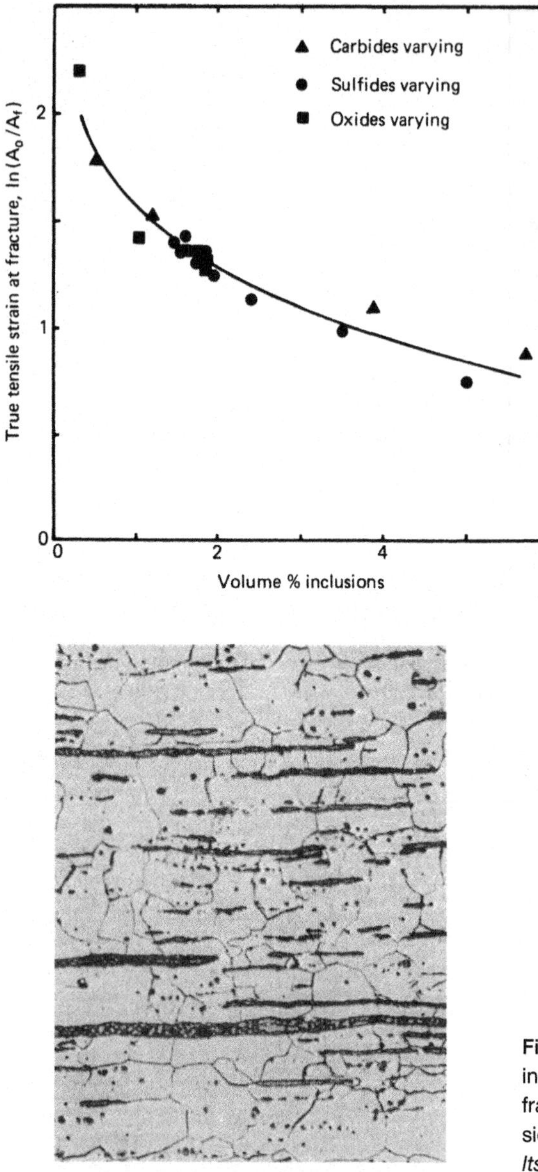

Figure 12.3. Effect of volume fraction natural inclusions on the tensile ductility of steel. Data from F. B. Pickering, *Physical Metallurgy and Design of Steels*, London: Applied Science Pub, 1978, p. 51.

Figure 12.4. Microstructure of wrought iron (top) showing elongated glassy silicates inclusions and the woody fracture (bottom) caused by cracking along the inclusions. From J. Aston and E. B. Story, *Wrought Iron and Its Manufacture, Characteristics and Applications*, Byer Co. Pittsburg, 1939.

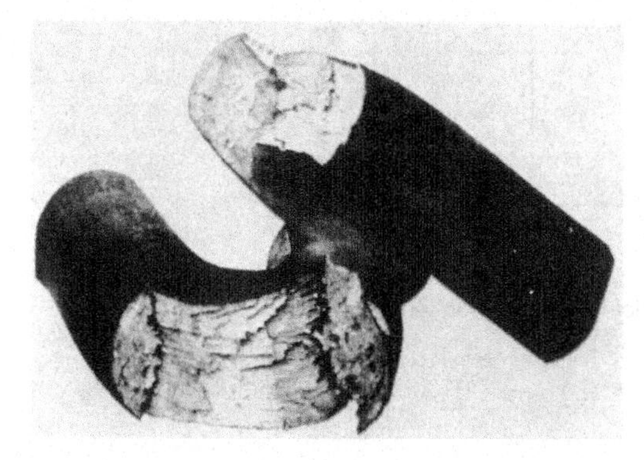

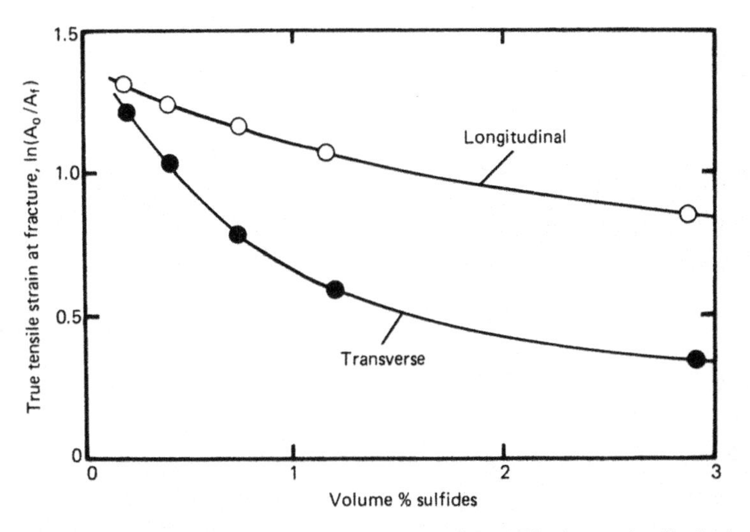

Figure 12.5. Effect of sulfides on the ductility of steel. The lower ductility in the transverse direction is cause by the inclusions being elongated during hot rolling. Data from F. B. Pickering, *ibid*.

to the working direction. Wrought iron, though no longer a commercial product, is an extreme example. Crack paths follow glassy silicate inclusions producing woody fractures (see Figure 12.4).

In steels, MnS inclusions, elongated during hot working are a major cause of fracture anisotropy as illustrated in Figure 12.5. Transverse ductility can be markedly improved by *inclusion shape control* (Figure 12.6). This is a practice of adding small amounts of Ca, Ce, Ti, or rare earths elements to react with the sulfur to form hard inclusions that remain spherical during working. If the added cost is justified, the inclusion content can be reduced by desulfurization of vacuum melting.

The formability of medium and high carbon steels can be improved by spheroidizing the carbides. For example, the manufacture of bolts requires upsetting their ends.

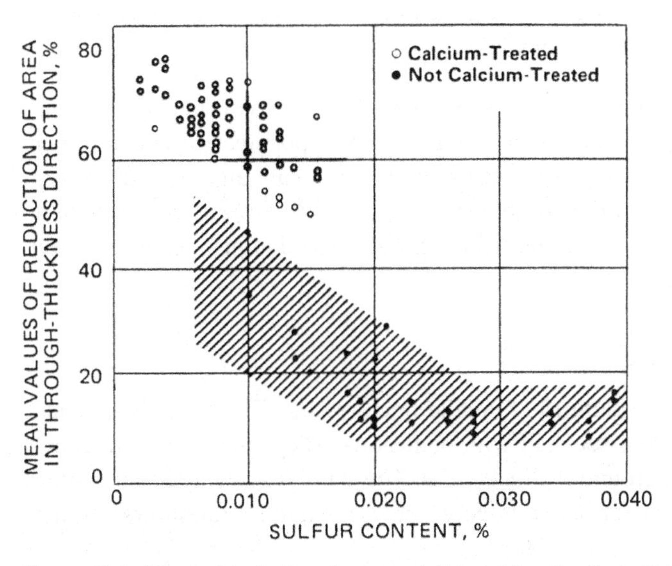

Figure 12.6. Effect of inclusion shape control in raising the through-thickness ductility of an HSLA steel. From H. Pircher and W. Klapner in *Micro Alloying 75*, Union Carbide, 1977.

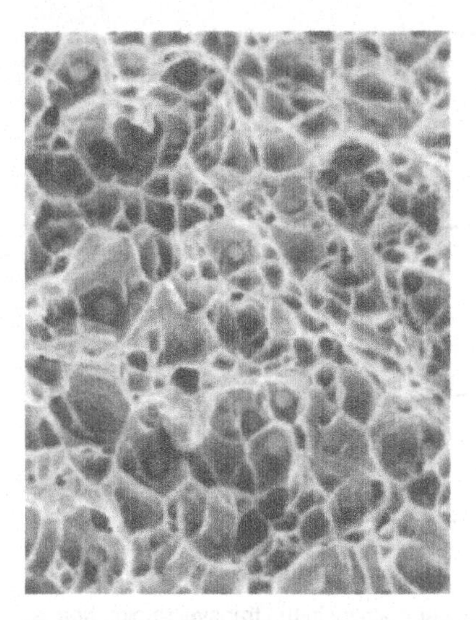

Figure 12.7. Dimpled fracture surface of steel. Note that the microvoids are associated with inclusions. Courtesy of J. W. Jones.

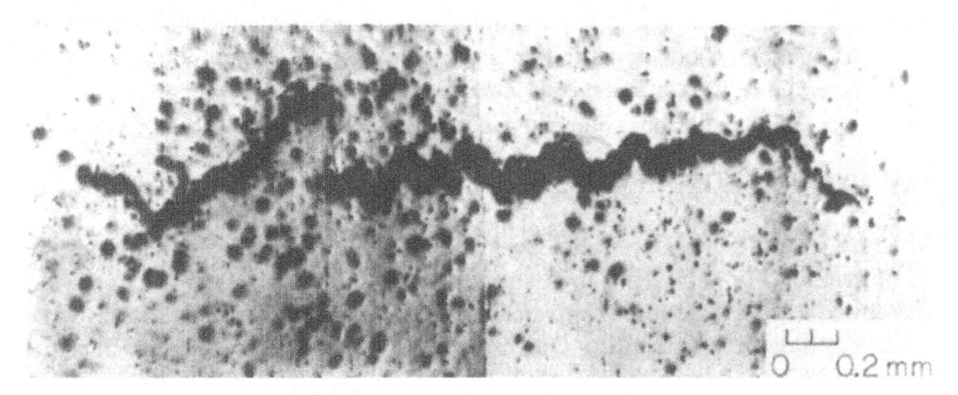

Figure 12.8. Crack in the center of a neck in a tensile specimen of 1018 steel. Oil stain from diamond polishing exaggerates the hole size. From F. McClintock in *Ductility* (ASM, 1968).

Bolts made from medium carbon steels must be spheroidized to withstand the circumferential strains in upsetting. Although spheroidization of hypoeutectoid steels is often accomplished by heating the steel above the lower critical temperature into the $\alpha + \gamma$ phase region and then slowly cooling just below the critical into the $\alpha +$ carbide region, spheroidizing occurs much quicker if the steel is simply heated to just below the lower critical*.

12.3 DUCTILE FRACTURE

Scanning electron microscope pictures of ductile fracture (e.g., Figure 12.7) are characterized by dimples that are associated with inclusions. They suggest that fractures initiate by opening of holes around the inclusions. These holes grow by plastic deformation until they link up to form macroscopic cracks. Figure 12.8 shows a crack

* J. O'Brien and W. Hosford, *Met. and Mat. Trans.*, v. 33 (2002).

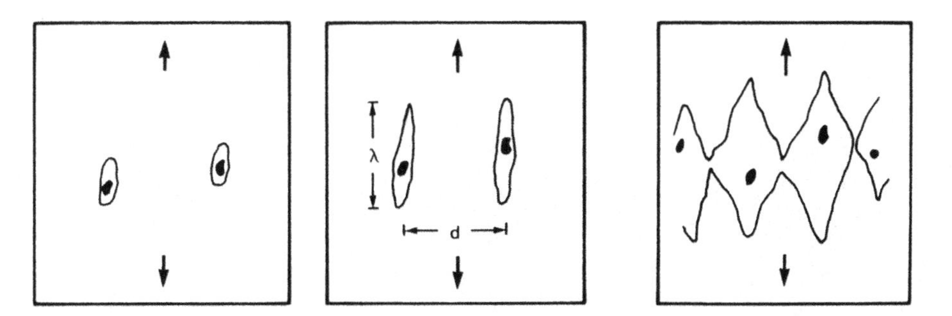

Figure 12.9. Schematic sketch showing how a ductile crack can form by necking between elongated voids around inclusions.

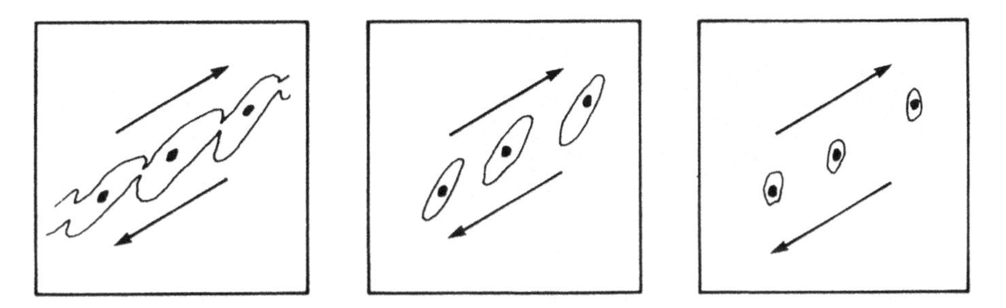

Figure 12.10. Schematic sketch showing how a ductile crack can form in shear by linking of voids around inclusions.

near the center of the neck in a tensile bar of cold-rolled 1018 steel prior to final fracture.

The joining of voids may be by their preferential growth parallel to the direction of highest principal stress until their length, λ, approximately equals their separation distance, d. Then the ligaments between them may neck on a microscopic scale as suggested in Figure 12.9.

Ductile failure may also occur by localized shear. One example is the shear lip formed on the cup-and-cone fractures in tension tests of ductile metals. Shear fracture may also occur in rolling and other forming processes. Void formation around inclusions also plays a dominant role in shear fractures as indicated in Figures 12.10 and 12.11.

12.4 HYDROSTATIC STRESS

Fracture strains depend on the level of hydrostatic stress during deformation as shown in Figure 12.12. High hydrostatic pressure suppresses void growth, thereby delaying fracture. Conversely, hydrostatic tension accelerates void growth.

Bridgman* made a series of tension tests on a mild steel under superimposed pressure. He found that the pressure suppressed final fracture, greatly increasing the reductions of area. Figure 12.13 shows his specimens.

* P. W. Bridgman, *Studies in Large Plastic Flow and Fracture*, Wiley, 1952.

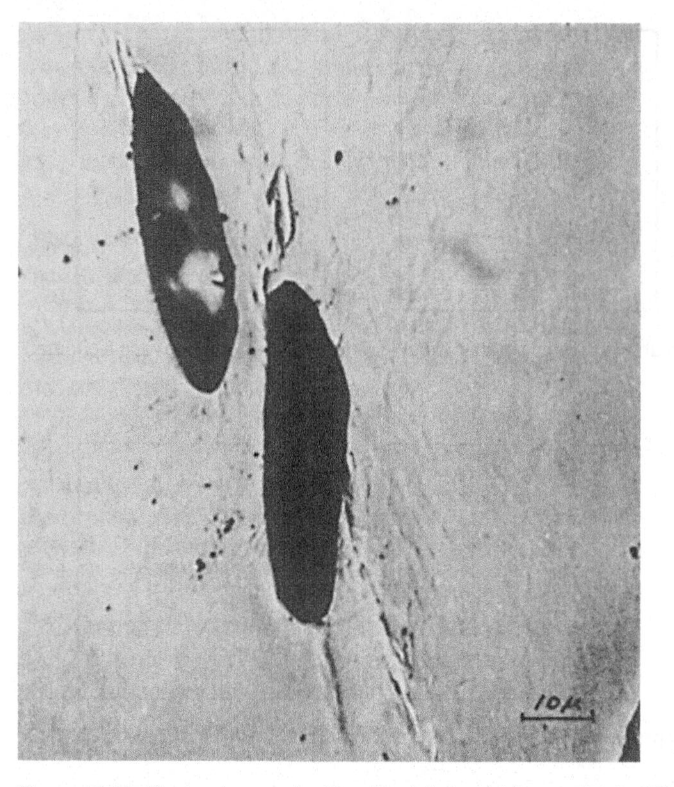

Figure 12.11. Photomicrograph of localized shear with large voids in OFHC copper. From H. C. Rogers, *Trans. TMS-AIME* v. 218 (1960).

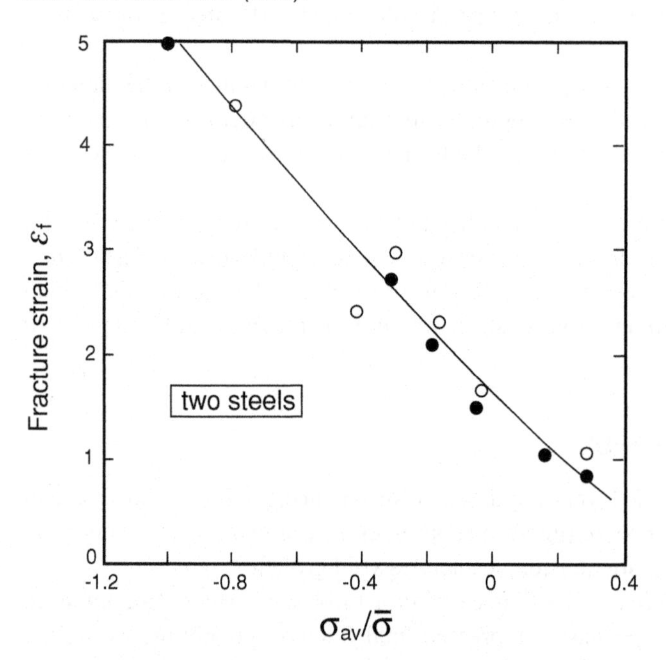

Figure 12.12. Correlation of fracture strain with the ratio of hydrostatic stress to effective stress for two steels. Adapted from A. L. Hoffmanner, *Interim Report, Air Force Contract F 33615-67-1466* TRW (1967).

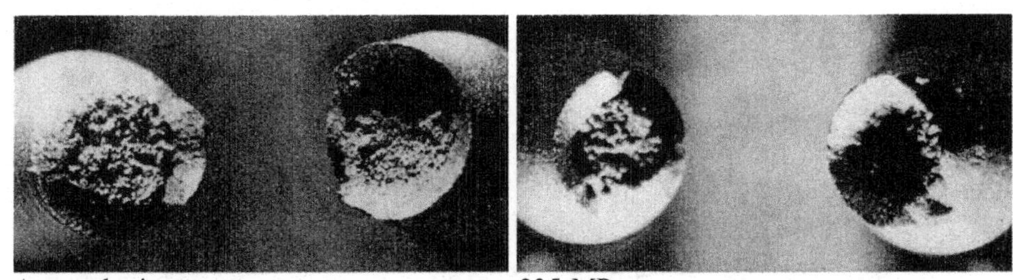

Atmospheric pressure 235 MPa

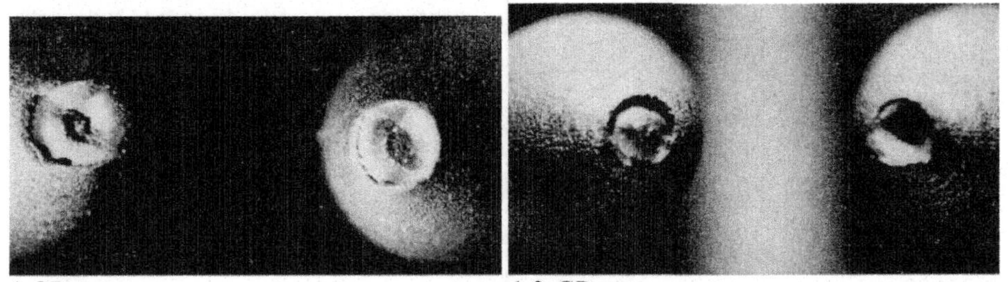

1 GPa 1.3 GPa

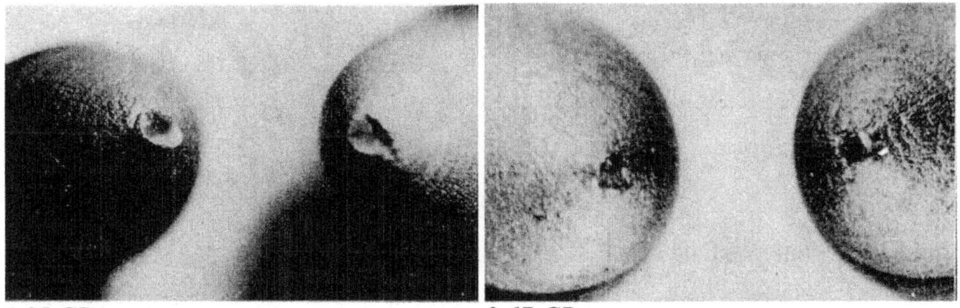

1.85 GPa 2.67 GPa

Figure 12.13. Increase of reduction of area with increased pressure. From P. W. Bridgman, *Studies in Large Plastic Flow and Fracture* (Wiley, 1952).

One measure of the extent of void growth is the decrease in density discussed in Chapter 11. Figure 12.14 shows the effect of hydrostatic pressure on density changes during strip drawing of 6061-T6 aluminum.

Formability problems would be minimized if all stress components could be maintained compressive. Materials of very limited ductility can be extruded if the exit region is under high hydrostatic pressure. Without such special conditions, tensile stresses are apt to arise in processes that are nominally compressive. Edge cracking during rolling and formation of center bursts in extrusion are examples. During upset forging, barreling causes hoop tensile stresses.

Although it has been suggested that failure strains depend on the level of the largest principal stress, σ_1, rather than on the hydrostatic stress, σ_m, the distinction is not great as can be seen from the extreme possibilities of axisymmetric elongation, $\sigma_m = \sigma_1 - (2/3)\bar{\sigma}$ and axisymmetric compression, $\sigma_m = \sigma_1 - (1/3)\bar{\sigma}$. All other

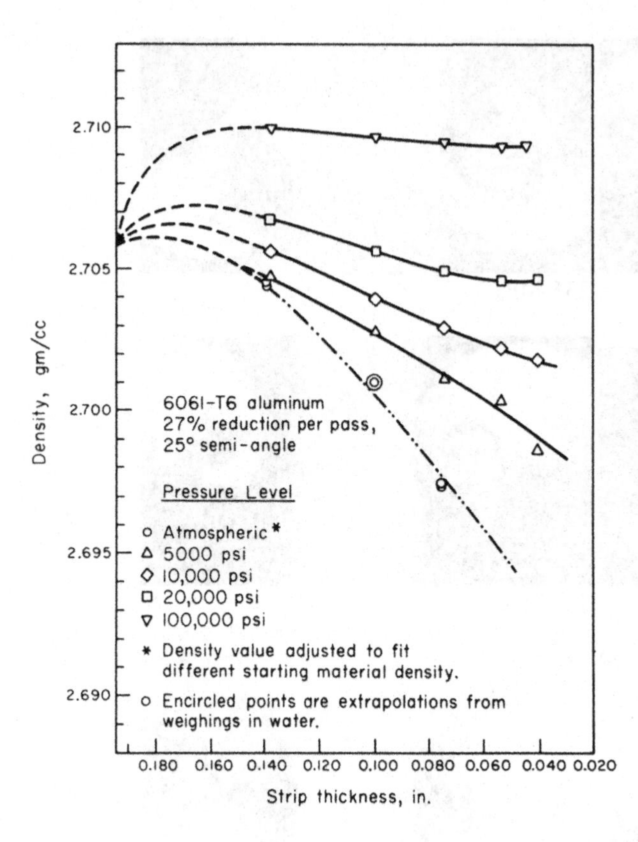

Figure 12.14. Loss of density during strip drawing with superimposed hydrostatic pressure. Note that drawing at the highest pressure increased the density by closing up pre-existing pores. From H. C. Rogers and L. F. Coffin Jr. *Final Rept. Contract NO w-66-0546-d* (June 1996) Naval Air Systems Command. See also H. C. Rogers in *Ductility* (ASM 1968).

conditions are intermediate so

$$\sigma_m - (1/3)\bar{\sigma} \leq \sigma_1 \leq \sigma_m + (2/3)\bar{\sigma}. \tag{12.1}$$

Cockroft and Latham* suggested a fracture criterion

$$\int_0^{\bar{\varepsilon}_f} \sigma_1 \, d\bar{\varepsilon} = C. \tag{12.2}$$

The concept is that damage accumulates during processing until the integral reaches a critical value, C. This criterion recognizes the dependence of the fracture strain on σ_1 (or almost equivalently on σ_m). However if σ_1 is even temporarily compressive, it suggests an accumulation of negative damage. Other criteria have been proposed.[†]

Gurson[‡] has quantitatively treated the role of porosity in ductile fracture. He developed a yield criterion for porous materials that is sensitive to hydrostatic pressure and can be used to predict void growth.

* M. G. Cockroft and D. J. Latham, *J. Inst. Metals,* v 96 (1968), pp. 33–39.

† See for example A. G. Atkins, *Metal Sci.* (Feb. 1981), pp. 81–3; A. K. Ghosh, *Met. Trans.*, 7A (1976), pp. 523–33; F. A. McClintock, *J. Appl. Mech., Trans. ASME*, 35 (1968), pp. 363–71.

‡ A. L. Gurson, *J. Engr. Mater. Tech*, v. 99 (1977), pp. 2–15.

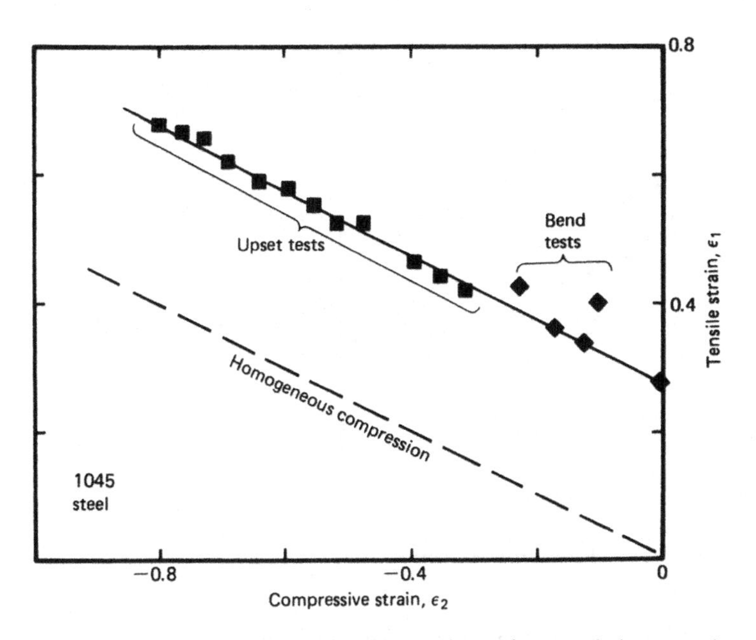

Figure 12.15. Forming limits defined by strains at fracture during upsetting. Data from H. A. Kuhn, *Formability Topics – Metallic Materials* (ASTM STP 647 1978).

12.5 BULK FORMABILITY TESTS

Formability does not always correlate with ductility measured in a tension test, especially if cracking starts on the surface. For this reason, other tests are often used. Kuhn* proposed a simple procedure for evaluating formability with upset compression tests. Barreling during upsetting causes hoop tensile stresses that may lead to cracking. He plotted the hoop strain, ε_1, when the first cracks were observed as a function of the compressive strain, ε_2. The ratio of ε_1 to ε_2 was varied by changing the lubrication and height-to-diameter ratio of the specimens. Figure 12.15 is a plot for a 1045 steel. Additional points were obtained from bend tests on wide specimens. The straight line indicates that

$$\varepsilon_{1f} = C - (1/2)\varepsilon_{2f}, \tag{12.3}$$

where C is the value of $\varepsilon_{1f} =$ for plane-strain. This line parallels $\varepsilon_1 = -(1/2)\varepsilon_2$ for homogeneous compression. Fracture strains vary with orientation as shown in Figure 12.16.

A simple test to measure the effect of spheroidization involves measuring the circumferential fracture strain when a hollowed tube of the work material is expanded by pushing it onto a conical tool as shown in Figure 12.17.†

* H. A. Kuhn, *Formability Topics – Metallic Materials*, ASTM STP 647 (1978).

† J. O'Brien and W. Hosford, *ibid.*

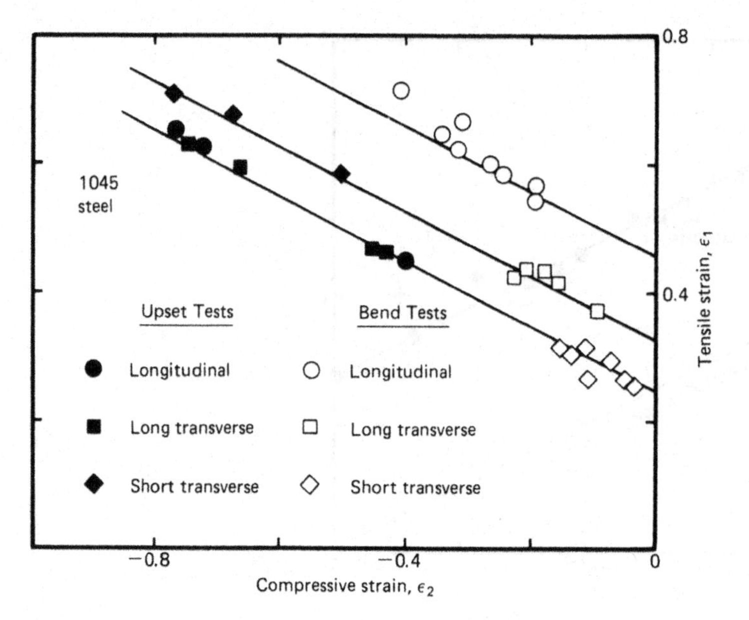

Figure 12.16. Effect of specimen orientation of forming limits. Data from Kuhn, *ibid.*

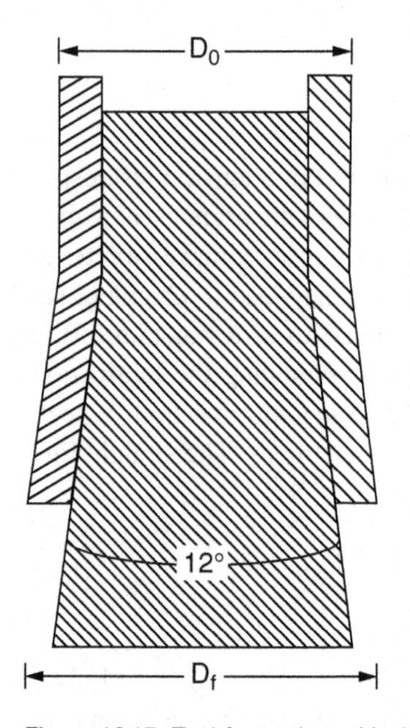

Figure 12.17. Tool for testing critical hoop expansion strain, $\varepsilon_h = \ln(D_0/D_f)$ when the bottom of the tubular specimen splits.

12.6 FORMABILITY IN HOT WORKING

Formability during hot working is usually better than in cold working because of the lower stresses. However sometimes failures occur at very low strains during hot working. This is called *hot shortness*. The usual cause is the presence of a liquid phase at the grain boundaries. The hot-working temperatures for aluminum alloys are very close to the eutectic temperature. Sometimes the liquid phase is nonmetallic. In the absence of manganese, sulfur would react with iron to form iron sulfide, which has a low melting temperature and wets the grain boundaries. Fortunately if manganese is added to the steel, the sulfur preferentially combines with the manganese to form MnS, which is solid at hot working temperatures and does not wet the grain boundaries.

Small amounts of copper or tin picked up by melting scrap can cause surface cracks during hot rolling of steel. These "tramp" elements don't oxidize easily so as the iron at the surface is oxidized their concentrations at the surface may reach a point where they form a liquid phase, which wet the grain boundaries.

NOTE OF INTEREST

Percy W. Bridgman (1882–1961) was born in Cambridge, Massachusetts. He studied at Harvard where he received a doctorate in 1908. He joined the faculty there in 1910. His research was mainly on the physics of high pressure, for which he received the Nobel Prize in Physics in 1946.

REFERENCES

W. F. Hosford, *Mechanical Behavior of Materials*, 2nd ed., Cambridge University Press, 2010.

E. M. Meilnik, *Metalworking Science and Engineering*, McGraw-Hill, 1991.

PROBLEMS

12.1. **a)** Explain why inclusion-shape control is of much greater importance in high-strength steels than in low-carbon steels.

b) Explain why inclusion-shape control improves the transverse and through-thickness properties, but has little effect on the longitudinal ones.

12.2. Figure 12.1 shows that, for a given sheet material, greater reductions are possible before edge cracking if square edges are maintained than if the edges are round. Explain why.

12.3. **a)** Wrought iron has a high toughness when stressed parallel to the prior working direction, but a very low toughness when stressed perpendicular to the prior working direction. Explain why.

b) When wrought iron was a commercial product, producers claimed that its corrosion resistance was superior to that of steel. What is the basis of this claim?

12.4. **a)** With the same material, die angle, and reduction, central bursts may occur in drawing but not in extrusion. Explain why this may be so.

b) Why may a given material be rolled to strains much higher than the fracture strain in a tension test?

12.5. Figure 12.18 is a Kuhn-type forming limit diagram for upsetting a certain grade of steel. The line gives the combination of strains that cause cracking.

a) Superimpose on this diagram the strain path that leads to failure at point *P*.
b) What differences in test variables would lead to cracking at point *S*?

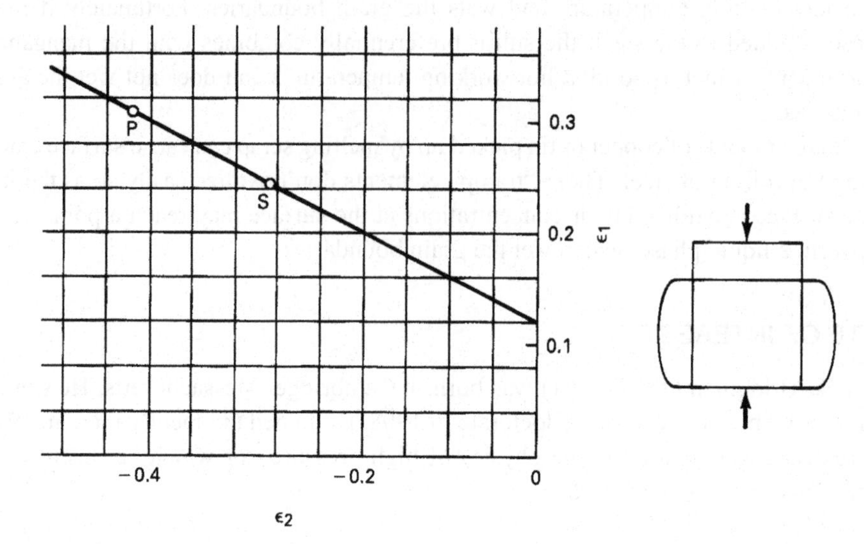

Figure 12.18. Figure for Problem 12.5.

13 Bending

All sheet-forming operations involve some bending and bending is a major feature. Springback occurs after the bending forces are removed and the material will have residual stresses. If the bend radius is too sharp, there may be tensile failure on the outside of the bend or buckling on the inside.

13.1 SHEET BENDING

Consider the bending of a flat sheet of a non-strain-hardening material subjected to a pure bending moment. Figure 13.1 shows the coordinate system. Let r be the radius of curvature measured at the mid-plane and let z be the distance of an element from the mid-plane. The engineering strain at z can be derived by considering the arc lengths, L, measured parallel to the x-axis. The arc length at the mid-plane, L_0, doesn't change during bending and may be expressed as $L_0 = r\theta$, where θ is the bend angle. At z, the arc length is $L = (r + z)\theta$. Before bending both arc lengths were equal so the engineering strain, e, is

$$e_x = (L - L_0)/L_0 = z\theta/r\theta = z/r. \tag{13.1}$$

The true strain, ε_x, is

$$\varepsilon_x = \ln(1 + z/r), \tag{13.2a}$$

For many bends the strains are low enough so we can approximate

$$\varepsilon_x = z/r. \tag{13.2b}$$

For wide sheets ($w \gg t$) the width strain, ε_y, is negligible so sheet bending can be considered to be a plane-strain operation, $\varepsilon_z = -\varepsilon_x$. The value of e_x varies from $-t/(2r)$ on the inside ($z = -t/2$) to zero at the mid-plane and $+t/(2r)$ on the outside surface. For small strains, $\varepsilon_x \approx e_x$. The internal stress distribution can be found from the strain distribution and the stress-strain curve. Assume that the material is elastic-ideally plastic (no strain hardening). Let the tensile yield strength in plane-strain be Y.

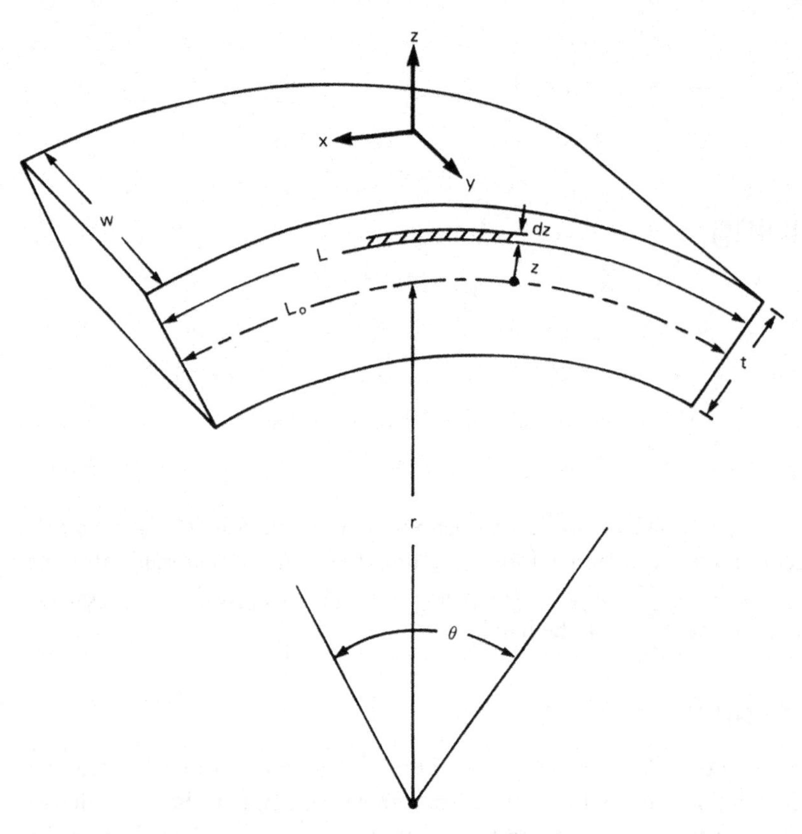

Figure 13.1. Coordinate system for analyzing sheet bending.

Figure 13.2 shows the strain and stress distribution. The entire cross section will yield at a stress $\sigma_x = Y$ or $-Y$ except for an elastic core near the mid-plane. For most bending operations this elastic core can be neglected.

The bending moment can be calculated assuming that there is no net external force, F_x. However there are internal forces, $\mathrm{d}F_x = \sigma_x w\,\mathrm{d}z$, acting on incremental elements of the cross section. The contribution of stress, σ_x, in every element to the total bending

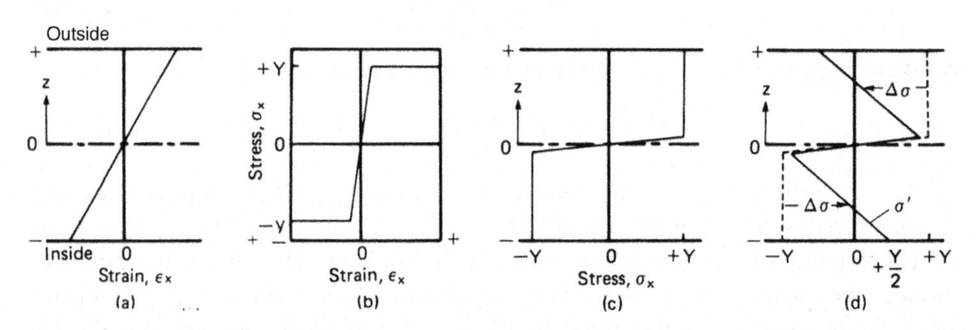

Figure 13.2. Stress and strain distribution across a sheet thickness. During bending the strain varies linearly across the section (a). With a non-strain-hardening material (b) the bending causes the stress distribution in (c). Elastic unloading results in the residual stresses in (d).

moment is the product of this incremental force times its lever arm, so $\mathrm{d}M = z\sigma_x w\,\mathrm{d}z$. The total bending moment is then

$$M = \int_{-t/2}^{+t/2} w\sigma_x z\,\mathrm{d}z = 2\int_{0}^{+t/2} w\sigma_x z\,\mathrm{d}z. \tag{13.3}$$

Integration of equation 13.3 is simplified by taking twice the integral between 0 and $t/2$ because the sign of σ_x changes at the mid-plane. For an ideally plastic material

$$M = 2wY\int_{0}^{t/2} z\,\mathrm{d}z = wYt^2/4. \tag{13.4}$$

EXAMPLE 13.1: A steel sheet, 1 mm thick, is bent to a radius of curvature of 12 cm. The flow stress in plane-strain is 210 MPa and $E' = 224$ GPa.

a) What fraction of the cross section doesn't yield?
b) In calculating the bending moment, what percent error would be caused by neglecting the fact that the center hadn't yielded?

SOLUTION:

a) The yielding strain $= Y/E' = 210/\text{MPa}/224\text{ GPa} = 0.00094$. This strain occurs at a position $z = \rho\varepsilon = (0.00094)(120) = \pm 0.11$ mm. The fraction not yielding $= 0.11/0.5 = 22\%$
b) To calculate the exact moment, divide the integral in equation 13.3 into two parts,

$$= 2\int_{0}^{0.00012} \frac{E'}{r} z^2\,\mathrm{d}z + 2\int_{0.00012}^{0.0005} w\sigma_0 z\,\mathrm{d}z$$

$$= (2/3)\frac{224\times 10^9}{0.12} 0.00012^3 + 210\times 10^6(0.0005^2 - 0.00012^2) = 51.6\,\mathrm{Nm}$$

An approximate solution, neglecting the elastic core is $M/w = 2\int_0^{0.0005} \sigma_0 z\,\mathrm{d}z = +210\times 10^6(0.0005^2 - 0) = 52.5$ Nm. The error in neglecting the elastic core is only $0.9/51.6 = 1.7\%$

The bending moment applied by the tools must equal the internal mending moment. When the external forces are removed, the bending moment must vanish, $M = 0$, as the material unloads elastically. The unloading is elastic so

$$\Delta\varepsilon_x = E'\Delta\varepsilon_x \tag{13.5}$$

where $E' = E/(1+\nu)$ and the change of strain, $\Delta\varepsilon_x$, is given by

$$\Delta\varepsilon_x = z/r - z/r' \tag{13.6}$$

where r' is the radius after springback. The change of bending moment is then

$$\Delta M = 2w \int_0^{t/2} \Delta\sigma_x z\,dz = 2w \int_0^{t/2} E(1/r - 1/r)z^2\,dz. \tag{13.7}$$

Integrating

$$\Delta M = (wE't^3/12)(1/r - 1/r'). \tag{13.8}$$

After springback the bending moment must be zero so $M - \Delta M = 0$. Combining equations 13.4 and 13.8, $(wE't^3/12)(1/r - 1/r') = wYt^2/4$ so

$$1/r - 1/r' = 3Y/(tE'). \tag{13.9}$$

The residual stress is,

$$\begin{aligned}\sigma'_x &= \sigma_x - \Delta\sigma_x = Y - \Delta\sigma_x = Y - E'\Delta e_x = Y - E'z(1/r - 1/r') \\ &= Y - E'z[3Y/(tE')],\end{aligned}$$

so

$$\sigma'_x/Y = 1 - (3z/t). \tag{13.10}$$

Figure 13.2d shows that on the inside surface $\sigma_x = +Y/2$ and on the outside surface

$$\sigma_x = -Y/2.$$

A similar treatment for work hardening ($\sigma_x = K'\varepsilon_x^n$), results in the moment under load being

$$M = 2(wK'/r^n)(t/2)^{2+n}/(2+n). \tag{13.11}$$

Substituting ΔM from equation 13.8 and M from equation 13.3 into $M - \Delta M = 0$,

$$\left(\frac{1}{r} - \frac{1}{r'}\right) = \left(\frac{6}{2+n}\right)\left(\frac{K'}{E'}\right)\left(\frac{t}{2r}\right)^n\left(\frac{1}{t}\right) \tag{13.12}$$

The magnitude of the springback predicted by equation 13.12 can be very large. The residual stress pattern can be found by combining equations 13.5, 13.6, and 13.12

$$\sigma'_x = K'\left(\frac{z}{r}\right)^n\left[1 - \left(\frac{3}{2+n}\right)\left(\frac{2z}{t}\right)^{1-n}\right] \tag{13.13}$$

Figure 13.3 shows how σ_x and σ'_x vary through the thickness of the sheet.

EXAMPLE 13.2: Find the tool radius that will produce a part of final radius, $r' = 25$ cm in a steel of thickness 1.2 mm. Assume a flow stress of 300 MPa and a plane-strain modulus of 224 GPa.

SOLUTION: Substituting $r' = 250$ mm, $t = 0.0012$ m, $\sigma_0 = 300$ MPa into equation 13.7, $r = 136$ mm or 13.6 cm. If this were part of a complex stamping, it would be impossible to build this correction into the tooling of a complex part.

13.2 BENDING WITH SUPER-IMPOSED TENSION

The amount of springback predicted by equation 13.12 is so large that it cannot be compensated for by over-bending. However, springback can be minimized or even eliminated by applying tension while bending. This greatly reduces springback. Tension

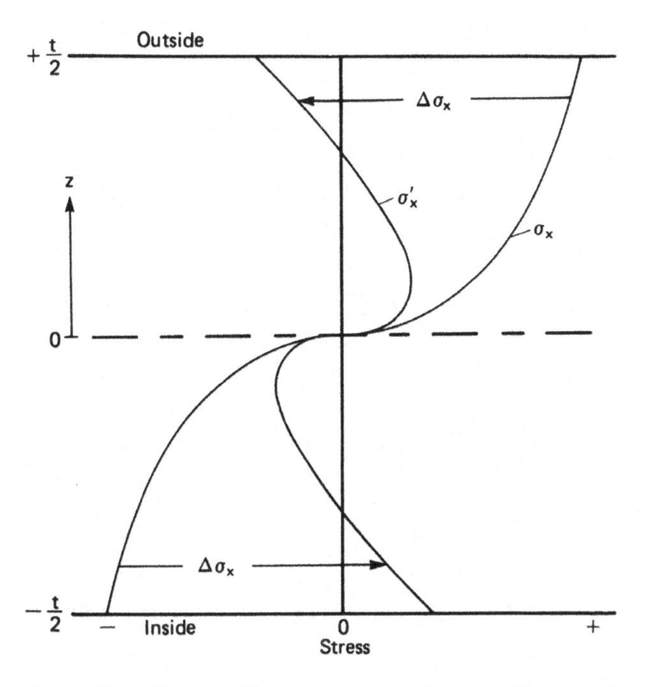

Figure 13.3. Stress distribution during bending and after unloading for a strain-hardening material.

shifts the neutral plane toward the inside of the bend and if sufficient, can shift it completely out of the sheet. In this case the entire cross section yields in tension. This is illustrated in Figure 13.4. In an ideally plastic material there would be no springback because the stress, $\sigma_x = Y$, is the same throughout the cross section. However, if the material strain hardens, the springback won't completely vanish.

Consider a simple case in which the stress-strain curve is approximated by $\sigma_x = Y + (d\sigma_x/d\varepsilon_x)\varepsilon_x$ where $d\sigma_x/d\varepsilon_x$ is constant. The stress can be divided into two parts: a bending stress, $\sigma_{xb} = \varepsilon_{xb}(d\sigma_{xb}/d\varepsilon_{xb}) = (z/r)(d\sigma_{xb}/d\varepsilon_{xb})$ and a uniform tensile stress, σ_{xu}. The springback is entirely caused by changes of σ_{xb} on unloading. Relaxation will cause a change of the bending strain, $\Delta\varepsilon_{xb} = \sigma_{xb}/E' = (z/r)(d\sigma_{xb}/d\varepsilon_{xb})/E'$. The relaxed bending strain is then $\varepsilon'_{xb} = \varepsilon_{xb} - \Delta\varepsilon_{xb} = (z/r)[1 - (d\sigma_{xb}/d\varepsilon_{xb})/E']$. Since

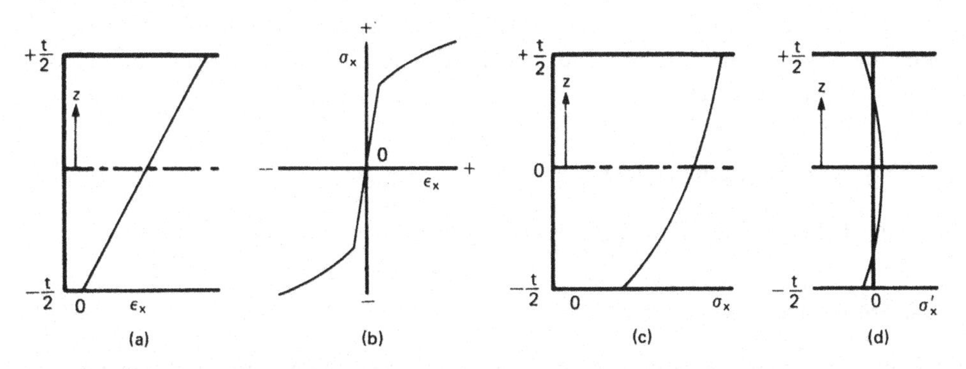

Figure 13.4. Bending with superimposed tension. With sufficient tension, the neutral axis moves out of the sheet so the entire thickness is strained in tension (a). With the stress-strain curve shown in (b) the stress distribution in (c) results. On unloading there are only minor residual stresses.

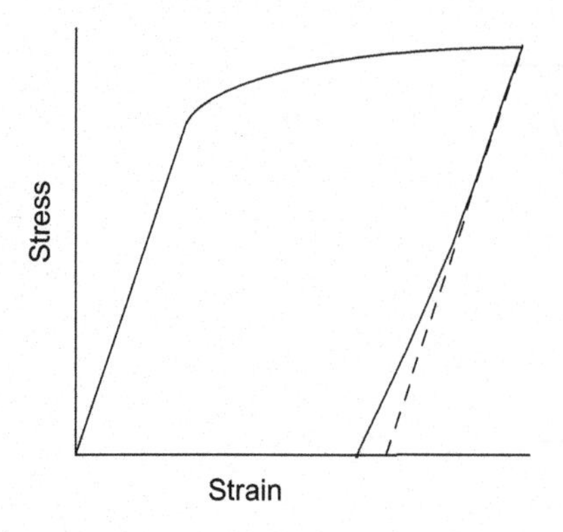

Figure 13.5. Young's modulus on unloading is lower than for loading.

$\varepsilon'_{xb} = z/r'$, the relaxed bending radius, r', is given by $z/r' = (z/r)[1 - (\mathrm{d}\sigma_x/\mathrm{d}\varepsilon_x)/E']$ so

$$r'/r = \left[1 - \frac{(\mathrm{d}\sigma_x/\mathrm{d}\varepsilon_x)}{E'}\right]^{-1}. \tag{13.14}$$

Springback is very much reduced by applied tension because $\mathrm{d}\sigma_x/\mathrm{d}\varepsilon_x$ is much lower than E'.*

EXAMPLE 13.3: Reconsider Example 13.2 with sufficient tension to cause a net tensile strain of 0.02 at the centerline. Let the stress-strain curve be approximated by $\sigma = 656\varepsilon^{0.2}$. (This corresponds to a flow stress of 300 MPa at the centerline.)

SOLUTION: Substituting $\mathrm{d}\sigma/\mathrm{d}\varepsilon = nK\varepsilon^{n-1} = 0.2(300)0.02^{-0.8} = 1372\,\mathrm{MPa}$ and $E' = 224$ GPa into equation 13.14, $r = 25\,\mathrm{cm}(1 - 1372/224{,}000) = 24.85$ cm. Compare this with 13.6 cm found in Example 13.2. For the small predicted springback (24.85 cm vs. 25 cm), it is possible to correct for the springback in the tooling design.*

13.3 YOUNG'S MODULUS ON UNLOADING

The appropriate value of Young's modulus to use in springback calculations is that measured during unloading. This is almost always lower than that measured during loading as illustrated in Figure 13.5 because the springback is larger than that calculated using the value of E for loading. The decreased slope can be thought of as a precursor to the Bauschinger effect.

13.4 REDUCING SPRINGBACK

Springback can be reduced by coining the corner as shown in Figure 13.6.

13.5 NEUTRAL AXIS SHIFT

The analyses above are based on the neutral plane remaining at the mid-plane of the sheet. Actually the neutral plane moves toward the inside of the bend. There are two

* See J. L. Duncan and J. E. Bird, *Sheet Metal Industries* (Sept. 1978).

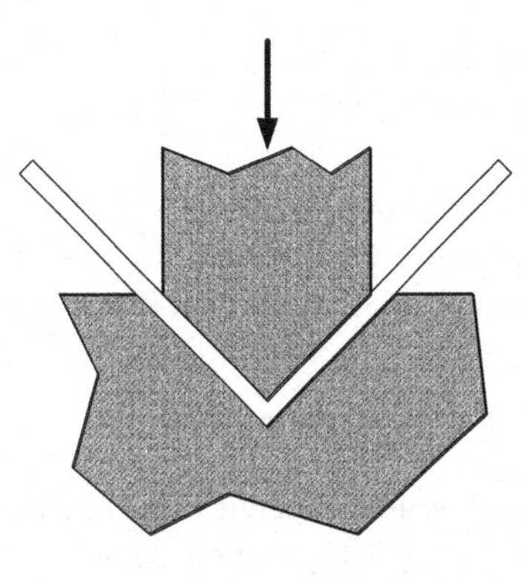

Figure 13.6. Springback can be greatly reduced by coining the corner.

reasons for this: On the inside of the bend, elements thicken. Also the true compressive strains on the inside are greater in magnitude than the strains on the outside so the stresses are greater in magnitude.

The thickness of element i, is $t_i = t_0/(1+e_i) = t_0/(1+z_i/\rho)$ and the true strain, ε_i, at element i is $\ln(1+z_i/\rho)$. If e_i is positive, the true stress, σ_i, on element i is $\sigma_i = K\varepsilon_i^n = K[\ln(1+z_i/\rho]^n$. If e_i is negative, the true stress, σ_i, on element i is $\sigma_i = -K(\ln|(1+z_i/\rho)^n|$.

The force on element i, on the tensile side of the bend is then

$$\mathrm{F}_0 = [wt_0/(1+z_i/\rho)]K[\ln(1+z_i/\rho)]^n \tag{13.15}$$

and on the compressive side

$$\mathrm{F}_I = -[wt_0/(1+z_i/\rho)]K|\ln(1+z_i/\rho)|^n, \tag{13.16}$$

where z_i is negative on the inside of the bend and positive on the outside. The ratio of the two at equal displacements from the original centerline is

$$F_0/F_I = \left[\frac{\ln(1+z_{i0}/\rho)}{|\ln(1+z_{iI}/\rho)|}\right]^n \frac{(1+z_{iI}/\rho)}{(1+z_{i0}/\rho)}, \tag{13.17}$$

where z_{i0} refers to an element on the outside of the bend and z_{iI} refers to an element on the inside of the bend. Figure 13.7 is a plot of equation 13.17.

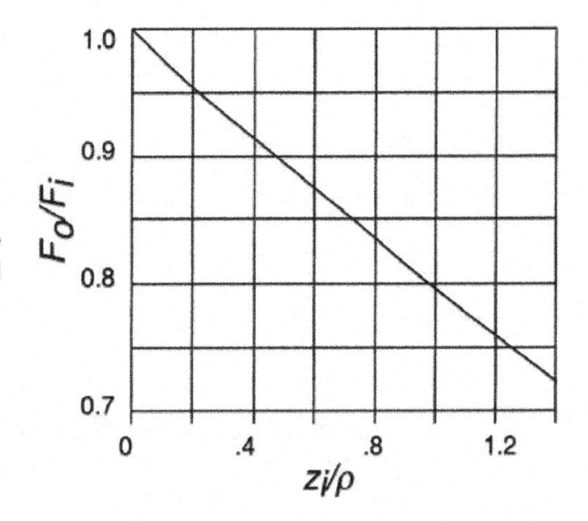

Figure 13.7. The ratio of the forces on elements on the outside and inside of the bend, both initially displaced by z from the middle of the sheet.

EXAMPLE 13.4: Consider bending of a 1.0 mm thick sheet to a radius of 2.5 cm. Calculate the ratio of the forces on elements at the outside and inside surfaces.

SOLUTION: Substituting $z/\rho = .02$ into equation 13.17, $F_0/F_i = 0.96$. The average ratio for elements on the outside of the bend ($z/\rho = .01$) is 0.98. This illustrates that the shift of the neutral plane can be neglected for most bends.

13.6 BENDABILITY

Cracking can occur if the tensile strains on the outside of a bend are great enough. The limiting values of r/t depend on the material (see Figure 13.8). The bend radius, R, in this figure, as is usual in tables of permissible bend radii, is the radius on the inside of the bend rather than at the mid-plane. Therefore $R/t = r/ - 1/2$. The solid lines in Figures 13.8a and b approximate failure when the tensile strain is equal to the elongation, El, measured in a tension test.

$$R/t = 2El - 1 \tag{13.18}$$

For Figure 13.8a where El is the tensile elongation and for Figure 13.7b, where A_r is the reduction in area in a tension test,

$$R/t = 1/(2A_r) - 1. \tag{13.19}$$

These approximations are reasonable for materials with a low ductility (low r/t), but are not accurate for sharp bends (high R/t) because the neutral axis shifts with sharp bends and the shift depends on the friction and applied tension. Furthermore, strains are not limited to the nominally bent region, but spread out. With low bend angles the maximum strains are less than predicted by equation 13.1.

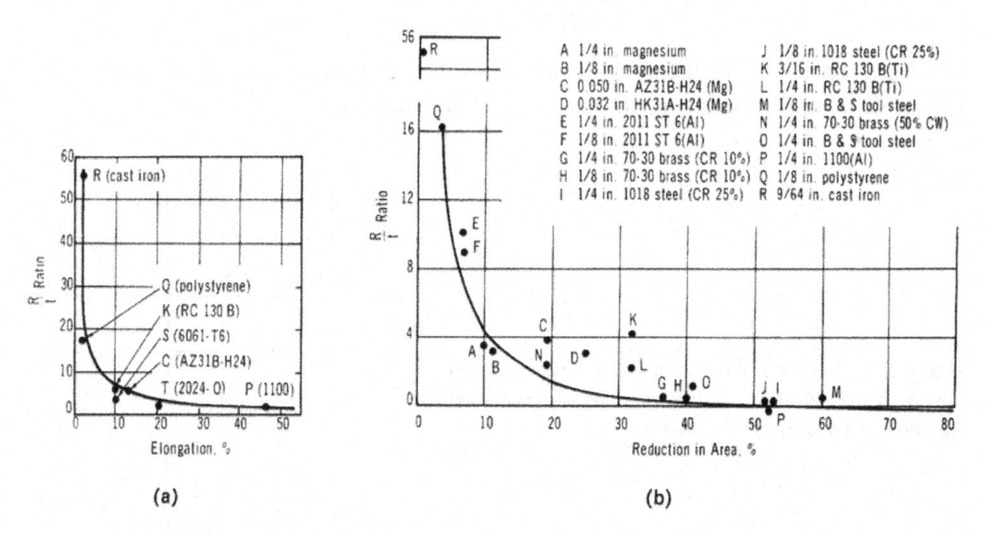

Figure 13.8. Correlation of the limiting bending severity = r/t with tensile ductility. Note that here R is the inside radius. From J. Datsko, *Material Properties and Manufacturing Processes*, John Wiley 1966.

13.7 SHAPE BENDING

The bending of various shapes such as T-, L-, and I-sections and square and circular tubes as illustrated in Figure 13.9 can be analyzed in a similar manner. Again it is assumed for simplicity that there is no strain hardening, and there is no net x-direction force. It will be assumed that the condition in each element is uniaxial tension or compression in the x-direction rather than plane-strain. Unlike the analysis for sheet bending, the width, w, is a function of z. For shapes symmetrical about the mid-plane, (Figures 13.8a and 13.8b), $M = 2Y\int_0^{h/2} wz\,\mathrm{d}z$ and $\Delta M = 2E(\frac{1}{r} - \frac{1}{r'})\int_0^{h/2} wz^2\,\mathrm{d}z$ so the springback is given by

$$\left(\frac{1}{r} - \frac{1}{r'}\right) = \left(\frac{Y}{E}\right)\left(\frac{2Q}{h}\right), \tag{13.20}$$

where $Q = [(h/2)\int_0^{h/2} wz\,\mathrm{d}z]/[\int_0^{h/2} wz^2\,\mathrm{d}z]$.

The residual stress distribution is given by

$$\sigma'_x = Y\left(1 - \frac{2Qz}{h}\right). \tag{13.21}$$

For a flat sheet in which w is not a function of z, equation 13.20 reduces to $Q = 3/2$. For cross sections in which w increases with z, $Q < 3/2$ so there is less springback than with flat sheets with the same value of h, whereas if w decreases with z, $Q > 3/2$ so there is more springback. As with flat sheets, the amount of springback decreases with applied tension.

There may be appreciable movement of the neutral plane during bending if the bends are tight (if r/h is small). With cross sections that are not symmetric about the neutral plane moves to the inside of the bend, because the cross-sectional area

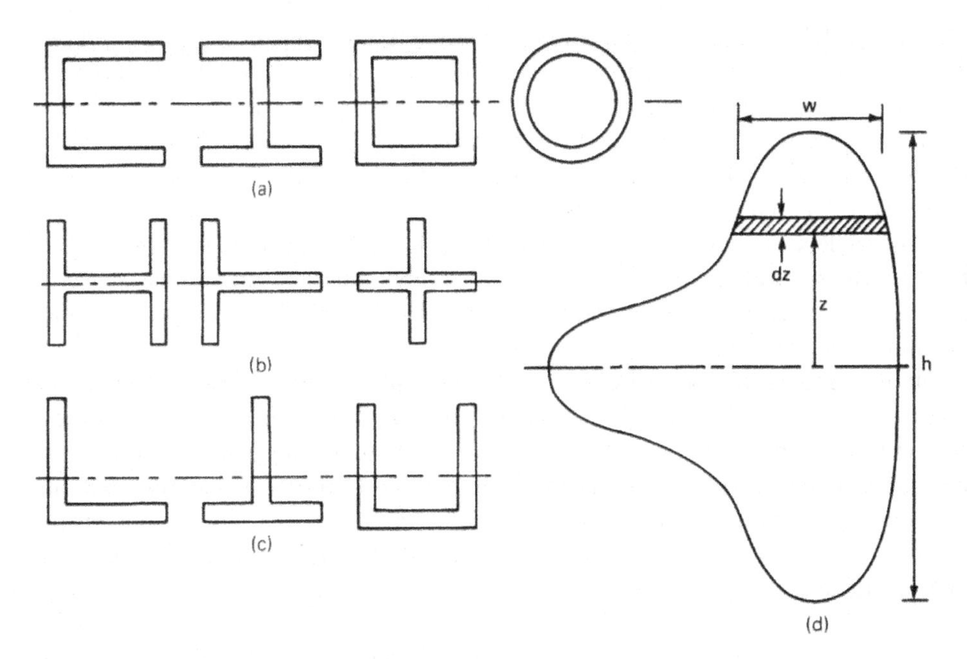

Figure 13.9. Cross sections of various shapes.

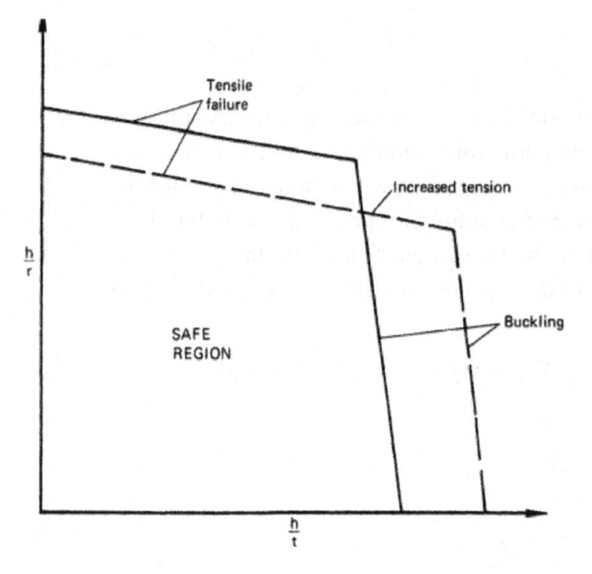

Figure 13.10. Schematic forming limits for shape bending. Critical levels of values of h/r and h/t depend on the shape and the material. Increasing axial tension lowers the critical h/r values and raises the critical h/t values.

on the outside decreases and the cross-sectional area on the inside increases. With nonsymmetric cross sections, yielding will start on one side before the other and this will cause a shift of the neutral plane.

13.8 FORMING LIMITS IN BENDING

There are two possible failure modes in bending of shapes: There may be a tensile failure on the outside of the bend or buckling on the inside. If the movement of the neutral axis is neglected, the tensile strain on the outside of the bend is $e = h/(2r)$ and failure occurs when this strain reaches a critical value. For brittle materials, this strain is approximately the percent elongation in a tension test. For ductile materials necking occurs at a strain a little greater than the uniform elongation in a tension test because of support from elements nearer the neutral plane. This support varies with the section shape, which can be characterized by h/t, where t is the thickness of the section. In slender sections (h/t is large) there is less support to the outer sections than if h/t is small. Therefore the critical value of h/r decreases with h/t.

The tendency to buckling on the inside of the bend depends primarily on h/t, but it also increases with bend severity. Wood* determined the forming limits for many materials and shapes in stretch forming. Figure 13.10 is a schematic forming limit diagram for shape bending.

In the bending of tubes the outer fibers tend to move inward, distorting the cross section or even causing a collapse of the tube. In low production items, this tendency can be lessened by filling the tube with sand or a low melting point metal. In high production items, a ball or plug mandrel can be used to preserve the cross section. This internal support will however lower the forming limits by causing greater tensile strains in the outside fibers. Figure 13.11 illustrates the use of a mandrel in the bending of a square tube.

* W. W. Wood, *Final Report on Sheet Metal Forming Technology*, v. II, SD-TDR-63-7-871 (July 1963).

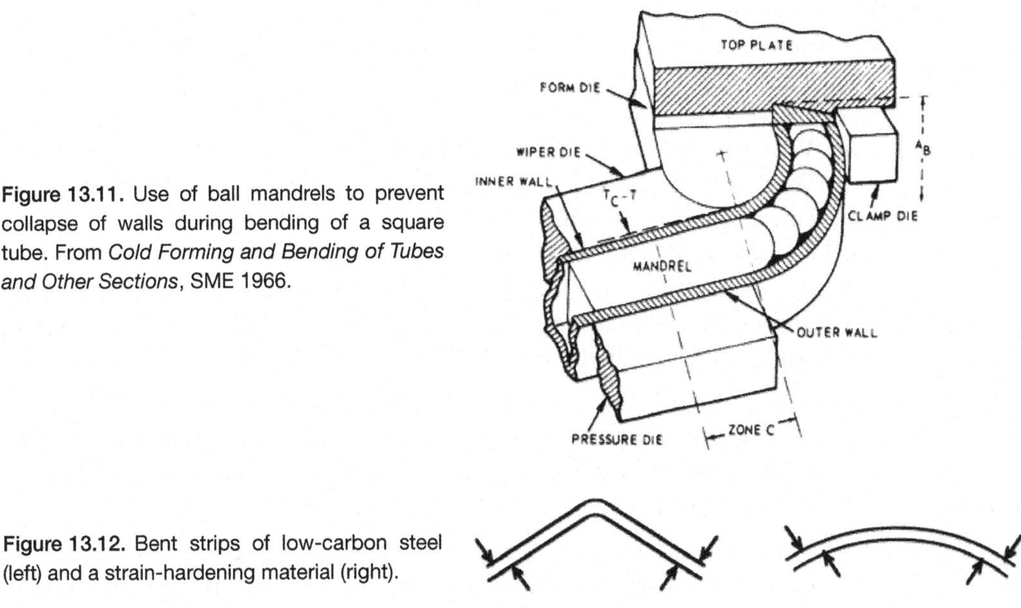

Figure 13.11. Use of ball mandrels to prevent collapse of walls during bending of a square tube. From *Cold Forming and Bending of Tubes and Other Sections*, SME 1966.

Figure 13.12. Bent strips of low-carbon steel (left) and a strain-hardening material (right).

NOTE OF INTEREST

If a strip of a low-carbon steel with a yield point is bent, the bend tends to concentrate at one point because once one region yields it is softer than other regions. In contrast a strip of a strain-hardening metal will spread the bend gradually over the entire surface. See Figure 13.12.

REFERENCES

R. J. Kervick and R. K. Springborn, eds., *Cold Forming and Bending of Tubes and Other Sections*, ASTME, 1966.

Z. Marciniak, J. L. Duncan, and S. J. Hu, *Mechanics of Sheet Metal Forming*, Butterworth-Heinemann, 2002.

PROBLEMS

13.1. An old shop-hand has developed a simple method of estimating the yield strength of steel. He carefully bends the strip with his hands and to a given radius, releases it, and wonders whether it has taken a permanent set. He repeats the process until it does take a permanent set. A strip 0.25-in thick, 1-in wide, and 10-in long first takes a permanent set at a radius of 10 in. Estimate the yield strength.

13.2. A coiler is being designed for a cold-rolling line of a steel mill. The coil diameter should be large enough so that coiling involves only elastic deformation. The sheet to be coiled is 1-mm thick and 2-m wide, and has a yield strength of 275 MPa.

a) What is the minimum diameter of the coiler?

b) For this diameter, find the horsepower consumed by coiling at 30 m/s.

13.3. In many applications, the minimum thickness of a sheet is determined by its stiffness in bending. If aluminum is substituted for steel to save weight, its thickness must be greater.

a) By what factor must the thickness be increased?

b) What weight saving would be achieved?

c) Would the weight saving be greater, less, or unchanged if both sheets were corrugated instead of being flat?

Hint: For elastic bending, the deflection, δ, is given by $\delta = A(FL^3/E'wt)$, where F is the force, L is the span length, E' is the plane-strain modulus, w the sheet width, and t is the sheet thickness. For steel, $E' = 220$ GPa and $\rho = 7.9$ Mg/m^3. For aluminum, $E' = 73$ GPa and $\rho = 2.7$ Mg/m^3.

13.4. For some designs, the minimum sheet thickness is controlled by the ability to absorb energy elastically in bending without any plastic deformation. In this case, what weight saving can be achieved by substituting aluminum (Y = 25ksi) for steel (Y = 35ksi)? Use the data in Problem 13.3.

13.5. An aluminum sheet, 1-mm thick, is to be bent to a final radius of curvature of 75 mm. The plastic portion of the stress-strain curve is approximated by $175 + 175\varepsilon$ MPa. Accounting for springback, what radius of curvature must be designed into the tools if the loading is:

a) pure bending?

b) tensile enough so that the mid-plane is stretched 2% in tension?

13.6. What fraction of the cross section remains elastic in Problem 13.5a?

13.7. It has been suggested that the residual hoop stress in a tube can be found by slitting a short length of tube longitudinally and measuring the diameter, d, after slitting and comparing this with the original diameter, d_0. A stress distribution must be assumed. Two simple stress distributions are suggested by Figure 13.13. For a copper tube, $d_0 = 25$ mm, d = 25.12 mm, and $t = 0.5$ mm. Assume that $E = 110$ GPa and $\nu = 0.30$. Find the residual stress at the surface using both assumptions about the stress distribution.

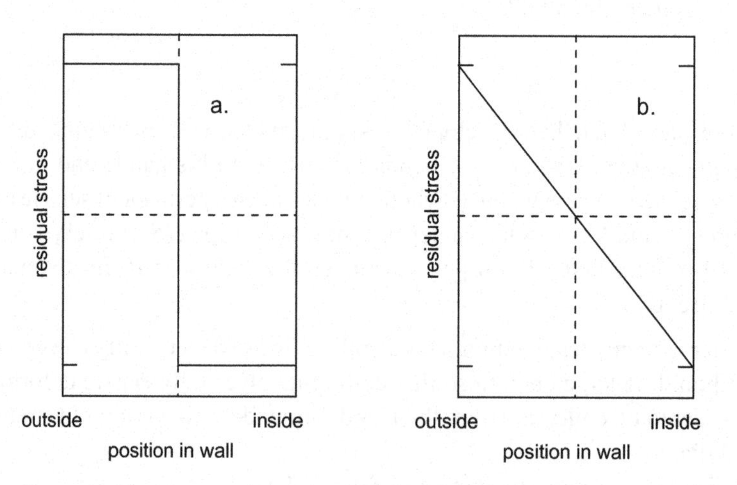

Figure 13.13. Assumed stress distributions for Problem 13.7.

13.8. A round bar (radius R and length L) was plastically deformed in torsion until the entire cross section yielded. Assume an ideally plastic material with a shear strength, k, and a shear modulus, G. When unloaded, it untwisted by an amount $\Delta\theta$ (radians).

a) Derive an expression for the level of residual stress, τ', as a function of the radial position, r, and R, G, and k.

b) Find the relative springback, $\Delta\theta/L$, in terms of G, k, and R.

13.9. A plate was bent to a radius of curvature, R. After springback, the radius was R'. Later, the plate was etched, removing the outer surface. How would you expect the last radius, R', to compare with R and R'?

13.10. Consider the bending of a strip, 80-mm wide and 1.0-mm thick. The stress-strain relation in the elastic region is $\sigma = 210\varepsilon$ GPa, and in the plastic region it is $\sigma = 250\varepsilon$ MPa.

a) What is the limiting curvature to which the strip can be bent without yielding?

b) If the strip is bent to a radius of curvature of 500 mm, what is the radius when it is released? Assume bending by a pure bending moment.

13.11. A steel sheet, 1.5-m wide and 1.0-mm thick, is bent to a radius of 100 mm. Assume no strain hardening, no friction, and no tension. The effective stress-strain relation is $\bar{\sigma} = 650(0.015 + \bar{\varepsilon})^{0.20}$. E = 210 GPa, and $\nu = 0.29$. Find the radius of curvature after unloading.

14 Plastic Anisotropy

14.1 CRYSTALLOGRAPHIC BASIS

The primary cause of anisotropy of plastic properties is preferred orientation of grains (i.e., statistical tendency for grains to have certain orientations).* Mechanical working of metals produces preferred orientations or crystallographic textures. Recrystallization during annealing usually changes the crystallographic texture but doesn't cause randomness. (Repeated heating and cooling through the $\alpha \rightarrow \gamma \rightarrow \alpha$ transformation may be a possible exception.) How crystallographic textures develop and how texture influences the anisotropy are not treated in this text. The anisotropy of textured titanium sheet will be considered, however, because of the insight it gives to the subject.

Alpha-titanium alloys have a hexagonal-close packed crystal structure. Deformation occurs primarily by slip in close-packed $\langle 11\bar{2}0\rangle$ directions, which lie in the basal plane parallel to the edges of the hexagonal cell as shown in Figure 14.1. Whether slip occurs on the (0001) basal planes or on the $\{10\bar{1}0\}$ prism planes, there is no strain parallel to the c-axis because the slip direction is normal to the c-axis. Not even slip on the $\{10\bar{1}1\}$ pyramidal planes will cause c-axis strain.

Rolled sheets of α-titanium alloys develop strong textures that may be most simply described in terms of an ideal texture with the c-axis of most grains aligned with the normal to the rolling plane. There is some spread from this ideal orientation, particularly with a rocking of the c-axis $\pm$ up to 40° from the sheet normal toward the transverse direction.

When a tensile specimen cut from a sheet is extended, slip can easily occur on the $\{10\bar{1}0\}$ prism planes. Even though the yield strength may be almost unchanged with the orientation of the tensile specimen relative to the rolling direction, the material is not isotropic. Slip on the $\{10\bar{1}0\}$ prism planes causes contraction in the plane of the sheet, but no thinning. A useful parameter to describe this sort of anisotropy is the ratio, R, of the plastic strain in the width direction to that in the thickness direction,

$$R = \frac{\varepsilon_w}{\varepsilon_t} \tag{14.1}$$

* In contrast, anisotropy of fracture behavior is largely governed by mechanical alignment of inclusions, voids and grain boundaries.

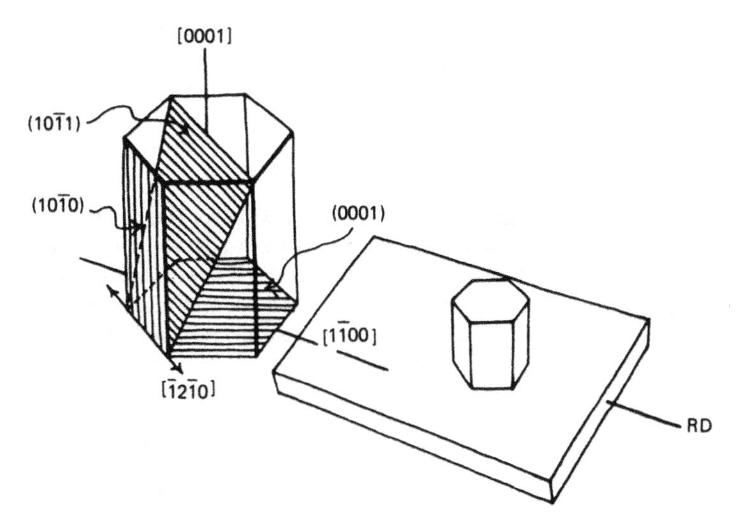

Figure 14.1. Idealized texture of an α-titanium sheet with the (0001) plane parallel to the plane of the sheet. Although there are several possible slip planes, slip is restricted to the <11$\bar{2}$0> family of slip directions, all of which lie in the (0001) plane so slip can cause no thickening or thinning of the sheet. From W. F. Hosford and W. A. Backofen in *Fundamentals of Deformation Processing,* Syracuse University Press, 1964.

where ε_w and ε_t, are the contractile strains in the width and thickness directions as illustrated in Figure 14.2. For an isotopic material, $R = 1$, but for ideally textured titanium, $\varepsilon_t = 0$, so $R = \infty$. In commercial titanium sheets, R-values of 3 to 7 are typical. A high R-value suggests that there is a high resistance to thinning and therefore a high strength in biaxial tension in the plane of the sheet and in through-thickness compression. Figure 14.3 illustrates schematically how the R-value affects the shape of the yield locus.

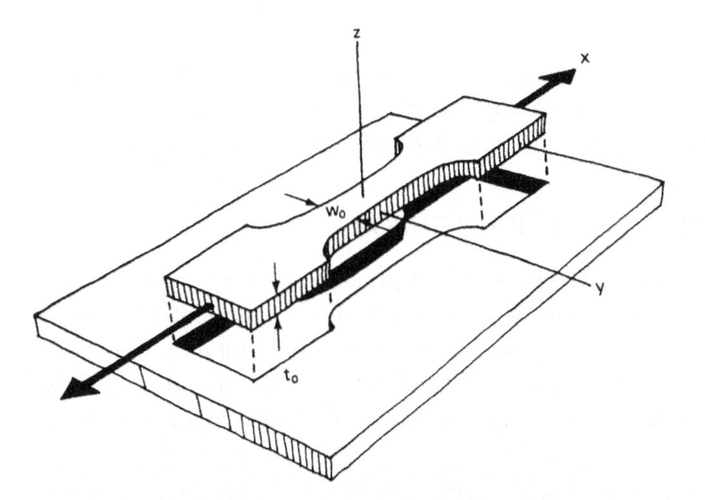

Figure 14.2. Strip tensile specimens cut from a sheet. The R-value is defined as the ratio of width-to-thickness strains, $\varepsilon_w/\varepsilon_t$ or $\varepsilon_y/\varepsilon_z$. From W. F. Hosford and W. A. Backofen, *ibid.*

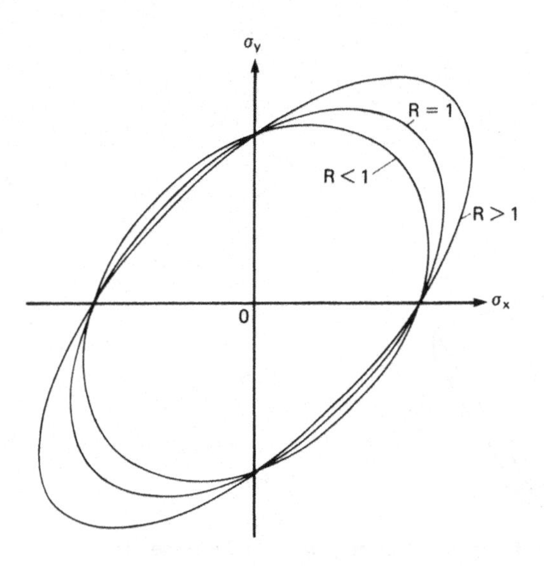

Figure 14.3. Schematic yield loci for textured materials with rotational symmetry. A high R-value implies resistance to thinning and hence a high yield strength under biaxial tension, while a low R-value indicates easy thinning and therefore a low yield strength under biaxial tension.

14.2 MEASUREMENT OF R

Although the R-value is defined as the ratio of width-to-thickness strains, the thickness strain of a thin sheet cannot be measured accurately. Instead the thickness strain is found from the width and length strains, $\varepsilon_t = -(\varepsilon_w + \varepsilon_l)$. The reduced section of the tensile specimen should be long enough so that the measurements can be made well away from the constraint of the shoulders. The R-value often varies somewhat with the strain, so the length strain at which the measurements are made should be specified. Often 15% elongation is used.

The R-values vary with the test direction in the plane of the sheet. It is customary to define an average value as

$$\bar{R} = \frac{R_0 + 2R_{45} + R_{90}}{4}. \tag{14.2}$$

For steels, the variation of the elastic modulus and the R-value with texture is usually similar. This correlation is neither exact nor fundamental, but is the basis for a commercial instrument. The instrument measures the velocity of sound in thin strips and is calibrated to give readings of the corresponding R-value.

14.3 HILL'S ANISOTROPIC PLASTICITY THEORY

In 1948, Hill* advanced a quantitative treatment of plastic anisotropy without regard to its crystallographic basis. He assumed materials with three orthogonal axes of anisotropy, x, y, and z about which the properties have two-fold symmetry. The yz, zx and xy planes are planes of mirror symmetry. In a rolled sheet it is conventional to take the x-, y-, and z-axes as the rolling direction, the transverse direction and the sheet-plane

* R. Hill, *Proc. Roy. Soc. London* 193A p. 281 and R. Hill, *Mathematical Theory of Plasticity*, Chapt XXXII, Oxford U. Press. See also W. F. Hosford and W. A. Backofen *in Fundamentals of Deformation Processing*, Syracuse U. Press 1964, p. 259.

normal. The theory also assumes equal the tensile and compressive yield strengths in every direction.

The proposed yield criterion is a generalization of the von Mises criterion:

$$F(\sigma_y - \sigma_z)^2 + G(\sigma_z - \sigma_x)^2 + H(\sigma_x - \sigma_y)^2 + 2L\tau_{yz}^2 + 2M\tau_{zx}^2 + 2N\tau_{xy}^2 = 2f(\sigma_{ij})^2, \tag{14.3}$$

where F, G, H, L, M, and N are constants that describe the anisotropy. Note that if $F = G = H = 1$ and $L = M = N = 3$, this reduces to the von Mises criterion. The constants F, G and H can be evaluated from simple tension tests. Let the x-direction yield strength be X. At yielding in an x-direction tension test, $\sigma_x = X$ and $\sigma_y = \sigma_z = \tau_{ij} = 0$. Substituting into equation 14.3, $(G + H)X^2 = 1$ or $X^2 = 1/(G + H)$. Similarly,

$$X^2 = \frac{1}{G+H}, \quad Y^2 = \frac{1}{H+F} \quad \text{and} \quad Z^2 = \frac{1}{F+G} \tag{14.4}$$

Solving simultaneously,

$$\begin{aligned} F &= (1/Y^2 + 1/Z^2 - 1/X^2)/2 \\ G &= (1/Z^2 + 1/X^2 - 1/Y^2)/2 \\ H &= (1/X^2 + 1/Y^2 - 1/Z^2)/2, \end{aligned} \tag{14.5}$$

where Y and Z are the y- and z-direction yield strengths. However measurement of Z is not feasible for sheets.

Using equation 2.18

$$\mathrm{d}\varepsilon_{ij} = \mathrm{d}\lambda \frac{\partial f(\sigma_{ij})}{\partial \sigma_{ij}}, \tag{14.6}$$

the flow rules are:

$$\begin{aligned} \mathrm{d}\varepsilon_x &= \mathrm{d}\lambda[H(\sigma_x - \sigma_y) + G(\sigma_x - \sigma_z)], \quad & \mathrm{d}\varepsilon_{yz} = \mathrm{d}\varepsilon_{zy} = \mathrm{d}\lambda L\tau_{yz} \\ \mathrm{d}\varepsilon_y &= \mathrm{d}\lambda[F(\sigma_y - \sigma_z) + H(\sigma_y - \sigma_x)], \quad & \mathrm{d}\varepsilon_{xz} = \mathrm{d}\varepsilon_{xz} = \mathrm{d}\lambda M\tau_{zx} \\ \mathrm{d}\varepsilon_z &= \mathrm{d}\lambda[G(\sigma_z - \sigma_x) + F(\sigma_z - \sigma_y)], \quad & \mathrm{d}\varepsilon_{xy} = \mathrm{d}\varepsilon_{yx} = \mathrm{d}\lambda N\tau_{xy} \end{aligned} \tag{14.7}$$

To derive these flow rules the yield criterion must be written with the shear stress terms appearing as $L(\tau_{yz}^2 + \tau_{zy}^2) + M(\tau_{zx}^2 + \tau_{xz}^2) + N(\tau_{xy}^2 + \tau_{yx}^2)$.

Note that in equation 14.7, $\mathrm{d}\varepsilon_x + \mathrm{d}\varepsilon_y + \mathrm{d}\varepsilon_z = 0$ which indicates constant volume.

The constant, N, can be found from a tension test made at an angle, θ, to the x-axis. In such a test yielding occurs when

$$F\sin^4\theta + G\cos^4\theta + H(\cos^4\theta - 2\cos^2\theta\sin^2\theta + \sin^4\theta) + 2N\cos^2\theta\sin^2\theta = Y_\theta^2, \tag{14.8}$$

where Y_θ is the yield strength in the θ-direction test. Equation 14.8 can be simplified to

$$Y_\theta = [H + F\sin^2\theta + Gs\cos^2\theta + (2N - F - 4H)\cos^2\theta\sin^2\theta)]^{-1/2}. \tag{14.9}$$

For a 45° test this becomes $F/2 + G/2 + N = Y_\theta^2$. Solving for N,

$$N = 2Y_{45}^2 - (F + G)/2. \tag{14.10}$$

By differentiating equation 14.8, it can be shown that there are minima and maxima in the value of Y_θ at $\theta = 0°$ and $90°$ and at

$$\theta^* = \arctan[(N - G - 2H/(N - F - 2H)]^{1/2}. \qquad (14.11)$$

There are four possibilities:

With $N > G + 2H$ and $N > F + 2H$ maxima occur at 0° and 90° and there is a minimum at θ^*,
With $N < G + 2H$ and $N < F + 2H$ minima occur at 0° and 90° and there is a maximum at θ^*,
If $G + 2H > N > F + 2H$, θ^* is imaginary and there is a maximum at 0° and a minimum at 90°,
If $G + 2H < N < F + 2H$, θ^* is imaginary and there is a minimum at 0° and a maximum at 90°.

An alternative way of evaluating N is from the R-values. From the flow rules,

$$\begin{aligned}
\varepsilon_x &= \lambda(\sigma_x'[(H + G)\cos^2\theta - H\sin^2\theta] \\
\varepsilon_y &= \lambda(\sigma_x'[(F + H)\cos^2\theta - H\sin^2\theta] \\
\varepsilon_z &= \lambda(\sigma_x'[-F\sin^2\theta - G\cos^2\theta] \\
\gamma_{xy} &= \lambda(\sigma_x' 2N\cos\theta\sin\theta. \qquad (14.12)
\end{aligned}$$

The strain ratio, R_θ, expressed in terms of the anisotropic parameters becomes $R_\theta = [H + (2N - F - 4H)\sin^2\theta\cos^2\theta]/(F\sin^2\theta + G\cos^2\theta)$. For $\theta = 45°$, $R_{45} = N/(F + G) - 1/2$. Solving for N, $N = (R_{45} + 1/2)(F + G)$. Substituting $R = H/G$, $P = H/F$ and $G + H = 1/X^2$.

$$N = (2R_{45} + 1)(R + P)/[2(R + 1)PX^2]. \qquad (14.13)$$

The R-value has a minimum or a maximum at 0° and 90° and there may be one minimum or one maximum between 0° and 90°.

For sheet metals, shear tests are necessary to evaluate L and M in equation 14.3. However γ_{yz} and γ_{zx} are normally zero in sheet forming so these parameters are not necessary.

14.4 SPECIAL CASES OF HILL'S YIELD CRITERION

For the special case in which x, y, and z are principal axes $\tau_{yz} = \tau_{zx} = \tau_{xy} = 0$), equation 14.3 can be expressed in terms of R and P. Substituting $(G + H)X^2 = 1$,

$$\left(\frac{F}{G}\right)(\sigma_y - \sigma_z)^2 + \left(\frac{G}{G}\right)(\sigma_z - \sigma_x)^2 + \left(\frac{H}{G}\right)(\sigma_x - \sigma_y)^2 = \left[\left(\frac{G}{G}\right) + \left(\frac{H}{G}\right)\right]X^2. \qquad (14.14)$$

Now substituting, $R = H/G$ and $R/P = F/G$ and multiplying by P,

$$R(\sigma_y - \sigma_z)^2 + P(\sigma_z - \sigma_x)^2 + RP(\sigma_x - \sigma_y)^2 = P(R + 1)X^2. \qquad (14.15)$$

The flow rules become

$$\begin{aligned}
\varepsilon_x : \varepsilon_y : \varepsilon_z = {} & R(\sigma_x - \sigma_y) + (\sigma_z - \sigma_z) : (R/P)(\sigma_y - \sigma_z) + R(\sigma_y - \sigma_x): \\
& (R/P)(\sigma_z - \sigma_y) + (\sigma_z - \sigma_x). \qquad (14.16)
\end{aligned}$$

Expressing the effective stress and effective strain for this criterion in a way that they reduce to σ_x and ε_x in a uniaxial tension test*:

$$\bar{\sigma} = \left[\frac{R(\sigma_y - \sigma_z)^2 + P(\sigma_z - \sigma_x)^2 + RP(\sigma_x - \sigma_y)^2}{R + P}\right]^{1/2}, \tag{14.17}$$

and

$$\bar{\varepsilon} = C\left[P(\varepsilon_y - R\varepsilon_z)^2 + R(\varepsilon_z - R\varepsilon_x)^2 + (R\varepsilon_x - P\varepsilon_y)^2\right]^{1/2}, \tag{14.18}$$

where $C = [(R+1)/R]^{1/2}/(R+P+1)$.

With rotational symmetry about the z-direction, $F = G = H = N/3$, $L = M$ and $R = P = R_{45}$. The x- and y-directions may be chosen to coincide with the principle stress axes so $\tau_{xy} = 0$. Substituting $P = R$ into equation 14.14 and the flow rules, equation 14.11, the yield criterion and flow rules become:

$$(\sigma_y - \sigma_z)^2 + (\sigma_z - \sigma_x)^2 + R(\sigma_x - \sigma_y)^2 = (R+1)X^2 \tag{14.19}$$

and

$$\varepsilon_x : \varepsilon_y : \varepsilon_z = (R+1)\sigma_x - R\sigma_y - \sigma_z : (R+1)\sigma_y - R\sigma_x - \sigma_z : 2\sigma_z - R\sigma_x - \sigma_y \tag{14.20}$$

Equation 14.19 plots in $\sigma_z = 0$ space as an ellipse as shown in Figure 14.4. The extension of the ellipse into the first quadrant increases with increasing R-value indicating that the strength in biaxial tension increases with R.

Equation 14.19 is often used with an average R, in analyses for materials that have properties that are not rotationally symmetric about the sheet normal. While this procedure is not strictly correct, it is often useful in assessing the role of normal anisotropy.

14.5 NONQUADRATIC YIELD CRITERIA

Calculations† of yield loci for textured fcc and bcc metals suggested that a nonquadratic yield criterion of the form

$$F|\sigma_y - \sigma_z|^a + G|\sigma_z - \sigma_x|^a + H|\sigma_x - \sigma y|^a = 1 \tag{14.21}$$

with an exponent much higher than 2 represents the anisotropy much better. With $a = 2$ this reduces to equation 14.15. Exponents of 8 for fcc metals and 6 for bcc metals were suggested. Although this criterion is a special case of Hill's 1979 criterion, it was suggested independently and is not one that Hill suggested would be useful.

For plane-stress conditions, this criterion reduces to

$$P|\sigma_x|^a + R|\sigma_y|^a + RP|\sigma_x - \sigma y|^a = P(R+1)X^a \tag{14.22}$$

* In Hill's papers the effective stress and strain functions are defined in a way that $\bar{\sigma}$ and $\bar{\varepsilon}$ do not reduce to σ_x and ε_x in a tension test.

† W. F. Hosford, "On Yield Loci of Anisotropic Cubic Metals," *7th North American Metalworking Res. Conf.* SME, Dearborn MI (1979) and R. Logan and W. F. Hosford, *Int. J. Mech. Sci.*, 22 (1980).

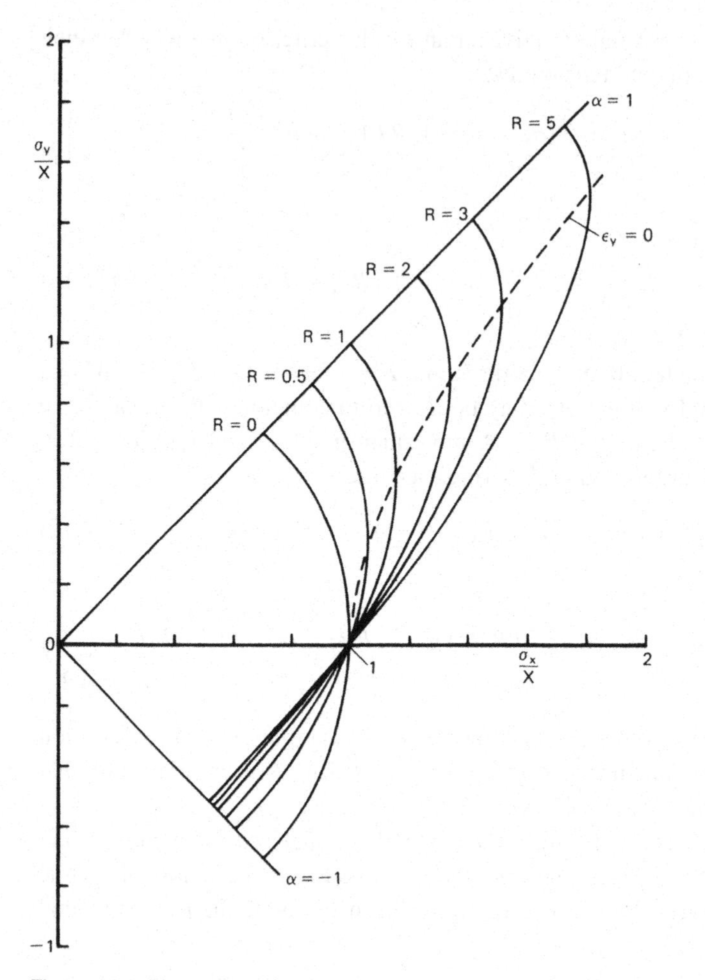

Figure 14.4. Plane-stress ($\sigma_z = 0$) yield locus for rotational symmetry about z according to the Hill criterion (equation 14.18). The dashed line indicates the locus of stress states for plane-strain, $\varepsilon_y = 0$.

If the exponent is an even integer, the absolute magnitude signs in equation 14.21 are unnecessary. With rotational symmetry about z, $R = P$ and the criterion reduces to

$$\sigma_x^a + \sigma_y^a + R(\sigma_x - \sigma_y)^a = (R+1)Y^a \tag{14.23}$$

The yield locus for equation 14.23 plots between the Tresca and Hill criteria (Figure 14.5).

Note that as a increases, the criterion approaches Tresca. The flow rules for equation 14.21 are:

$$\begin{aligned} \varepsilon_x &= \lambda\left[P\sigma_x^{a-1} + RP(\sigma_x - \sigma_y)^{a-1}\right], \\ \varepsilon_y &= \lambda\left[R\sigma_y^{a-1} + RP(\sigma_y - \sigma_x)^{a-1}\right], \\ \varepsilon_z &= -\lambda\left(P\sigma_x^{a-1} + R\sigma_y^{a-1}\right). \end{aligned} \tag{14.24}$$

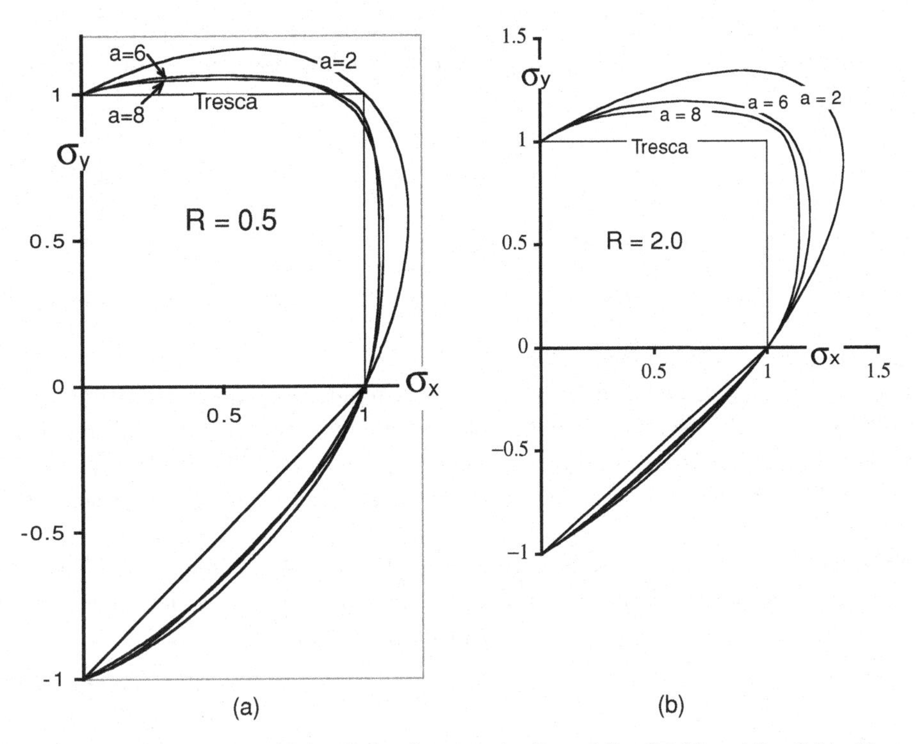

Figure 14.5. Shows plots of this criterion for several values of $R = 0.5$ (a) and $R = 2$ (b) with several values of the exponent a. Note as a increases, the loci approach the Tresca locus.

The effective stress and strain functions are:

$$\bar{\sigma} = \left[\frac{P|\sigma_x|^a + R|\sigma_y|^a + RP|\sigma_x - \sigma_y|^a}{P(R+1)}\right]^{1/a} \tag{14.25}$$

and

$$\bar{\varepsilon} = \varepsilon_x(1 + \alpha\rho)\frac{\sigma_x}{\bar{\sigma}}, \tag{14.26}$$

where

$$\rho = \frac{d\varepsilon_y}{d\varepsilon_x} = \frac{R[\alpha^{a-1} - P(1 - \alpha^{a-1})]}{P[1 + R(1 - \alpha)^{a-1}]}. \tag{14.27}$$

In 1979, Hill* proposed a generalized nonquadratic criterion to account for an "anomalous" observation that in some aluminum alloys with $R > 1$, $P > 1$ and $R_{45} > 1$, yield strengths in biaxial tension were found to be higher than the yield strength in uniaxial tension. This is not permitted with Hill's 1948 criterion.

$$\begin{aligned} f|\sigma_2 - \sigma_3|^m + g|\sigma_3 - \sigma_1|^m + h|\sigma_1 - \sigma_2|^m + a|2\sigma_1 - \sigma_2 - \sigma_3|^m \\ + b|2\sigma_2 - \sigma_3 - \sigma_1|^m + c|2\sigma_3 - \sigma_1 - \sigma_2|^m = 1, \end{aligned} \tag{14.28}$$

* R. Hill, *Math. Proc. Camb. Soc.* v. 75 (1979).

where the exponent, m, depends on the material. Hill suggested four special cases with planar isotropy ($a = b$ and $g = h$). Using the corresponding flow rules in these cases to express R

$$R = \frac{2^{m-1}a + h + 2b - c}{2^{m-1}a + g - b + 2c}. \tag{14.29}$$

Of Hill's special cases, only the 4th case, which can be expressed as

$$|\sigma_1 + \sigma_2|^m + (2R+1)|\sigma_1 - \sigma_2|^m = 2(R+1)Y^m \tag{14.30}$$

is free from concavity problems. Values of $1.7 < m < 2.2$ have been required to fit this to experimental data and different exponents are required for different R-values so this criterion cannot be used to predict the effect of R on forming operations. It should be noted that equation 14.21 is a special case of equation 14.22. However, it cannot account for the anomaly. It was suggested independently and is not one Hill's special cases.

The high exponent yield criteria, equations 14.21, and 14.31, do not provide any way of treating shear stress terms, τ_{yz}, τ_{zx}, or τ_{xy}. In 1989, Barlat and Lian* proposed the plane-stress criterion which accounts for the in-plane shear stress, σ_{xy}.

$$a|K_1 - K_2|^m + a|K_1 + K_2|^m + (a-2)|2K_2|^m = 2Y^m \tag{14.31}$$

where $K_1 = (\sigma_x + h\sigma_y)/2$ and $K_2 = \{[(\sigma_x - h\sigma_y)/2]^2 + p\tau_{xy}^2\}^{1/2}$. Here a, p, h and m are material constants. The exponent, m is approximately 8. It should be noted that this criterion reduces to equation 14.21 for planar isotropy.

Later Barlat and coworkers† proposed a criterion that allows for out of plane shear stresses, τ_{yz} and τ_{zx}. However this criterion requires six constants in addition to m. It will not be discussed further except to state that it does not reduce to equation 14.21 when τ_{yz} and $\tau_{zx} = 0$.

14.6 CALCULATION OF ANISOTROPY FROM CRYSTALLOGRAPHIC CONSIDERATIONS

Taylor‡ analyzed the deformation of a polycrystal by calculating the amount of slip, γ, in each grain necessary to achieve a fixed strain, ε. The ratio, $M = \gamma/\varepsilon$, is called the Taylor factor. He assumed that each grain of a polycrystal deforms in such a way that it undergoes the same shape change as the polycrystal as a whole. Later Bishop and Hill,§ using the same assumptions, showed that M can be calculated by considering the stress states capable of activating multiple slip. Taylor further assumed that the strain hardening could be described by

$$\tau = f(\gamma), \tag{14.32}$$

* F. Barlat and J. Lian, *Int. J. Mech. Sci.*, v. 5 (1989).
† F. Barlat, D. J. Lege and J. C. Brem, *Int. J. Plasticity*, v. 7 (1991).
‡ G. I. Taylor, *J. Inst. Met.* v. 62 (1938).
§ J. F. W. Bishop and R. Hill, *Phil. Mag.* v. 42.

where τ is the shear stress required to cause slip and γ is the total amount of slip on all slip systems. Using this assumption the stress-strain curve for polycrystals are related to that of a single crystal by

$$\sigma = \bar{M}\tau \quad \text{and} \quad \varepsilon = \gamma/\bar{M}, \tag{14.33}$$

where $\bar{M}$ is the average value of M for all orientations. These approaches have been used to calculate the yield loci for textured metals. Details are given elsewhere.

NOTE OF INTEREST

Rodney Hill (born 1921 in Leeds) earned his MA (1946), PhD (1948) and ScD (1959) from Cambridge University. He worked in the Armament Research Dept., Kent, British Iron & Steel Research Association, Bristol University, and Nottingham University before becoming a professor of Mechanics of Solids at Cambridge University. His book, *Mathematical Theory of Plasticity* (Oxford University Press, 1950) is a classic. It contains original work on applications of slip-line fields in addition to introducing the first complete theory of plastic anisotropy.

REFERENCES

R. Hill, *Mathematical Theory of Plasticity*, Oxford University Press, 1950.

W. F. Hosford, *Mechanics of Single Crystals and Textured Polycrystals*, Oxford University Press, 1992.

W. F. Hosford, *Mechanical Behavior of Materials*, 2nd ed. Cambridge University Press, 2010.

PROBLEMS

14.1. Show that the 1948 Hill criterion and flow rules predict an angular variation of R as:

$$R_\theta = \frac{H + (2N - F - G - 4H)\sin^2\theta\cos^2\theta}{F\sin^2\theta + G\cos^2\theta}.$$

14.2. Using the results of Problem 14.1, derive an expression for N/G in terms of R, $P(= R_{90})$, and R_{45}.

14.3. In strip tension tests, the strain ratios, $R_0 = 4.0$, $R_{90} = 2.0$, $R_{45} = 2.5$, and yield strengths, $Y_0 = 49$, $Y_{90} = 45$, were measured. Using the Hill 1948 criterion, calculate $Y_{22.5}$, Y_{45}, and $Y_{67.5}$ and plot Y as a function of θ.

14.4. A thin-wall tube was made from sheet metal by bending the sheet into a cylinder and welding. The prior rolling direction is the axial direction of the tube. The tube has a diameter of 5.0 in and a wall thickness of 0.025 in. The strain ratios are: $R_0 = 2.5$, $R_{90} = 0.8$, and $R_{45} = 1.8$, and the yield strength in the rolling direction is 350 MPa. Neglect elastic effects.

a) If the tube is capped and loaded under internal pressure, at what pressure will it yield?

b) Will the length increase, decrease, or remain constant?

c) If the tube is extended in tension, will the volume inside the tube increase, decrease, or remain constant?

d) Find the stress in the walls if the tube is capped and filled with water, and pulled in tension to yielding.

14.5. Consider a sheet with planar isotropy (equal properties in all directions in the sheet) loaded under plane-stress ($\sigma_z = 0$).

a) Express the ratio, $\rho = \varepsilon_y/\varepsilon_x$, as a function of the stress ratio, $\alpha = \sigma_y/\sigma_x$.

b) Write an expression for $\bar{\sigma}$ in terms of α, R, and σ_x.

c) Write an expression for $d\bar{\varepsilon}$ in terms of α, R, and $d\varepsilon_x$. Remember that $\bar{\sigma}\, d\bar{\varepsilon} = \sigma_x\, d\varepsilon_x + \sigma_y\, d\varepsilon_y + \sigma_z\, d\varepsilon_z$.

14.6. Using the 1948 Hill criterion for a sheet with planar isotropy,

a) Derive an expression for $\alpha = \sigma_y/\sigma_x$ for plane-strain ($\varepsilon_y = 0$) and plane-stress ($\sigma_z = 0$).

b) Find the stress for yielding under this form of loading in terms of X, R, and P.

c) For a material loaded such that $d\varepsilon_z = 0$ and $\sigma_z = 0$, calculate σ_x and σ_y at yielding.

d) For a material loaded such that $d\varepsilon_x = d\varepsilon_y$ and $\sigma_z = 0$, calculate σ_x and σ_y at yielding.

14.7. For a sheet with $X = 350$ MPa, $R = 1.6$, and $P = 2.0$, calculate:

a) The yield strength in the y-direction.

b) The value of σ_x at yielding with $\sigma_z = 0$ and $\varepsilon_y = 0$.

c) The values of σ_x and σ_y at yielding with $\sigma_z = 0$ and $\varepsilon_z = 0$.

d) The values of σ_x and σ_y at yielding with $\sigma_z = 0$ and $\varepsilon_x = \varepsilon_y$.

14.8. Consider a material with planar isotropy and $R = 0.5$. Calculate the stress ratio, $\alpha = \sigma_y/\sigma_x$, for plane-strain, $\varepsilon_y = 0$, using both the 1948 Hill criterion (equation 14.19) and the high-exponent criterion (equation 14.22) with $a = 8$ for

a) $R = 0.5$

b) $R = 2.0$.

14.9. Using equations 14.19 and 14.20 for planar isotropy with $\sigma_z = 0$, determine the plane-strain to biaxial strength ratio, $\chi = \sigma_{x(\sigma_x=\sigma_y)(\sigma_z=0)}/\sigma_{x(\sigma_y=\sigma_z=0)}$, as a function of R. Plot this ratio as a function of R from $R = 0.5$ to $R = 2$ for both the Hill criterion ($a = 2$) and for $a = 6$.

14.10. Marciniak proposed a method of biaxially stretching a sheet by using a cylindrical punch and a spacer sheet with a circular hole as sketched in Figure 14.6. It is often assumed that the stress state is balanced biaxial stress ($\sigma_x = \sigma_y$) if there is equal biaxial straining ($\varepsilon_x = \varepsilon_y$), but this isn't true unless $R_0 = R_{90} = R_{45}$. In experiments using this technique, it was found that $\varepsilon_y/\varepsilon_x = 1.035$ for an aluminum sheet with $R_0 = 0.55$ and $R_{90} = 0.89$.

a) Calculate the stress ratio, σ_y/σ_x, that would produce $\varepsilon_y/\varepsilon_x = 1.035$ using both the Hill criterion ($a = 2$) and for $a = 6$.

b) Calculate the ratio of strains, $\varepsilon_y/\varepsilon_x$, that would result from equal biaxial tension, $\sigma_x = \sigma_y$, using both the Hill criterion ($a = 2$) and for $a = 6$.

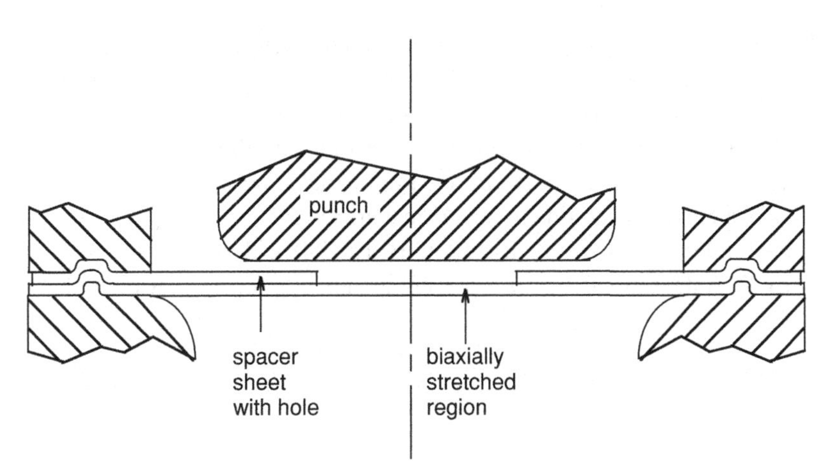

Figure 14.6. Set up for biaxial stretching.

14.11. W. Lode (*Z. Phys.* v. 36, 1926, pp. 913–30) proposed two variables for critically testing isotropic yield criteria and their flow rules:

$$\mu = \frac{2(\sigma_2 - \sigma_3)}{(\sigma_1 - \sigma_3)} - 1 \quad \text{and} \quad \nu = \frac{2(\varepsilon_2 - \varepsilon_3)}{(\varepsilon_1 - \varepsilon_3)} - 1.$$

Plot μ vs. ν for the Tresca and von Mises criteria and for equations 14.23 and 14.24 with $R = 1$ and $a = 6$. Compare your plot with Figure 14.7, which is based on experimental data (G. I. Taylor and H. Quinney, *Phil. Trans. Royal Soc. Ser. A* 203 (1931), pp. 323–62.

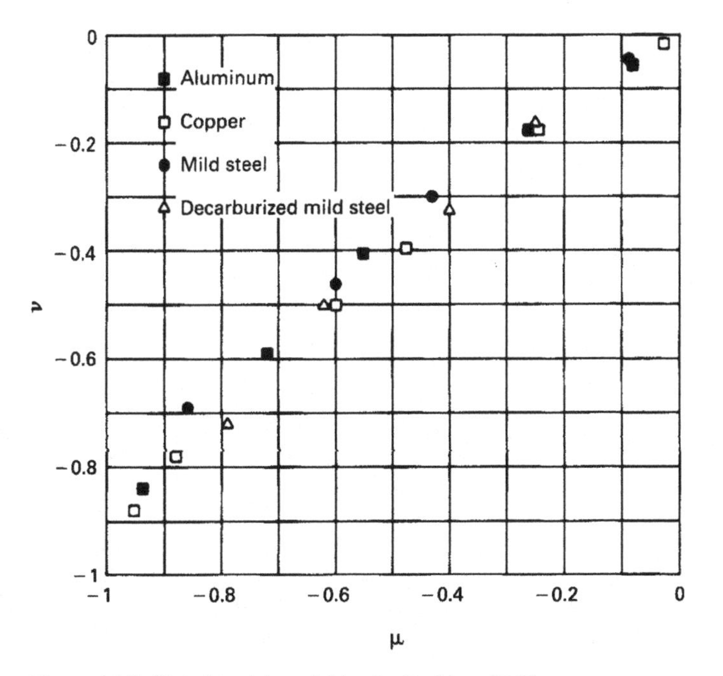

Figure 14.7. Plot of Lode's variables for Problem 14.11.

14.12. An automobile part is to be formed from a low-carbon steel. The steel has planar isotropy with R = 1.8. During tryouts, circles of 5-mm diameter were marked on the sheet before it was press formed. After forming, it was found that a circle in a critical region had become an ellipse with major and minor diameters of 4.87 and 6.11 mm. Assume that $\sigma_z = 0$.

a) Calculate the strains in the critical area.

b) Find the ratio, σ_y/σ_x, assuming Hill's 1948 criterion.

c) Find the ratio, σ_y/σ_x, assuming the high exponent criterion with $a = 6$.

15 Cupping, Redrawing, and Ironing

Sheet forming differs from bulk forming in several respects. In sheet forming, tension predominates whereas bulk forming operations are predominately compressive. In sheet forming operations at least one of the surfaces is free from contact with the tools. Useful formability is normally limited by localized necking, rather than by fracture as in bulk forming. There are instances of failure by fracture but these are unusual.

Sheet forming processes may be roughly classified by the state of stress. At one end of the spectrum is the deep drawing of flat-bottom cups. In this case, one of the principal stresses in the flange is tensile and the other is compressive. There is little thinning but wrinkling is of concern. At the other end of the spectrum are processes, usually called stamping, in which both of the principal stresses are tensile so thinning must occur. Rarely does the formability in sheet forming processes correlate well with the tensile ductility (either reduction in area or elongation at fracture).

15.1 CUP DRAWING

The deep drawing of flat-bottom cups is a relatively simple process. It is used to produce such items as cartridge cases, zinc dry cells, flashlights, aluminum and steel cans, and steel pressure vessels. The process is illustrated by Figure 15.1. There are two important regions: the flange where most of the deformation occurs and the wall, which must support the force necessary to cause the deformation in the flange. If the blank diameter is too large, the force in the wall may exceed its strength, thereby causing failure. Forming limits are described by a limiting drawing ratio (*LDR*), which is the largest ratio of d_0/d_1 that may be successfully drawn. In this respect deep drawing is similar to wire drawing. Indirect compression is induced by tensile forces in the drawn material. Each wedge-shaped element in the flange must undergo enough circumferential compression to permit it to flow over the die lip. A typical drawing failure is shown in Figure 15.2. The flanges must be held down to prevent wrinkling. Figure 15.3 shows what happens when the hold-down force is insufficient.

The following analysis, based on the work of Whiteley* uses the coordinate system illustrated in Figure 15.4. Among the simplifying assumptions are:

* R. Whiteley, *Trans ASM* v. 52 (1960). See also W. F. Hosford in *Formability: Analysis, Modeling and Experimentation*, TMS AIME (1978).

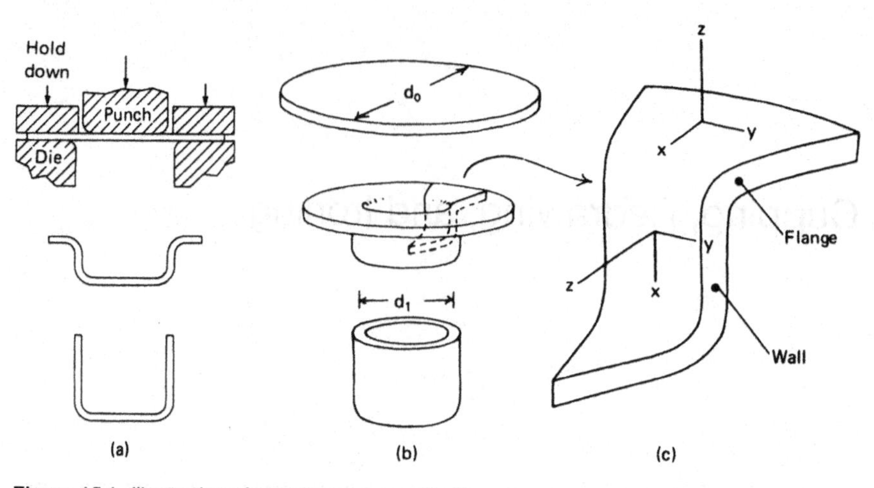

Figure 15.1. Illustration of cupping and coordinate axes.

Figure 15.2. Drawing failures by necking at the bottom of the cup wall. From D. J. Meuleman, *PhD thesis*, University of Michigan (1980).

Figure 15.3. Wrinkling of partially drawn cups due to insufficient hold-down. From D. J. Meuleman, *ibid*.

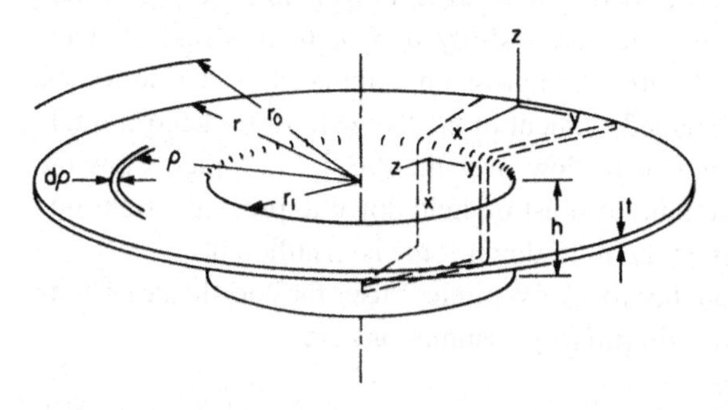

Figure 15.4. Schematic illustration of a partially drawn cup showing the coordinate system. From W. F. Hosford in *Formability: Analysis, Modeling and Experimentation*, TMS-AIME (1978).

1. All of the energy expended is used to deform the material in the flange. The work against friction and the work to bend and unbend the material as it flows over the die lip are initially neglected in the initial treatment, but will be accounted for later by an efficiency factor.
2. The material does not strain harden. It will be shown later that n has only a minor effect on the *LDR*.
3. Flow in the flange is characterized by plane-strain, $\varepsilon_z = 0$, so the thickness of the cup wall is the same as the thickness of the blank.

According to the assumption, $\varepsilon_z = 0$ in the flange, the total surface area is constant and the surface area inside an element is also constant. Therefore

$$\pi\rho^2 + 2\pi r_1 h = \pi\rho_0^2 \quad \text{so} \quad 2\rho\,\mathrm{d}\rho + 2\pi r_1 \mathrm{d}h = 0 \quad \text{or}$$
$$\mathrm{d}\rho = -r_1 \mathrm{d}h/\rho. \tag{15.1}$$

The circumferential strain, $\mathrm{d}\varepsilon_y = \mathrm{d}\rho/\rho$. Since $\mathrm{d}\varepsilon_z = 0$,

$$\mathrm{d}\varepsilon_x = -\mathrm{d}\varepsilon_y = -\mathrm{d}\rho/\rho = r_1 \mathrm{d}h/\rho^2, \tag{15.2}$$

where $\mathrm{d}h$ is the incremental movement of the punch. The incremental work done on the annular element between ρ and $\rho + \mathrm{d}\rho$ equals the volume of the element $(2\pi t\rho\,\mathrm{d}\rho)$ times the incremental work per volume, $\sigma_x \mathrm{d}\varepsilon_x + \sigma_y \mathrm{d}\varepsilon_y + \sigma_z \mathrm{d}\varepsilon_z = (\sigma_x - \sigma_y)\mathrm{d}\varepsilon_x$, so the total work on the element is $\mathrm{d}W = (2\pi t\rho\,\mathrm{d}\rho)(\sigma_x - \sigma_y)(r_1/\rho^2)\mathrm{d}h$. Although σ_x and σ_y vary throughout the flange the value of $(\sigma_x - \sigma_y)$ is constant and will be designated as σ_f. The total work for all such elements is

$$\frac{\mathrm{d}W}{\mathrm{d}h} = \int_{r_1}^{r} \frac{2\pi r_1 t\sigma_f \mathrm{d}\rho}{\rho} = 2\pi r_1 t\sigma_f \ln\left(\frac{r}{r_1}\right). \tag{15.3}$$

The drawing force, F_{d}, which equals $\mathrm{d}W/\mathrm{d}h$ has its largest value at the beginning of the draw when $r = r_0$, so

$$F_{\mathrm{d(max)}} = 2\pi r_1 t\sigma_f \ln(\mathrm{d}_0/\mathrm{d}_1), \tag{15.4}$$

where d_0 and d_1 are diameters of the blank and cup.

The axial stress that the wall must carry is then

$$\sigma_x = \frac{F_{\mathrm{d(max)}}}{2\pi r_1 t} = \sigma_f \ln(\mathrm{d}_0/\mathrm{d}_1). \tag{15.5}$$

The wall will begin to neck when $\sigma_x = \sigma_w$, the yield strength of the wall or

$$\sigma_w = \sigma_f \ln(\mathrm{d}_0/\mathrm{d}_1). \tag{15.6}$$

Both σ_w and σ_w are plane-strain yield strengths. They are equal for an isotropic material, so equation 15.6 predicts that $LDR = \exp(1) = 2.72$. This is much too high. The development should be modified by realizing that the calculated work in the flange and hence the calculated drawing force should be multiplied by $1/\eta$ to account for the work against friction and the work to cause bending. In this case $LDR = \exp(\eta)$. Usually the *LDR* is about 2, so the efficiency is about 0.70.

15.2 ANISOTROPY EFFECTS IN DRAWING

It has been found that the *LDR* increases with *R*. This is understandable because a high *R* implies a low resistance to in-plane contraction (deformation in the flange) and a high resistance to thinning (the failure mode of the wall.) If the material is not isotropic, the drawability will be given by

$$\ln(LDR) = \eta\beta, \tag{15.7}$$

where β is the ratio of the two plane-strain strengths,

$$\beta = \frac{\sigma_{w(\varepsilon_y=0)}}{\sigma_{f(\varepsilon_z=0)}}. \tag{15.8}$$

With Hill's 1948 criterion, it can be shown that

$$\beta = \sqrt{(R+1)/2}, \tag{15.9}$$

so

$$\ln(LDR) = \eta\sqrt{(R+1)/2}. \tag{15.10}$$

Although rotational symmetry about the sheet normal has been assumed in this development, most sheets don't have a single *R*-value. It has been common to use an average *R*-value in equation 15.10.

With the high-exponent criterion (equation 13.23) β in equation 15.8 is

$$\beta = (1/2)\left[\frac{2 + 2^a R}{\alpha^a + 1 + R(1-\alpha)^a}\right]^{1/a}, \tag{15.11}$$

where $\alpha = R^{1/(a-1)}/[1 + R^{1/(a-1)}]$.

Figure 15.5 compares the predictions of equation 15.10 (line labeled $a = 2$) with experimental data. It is obvious that equation 15.10 predicts too much dependence

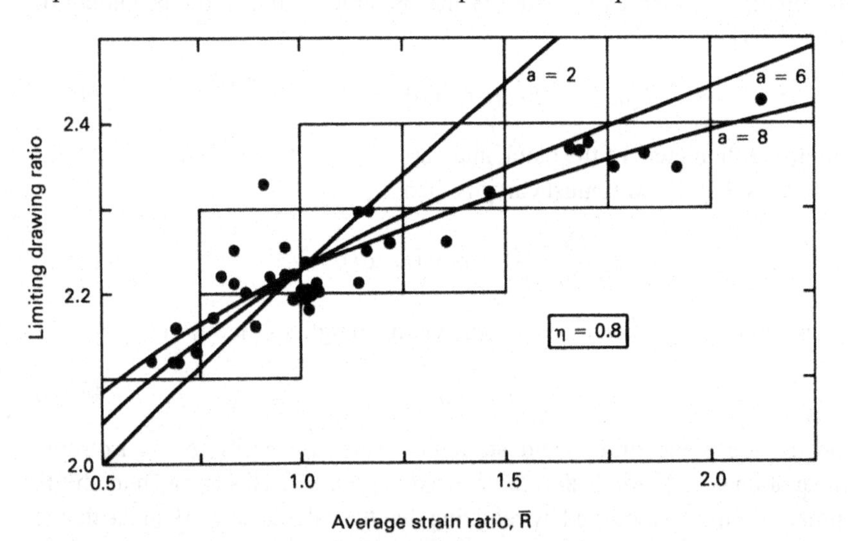

Figure 15.5. Dependence of the limiting drawing ratio on $\bar{R}$. The line labeled $a = 2$ is the prediction of equation 15.10, while the lines labeled $a = 6$ and $a = 8$ are the predictions of equation 15.7 using the high-exponent criterion. From R. W. Logan, D. J. Meuleman and W. F. Hosford in *Formability and Metallurgical Structure*, A. K. Sachdev and J. D. Embury, eds, The Metallurgical Society (1987).

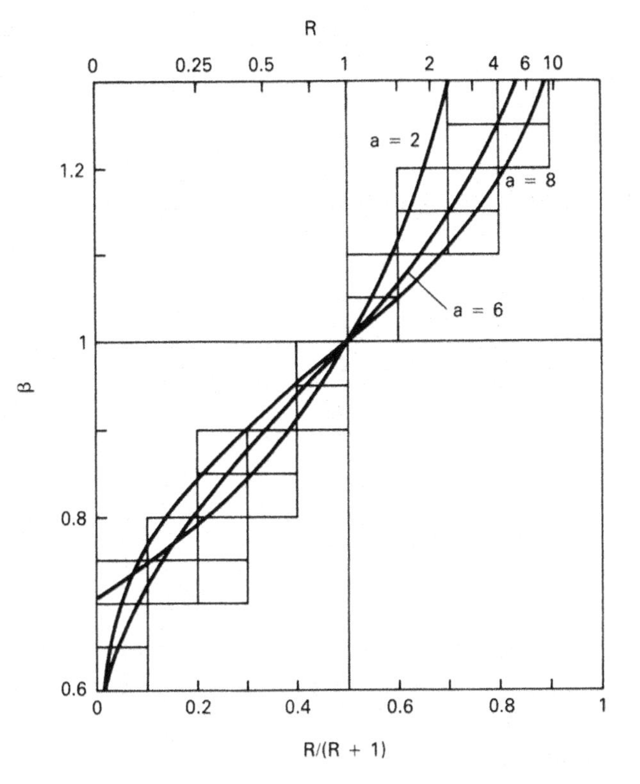

Figure 15.6. Relation between the ratio of plane-strain strengths, β, and the average R value.

of *LDR* on *R*. The lines $a = 6$ and $a = 8$ are the predictions of equation 15.7 when with β evaluated using the high-exponent criterion and its flow rules. In each case η was chosen as 0.8. Changing η shifts the level of the curves without affecting their shapes.

Figure 15.6 shows the results of calculations relating β to R. There is much less dependence for $a = 6$ and $a = 8$ than for $a = 2$.

15.3 EFFECTS OF STRAIN HARDENING IN DRAWING

Although work hardening was neglected in the treatment above, it can be treated analytically. The calculated effect of the strain-hardening exponent, n, on the *LDR* of isotropic material is shown in Figure 15.7. The dependence in the range $0.1 \leq n \leq 0.3$ is very small. The results of a more sophisticated analysis that allows for both strain hardening and thickness strains are shown in Figure 15.8 together with the same data as in Figure 15.5. Calculations were made for three values of n and the data are grouped according to the n-value of the material. Figure 15.9 summarized the effect of n from these experiments and calculations.

Work hardening also controls how the punch force changes during the stroke as shown in Figure 15.10. With increasing n, the maximum is reached later in the stroke. These calculations neglect the bending effect at the die lips early in the stroke. As a consequence the initial rise of the force is usually less rapid than shown here and the maximum occurs somewhat later.

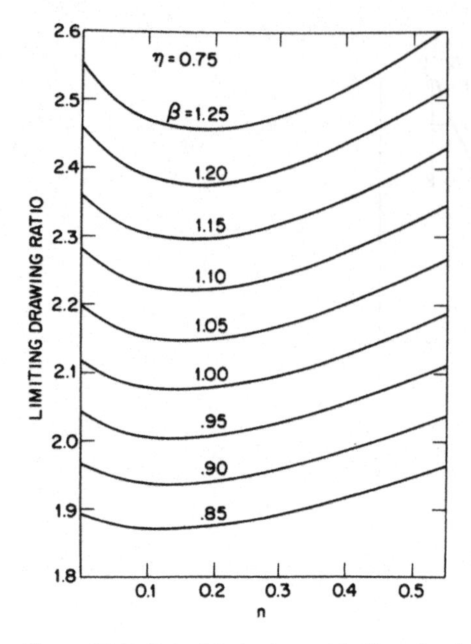

Figure 15.7. Calculated values of limiting drawing ratio for different strain-hardening exponents. From W. F. Hosford, *ibid.*

15.4 ANALYSIS OF ASSUMPTIONS

In the previous analyses, plane-strain was assumed in the flange so no thickness changes occur. This is not strictly true. The material that becomes bottom of the wall experiences more tension than compression while it is in the flange so therefore thins. In contrast, the material that becomes top of the wall experiences more compression than tension while it is in the flange so therefore thickens. Figure 15.11a shows the calculated thickness

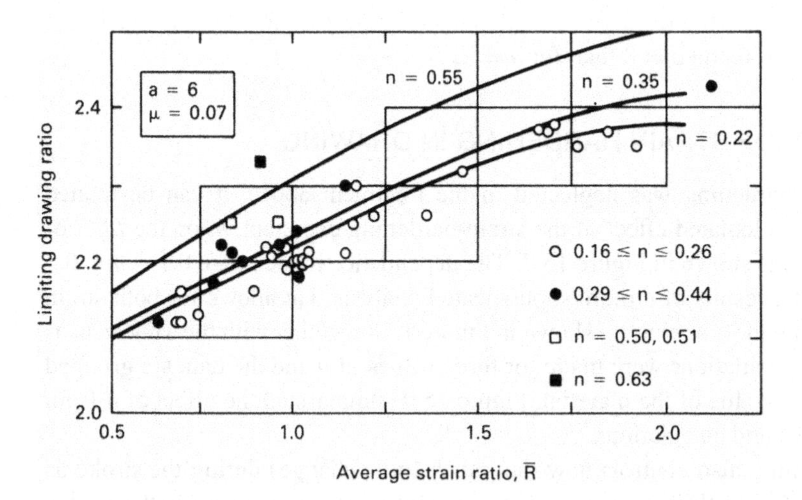

Figure 15.8. Effect of R and n on drawability. The solid lines are finite element calculations with $a = 6$. The experimental data are grouped according to their levels of n. From R. W. Logan, D. J. Meuleman and W. F. Hosford, *ibid.*

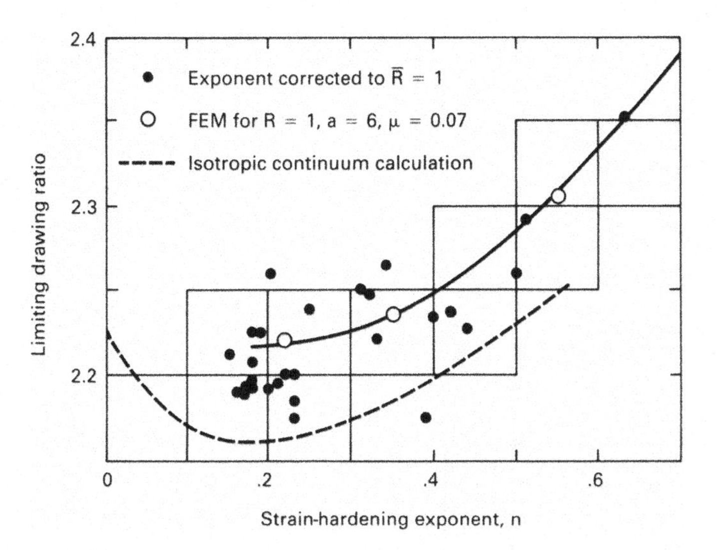

Figure 15.9. Summary of the effect of n on *LDR*. The experimental data are from Figure 15.5 corrected to $R = 1$. From R. W. Logan, D. J. Meuleman and W. F. Hosford, *ibid*.

changes for an isotropic non-work-hardening material and Figure 15.11b shows actual thickness measurements. The effects increase with drawing ratio.

The simplistic assumption of a deformation efficiency is equivalent to assuming that work in bending and unbending, $\mathrm{d}W_b/\mathrm{d}h$, and the work against friction, $\mathrm{d}W_f/\mathrm{d}h$, are proportional to the deformation work in the flange, $\mathrm{d}W/\mathrm{d}h$. For analysis of the *LDR*, it is important only that this is true at the maximum force. The bending and unbending is in plane-strain, $\varepsilon_y = 0$, so this contribution should be proportional to the work in the flange which also deforms with $\varepsilon_y = 0$. The friction originates in two places. One is at the die lip. Here the normal force is $N \approx \pi F_{\mathrm{d}}/2$, so the frictional contribution to the drawing force should be approximately equal to $\pi \mu F_{\mathrm{d}}/2$ (see Problem 15.5). The other place friction acts is in the flange. If the hold-down force is proportional to the

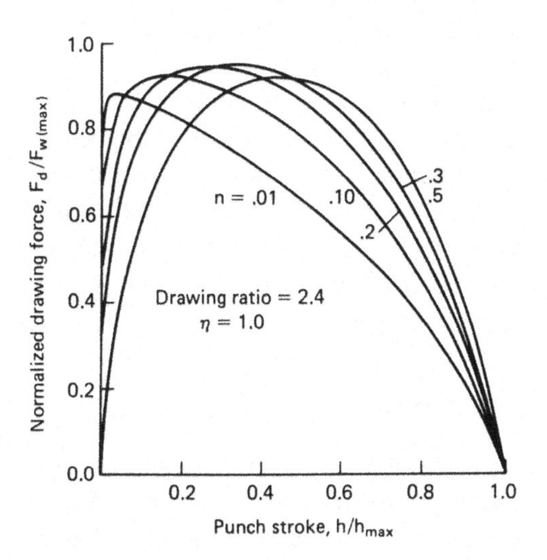

Figure 15.10. Calculated variation of punch force with stroke. From W. F. Hosford, *ibid*.

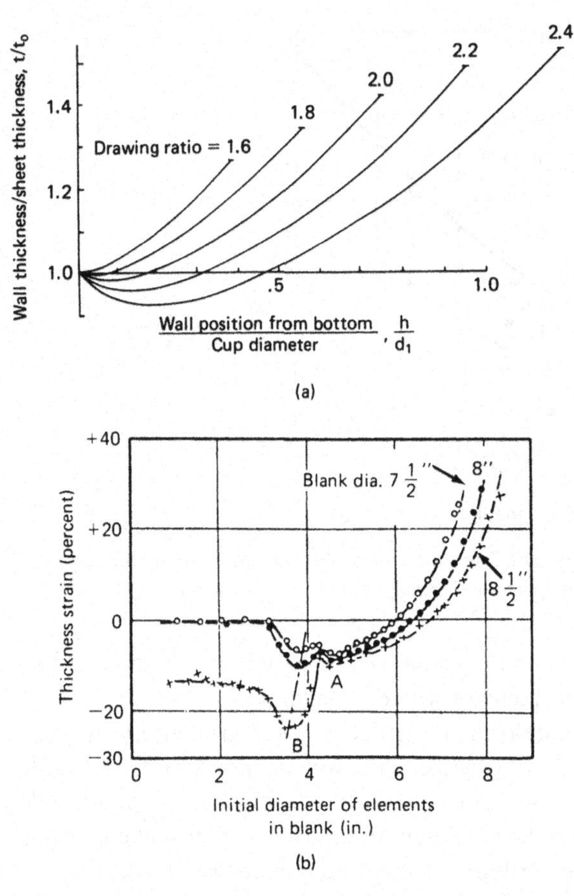

Figure 15.11. (a) Calculated variation of wall thickness and (b) experimental measurements by W. Johnson and P. B. Mellor, *Engineering Plasticity*, van Nostrand Reinhold, 1973.

drawing force the frictional force will be also. It has been suggested that the hold-down force be a constant fraction of the tensile strength, which is almost equivalent. Thus the efficiency should not vary significantly with material properties.

Several investigators have considered another possible mode of failure in materials of low n-value. They predict failure in the region between the die lip and punch early in the punch stroke before any wall has formed. Such failures would occur so early in the punch stroke that there would be little expenditure of energy against friction and bending so the efficiency at that point would be much higher than later in the stroke. Therefore these analyses, which are based on 100% efficiency, fail to account for the fact that the drawing force would be higher at a later stage.

15.5 EFFECTS OF TOOLING ON CUP DRAWING

Drawability depends on tooling and lubrication as well as on material properties. The work expended in bending and unbending as the material flows over the die lip increases with the ratio of the sheet thickness to the radius of the die lip. This causes a higher drawing force and lower *LDR*. However, if the die-lip radius is too large, wrinkling may occur in the unsupported region between the die and punch.

The radius of the punch nose is also important. When failures occur they are usually near the bottom of the wall where the material hasn't been work hardened. With a generous punch radius, the failure site is moved upward on the wall where the wall has been strengthened by more work hardening.

Lubrication has two opposing effects. Lubrication of the flange and die lip reduces the frictional work. However, high friction between the cup wall and the punch causes shear stresses to move the potential failure site upward into material that has been work hardened more. Abrasion to roughen the punch has been known to increase drawability. However, it is difficult to maintain such roughness in production. A similar increase of drawability has been achieved by drawing into a chamber of pressurized water.[*] The pressure increases the friction between the punch and the wall. Increased drawability can also be achieved by maintaining the punch temperature lower than the flange temperature.[†] The temperature difference keeps the flange softer than the wall.

Ideally the hold-down force should be adjusted to a level just sufficient to prevent wrinkling. Swift[‡] has recommended a hold-down pressure of ½ to 1% of the yield strength but the optimum pressure decreases with increasing sheet thickness. Above $t/d = 0.025$, no hold-down is necessary.[§]

15.6 EARING

The height of the walls of drawn cups usually have peaks and valleys as shown in Figure 15.12. This phenomenon is known as *earing*. There may be two, four or six ears, but four ears are most common. Earing results from planar anisotropy and ear height and angular position correlate well with the angular variation of *R*. For two or four ears, earing is described by the parameter

$$\Delta R = \frac{R_0 + R_{90} - 2R_{45}}{2}. \tag{15.12}$$

If $\Delta R > 1$, there are ears at 0° and 90° and if $\Delta R < 1$, ears form at 45°. This variation is shown in Figure 15.13. Earing is undesirable because the walls must be trimmed creating scrap. With earing, the full benefit of a high *R* on *LDR* is not realized.

Experimental correlations of earing with ΔR are shown in Figure 15.14.

The angular dependence of earing can be estimated by assuming that the state of stress at the outer edge of the blank is uniaxial compression and that the compressive strain is the same everywhere along the top of the final cup. The thickness strain then is

* W. G. Granzow, Paper No. F-1920. *Fabricating Machinery Assn.* Rockford IL, 1975.

† W. G. Granzow, *Sheet Metal Industries*, v. 55 (1979).

‡ Chung and Swift, *Proc. Inst. Mech. Eng.*, 165 (1951), p. 199.

§ D. F. Eary and E. A. Reed, *Techniques of Pressworking of Sheet Metal*, 2nd ed. Prentice Hall, 1974.

Figure 15.12. Ears in cups made from three different copper sheets. The arrow indicates the prior rolling direction. From D. V. Wilson and R. D. Butler, *J. Inst. Met*, v. 90 (1961–2).

$\varepsilon_{z\theta} = -\varepsilon_y/(R_{\theta+90}+1)$ where $R_{\theta+90}$ is the R-value measured in a tension test normal to θ. The thickness variation along the top of the cup is then described by

$$t_\theta = t_0 \exp \varepsilon_{z\theta} = t_0 \exp\left(\frac{-\varepsilon_y}{R_{\theta+90}+1}\right) = t_0\left(\frac{\mathrm{d}_1}{\mathrm{d}_0}\right)^{1/(R_{\theta+90}+1)}. \tag{15.13}$$

If it is further assumed that although the wall thickness varies linearly with height the value of

$$h_\theta(t_0+t_\theta)/2 = h_\theta t_0\left[1+(\mathrm{d}_1/\mathrm{d}_0)^{1/(R\theta+1)}\right]/2, \tag{15.14}$$

is the same everywhere along the top of the wall. Therefore

$$h_{45}\left[1+\left(\frac{\mathrm{d}_1}{\mathrm{d}_0}\right)^{1/(R_{45}+1)}\right]^{-1} = h_0\left[1+\left(\frac{\mathrm{d}_1}{\mathrm{d}_0}\right)^{1/(R_{90}+1)}\right]^{-1} = h_{90}\left[1+\left(\frac{\mathrm{d}_1}{\mathrm{d}_0}\right)^{1/(R_0+1)}\right]^{-1}. \tag{15.15}$$

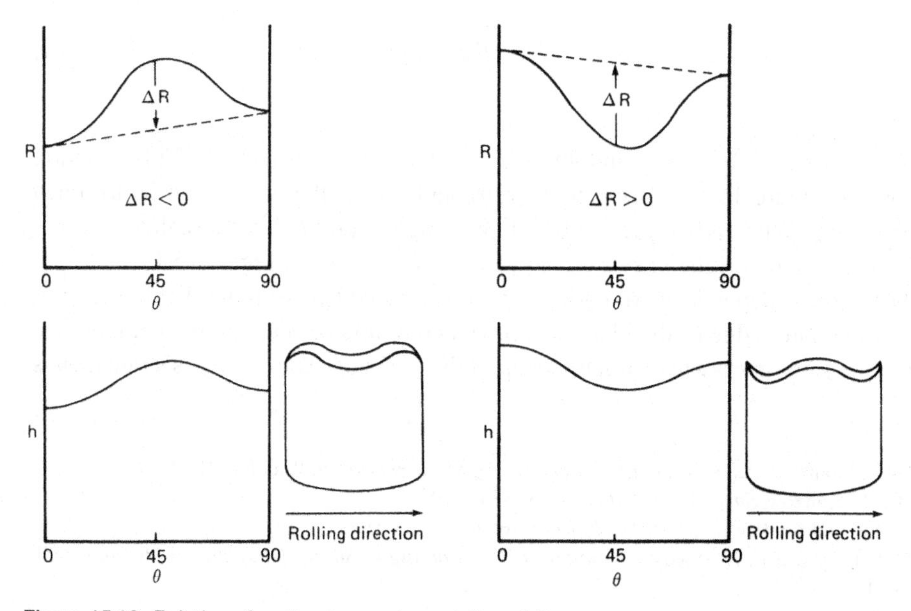

Figure 15.13. Relation of rearing to angular variation of *R*.

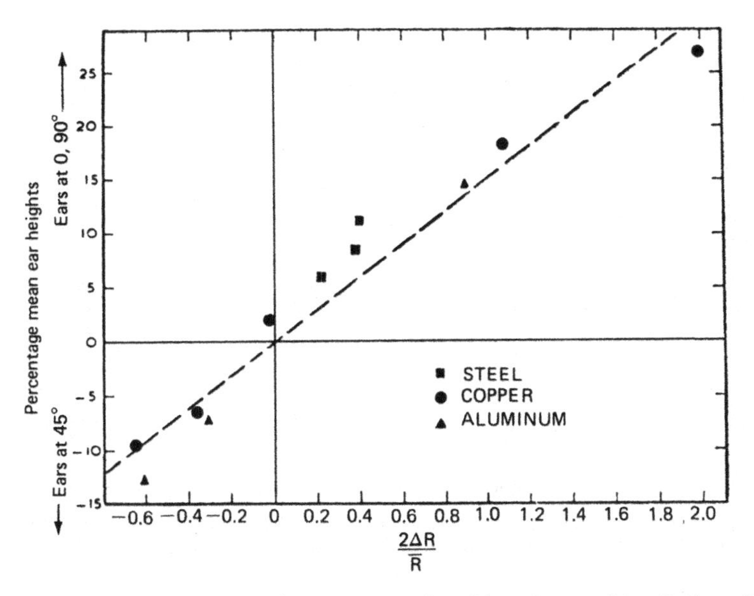

Figure 15.14. Correlation of the extent and position of ears with ΔR. From D. V. Wilson and R. D. Butler, *ibid*.

For a given drawing ratio, the relative ear height $2h_{45}/(h_0 + h_{90})$ can be found from equation 15.13 (See Problem 15.8).

15.7 REDRAWING

Only shallow cups can be made in a single drawing operation. However, redrawing can decrease the diameter and increase wall height. As in drawing, there is little thickness change so the surface area remains constant. Figure 15.15 illustrates direct and reverse redrawing. Hold-down is unnecessary in reverse redrawing.

There is a limiting diameter reduction as in drawing. If the redraw ratio is too high, the wall will fail in tension. Redrawing is almost a steady-state process. If the material were non-work-hardening and the wall thickness constant, the drawing force would remain constant except at the start and end. Because the cup walls are thicker at the top

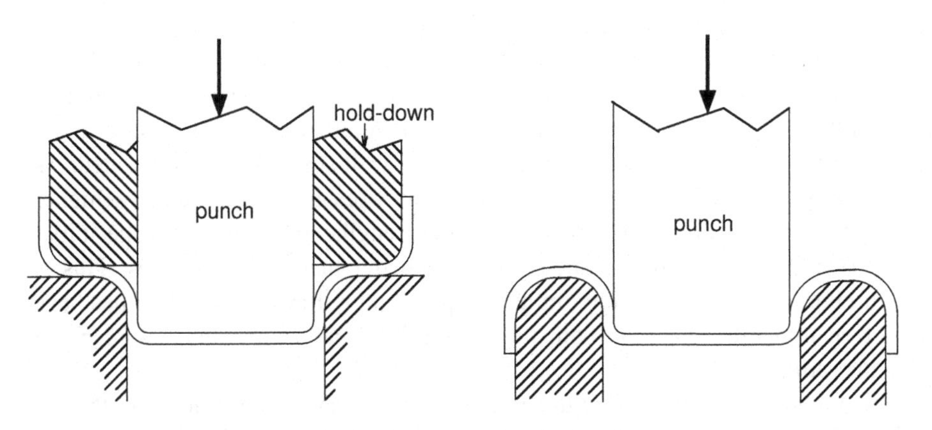

Figure 15.15. Direct redrawing (left) and reverse redrawing (right).

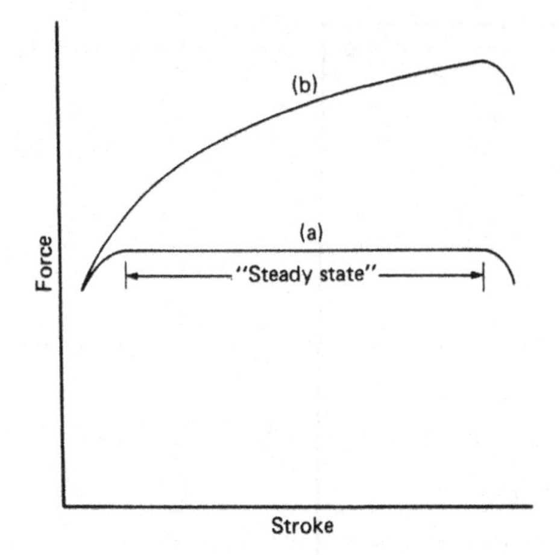

Figure 15.16. For a non-strain-hardening material the punch forces during redrawing are constant except at the very start and end (a). The redraw force for a strain-hardening material increases during the stroke because the initial drawing hardens the material at the top of the walls (b).

than the bottom, the redrawing force rises during the punch stroke. Furthermore, with a work-hardening material, the top of the wall will have been strengthened more during the initial draw so the punch force in redrawing must rise with stroke as the harder material is being deformed. This is shown schematically in Figure 15.16. The bottom of the cup wall, which must carry the redrawing force, is not strengthened during the initial draw. The limiting redraw ratio is greater for materials with low n-values (as cold-rolled sheets) than for those with high n-values (annealed sheets). High R-values are beneficial because the flow during the reduction of wall diameter and the potential necking failure of the wall are the same as in the initial cupping.

15.8 IRONING

Ironing is the forced reduction of wall thickness, which increases the wall height. Some ironing of the top of the wall may occur during the initial drawing. In this case the ironing adds to the total punch force as shown in Figure 15.17. Because the ironing occurs late in the draw, it may not affect the drawability.

Ironing is often used as a separate process after drawing and redrawing. In the manufacture of aluminum beverage cans, ironing is used to increase the wall height by a factor of three. Of course, there is a limiting reduction that can be made in a single ironing pass because high reductions may cause the wall strength to be exceeded. As with redrawing, the limiting reduction decreases with work-hardening capacity (higher n-values) because the bottom of the wall is not strengthened as much in the initial drawing. Heavily work-hardened sheets perform better than annealed ones. This is shown in Figure 15.18. Redrawing, unlike cupping and redrawing, the R-values have no effect on ironability because both the forced deformation and the failure mode involve the same plane-strain.

Ironing tends to greatly reduce the earing. This is because the valleys are thicker than the ears so are elongated more.

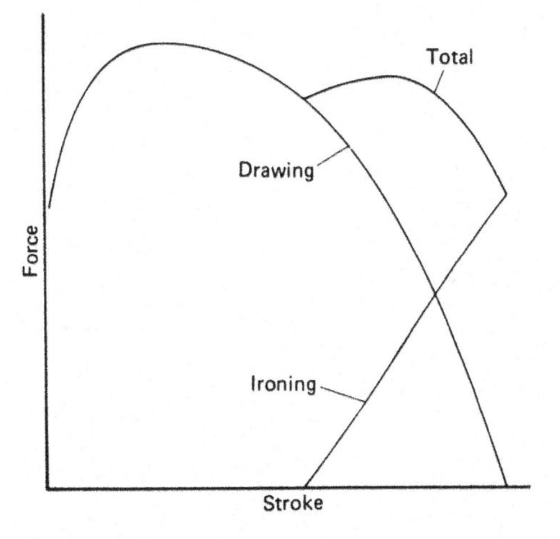

Figure 15.17. Any ironing during drawing adds to the drawing force near the end of the draw. The *LDR* is reduced only if the second peak is higher than the first.

Figure 15.19 shows the forces acting on the deforming metal during ironing. The deformation is much like strip drawing (Chapter 7) except for the action of friction. Friction between the cup wall and deforming material acts to help pull the cup through the die whereas friction between the ironing ring and the deforming material acts in the opposite direction. If friction is neglected, equation 7.11 is applicable.

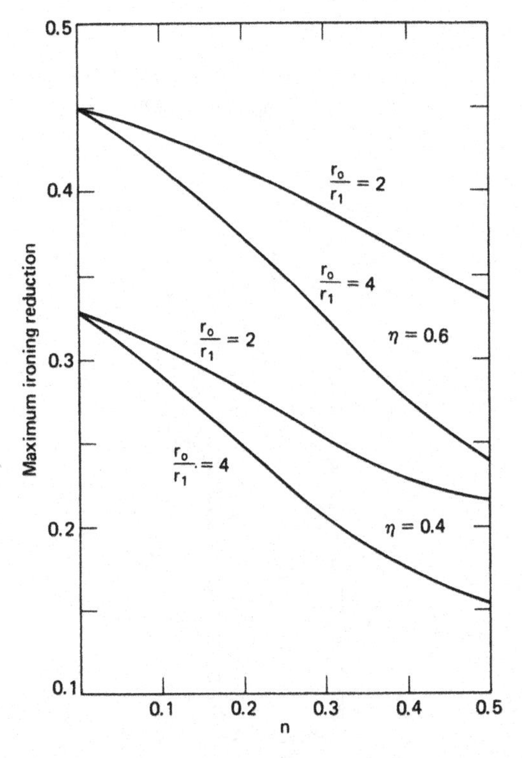

Figure 15.18. The calculated dependence of the limiting ironing reduction on the strain-hardening exponent, n, efficiency, η, and prior drawing and redrawing reduction, (r_0/r_1). From W. F. Hosford, *ibid.*

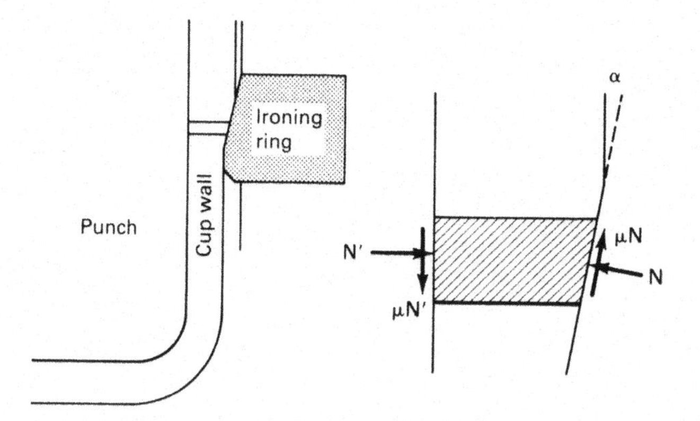

Figure 15.19. Frictional forces on opposite sides of the cup wall during ironing act in opposite directions.

It is possible to draw, redraw and iron in a single stroke of a punch as illustrated schematically in Figure 15.20.

15.9 PROGRESSIVE FORMING

Some shapes require progressive forming through several drawing steps. Figure 15.21 shows how a conical cup can be drawn by progressive forming.

15.10 TAPERED DIES

Drawing limits can be increased by eliminating the friction and the bending and unbending at the die radius and by eliminating the friction under a blank holder. Use of a conical

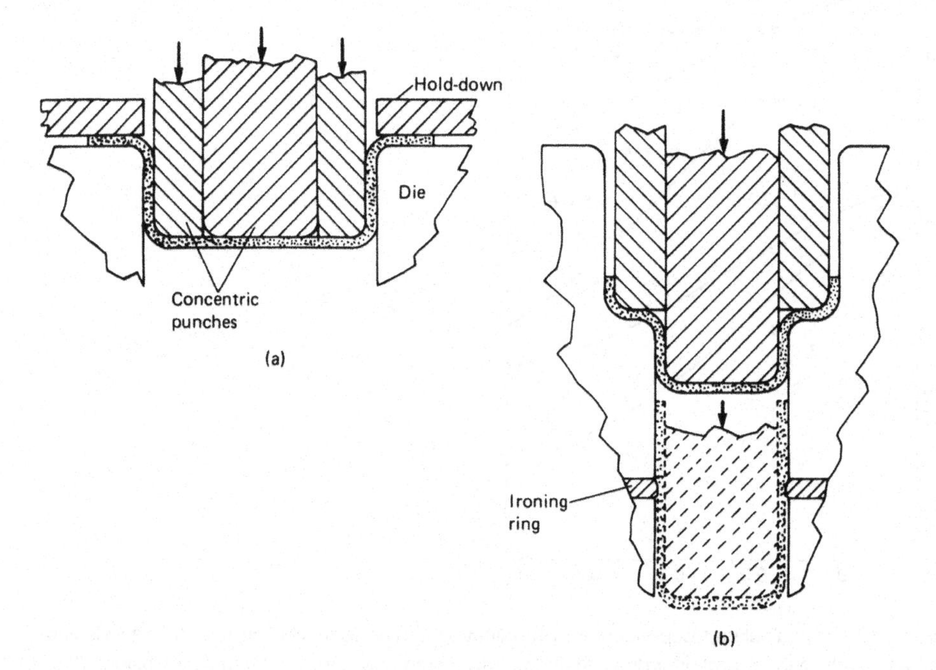

Figure 15.20. Illustration of tooling for cupping, redrawing and ironing in one stroke.

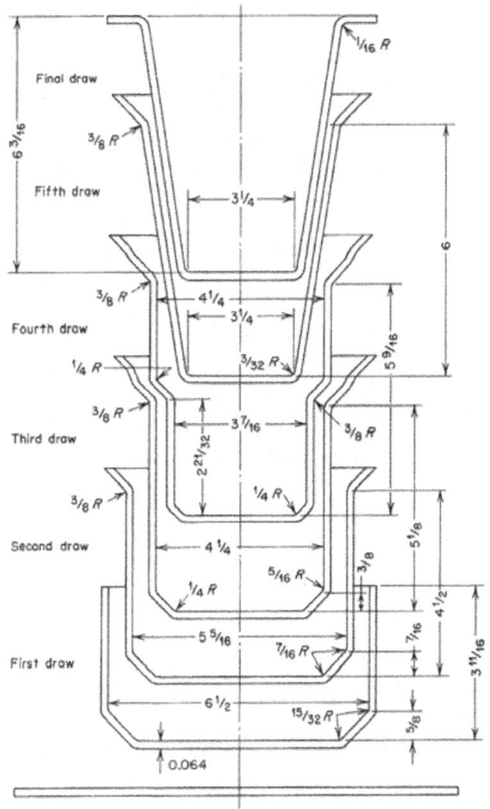

Figure 15.21. Progressive forming of a conical cup. From Serope Kalpakjian, *Mechanical Processing of Metals*, Van Nostrand (Courtesy of Alcoa).

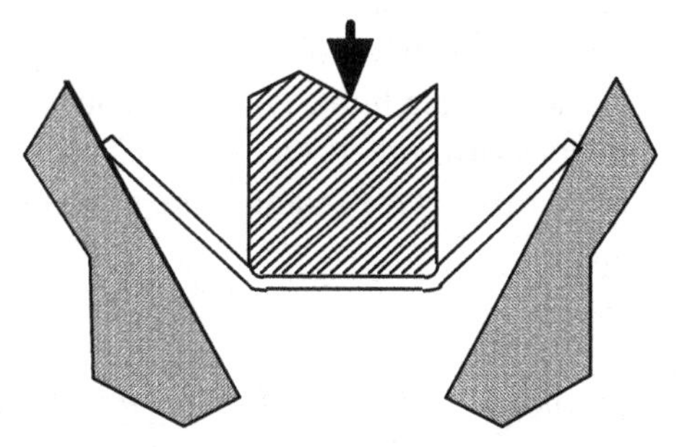

Figure 15.22. Conical die.

die (Figure 15.22) reduces the forces associated with these. Wrinkling of the flange won't occur if the punch diameter is less than 30 times the sheet thickness. The optimum die shape is the Huygen tractrix (Figure 15.23). With this, only the outer edge of the blank contacts the die and the flange remains conical throughout the draw.

It has been found* that tractrix dies make it possible to obtain drawing ratios of up to 2.8 if the ratio of punch diameter to sheet thickness <40. The disadvantages are wrinkling tendency, longer punch travel, and a greater variation of final wall thickness.

* G. E. Totten, K. Funatani, and L. Xie, *Handbook of Metallurgical Process Design.*

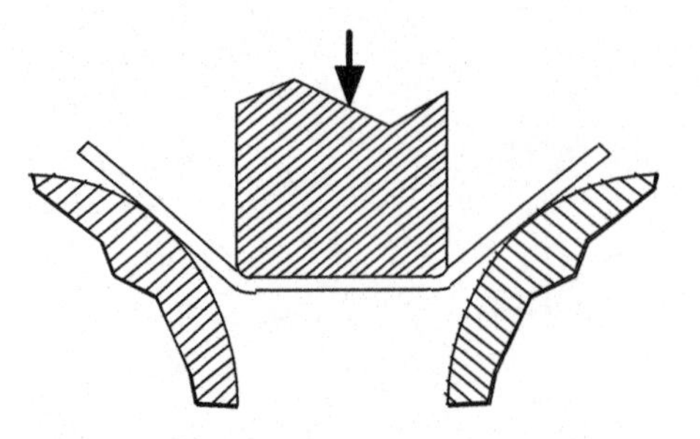

Figure 15.23. Tractrix die.

The tractrix is the curve along which a small object moves when pulled on a horizontal plane with a piece of thread by a puller, which moves rectilinearly. It is mathematically described by

$$x = 1/\cosh(t),$$
$$y = t - \tanh(t) \qquad (15.16)$$

15.11 DRAWING BOXES

Blank shape – shallow boxes are usually drawn from square blanks and trimmed after drawing. For deeper draws, this provides more metal than necessary at the corners. With deep draws, the ideal blank is nearly a circle with the same area. This leaves extra material at the corners (Figure 15.24). Slip-line field theory assuming plane-strain in the flange has been used to develop the ideal blank shapes for irregular cups. However there is usually thickening near the corners.

The drawing limit, H/W is about 0.75 as R/W increases from about 0.03 to 0.5. For very sharp corners ($R/W = 0.3$) failure occurs at the corners so the limiting H/W is very small. These relations are shown in Figure 15.25.

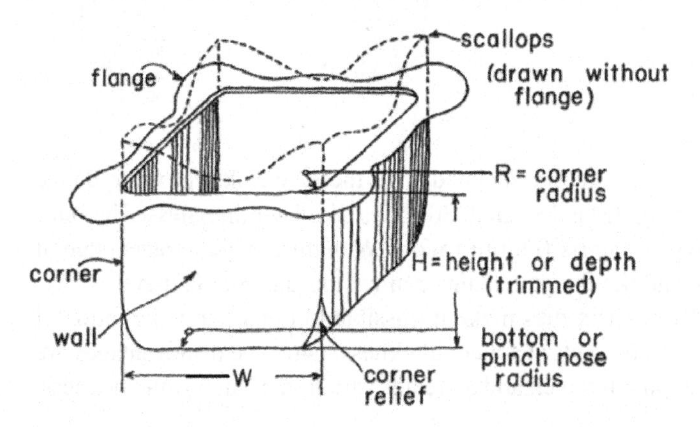

Figure 15.24. Square cup. From G. Sachs, *Principles and Methods of Sheet-Metal Fabrication*, Reinhold Pub (1951).

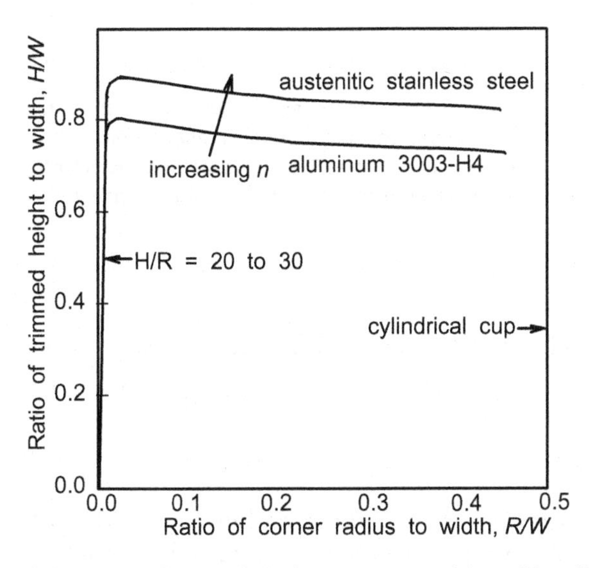

Figure 15.25. Drawing limits for square cups. Adapted from W. A. Backofen, *Deformation Processing*, Addison Wesley, 1972.

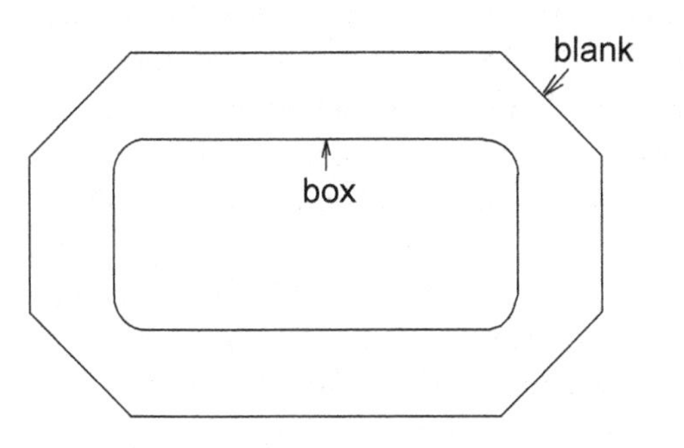

Figure 15.26. Blank for a rectangular box with truncated corners.

Sachs gives the limiting depth of draw for rectangular cups of length L to width, W, ratios between 1 and 3 as

$$H/W = \mathrm{C}\sqrt{(L/W)} \tag{15.17}$$

where C is a material-dependent constant, approximately 0.7 to 0.75. For $L/W > 3$, H/W is independent of length and between 1.2 and 1.35. Blanks for rectangular cups are often rectangles with truncated corners as shown in Figure 15.26.

15.12 RESIDUAL STRESSES

The walls have residual stresses after drawing. In the axial direction, there is residual tension in the outside of the walls and residual tension on the inside. These stresses are greatest near the top of the walls because there was little axial tension during the bending and unbending at the die lip. These residual stresses cause a bending moment that is resisted by a hoop tensile stress. Such residual stresses create a sensitivity to

Figure 15.27. Stainless steel cups which failed by stress corrosion in a laboratory atmosphere within 24 hrs. after drawing. The cracks are perpendicular to the hoop tensile stresses. From D. J. Meuleman, *ibid.*

stress corrosion in some materials. Figure 15.27 shows cup failures caused by stress corrosion.

NOTES OF INTEREST

The first aluminum two-piece beverage cans were produced in 1963. They replaced the earlier steel cans made from three separate pieces: a bottom, the wall, which was bent into a cylinder and welded, and a top. Today the typical beverage can is made from a circular blank, 5.5 inches in. diameter, by drawing it into a 3.5 inch diameter cup, redrawing to 2.625 in. diameter, and then ironing the walls to achieve the desired height. There are about 100 billion made in the United States each year. Beverage cans account for about one-fifth of the total usage of aluminum. This is enough for every man, woman, and child to consume one beer or soft drink every day.

Cartridge brass (Cu-35 Zn) is susceptible to stress-corrosion cracking in atmospheres containing ammonia. Cracks run along grain boundaries under tension. Stress relieving of brass was not common before the 1920's. During World War I, the U.S. Army commandeered barns in France to use as ammunition depots. Unfortunately, the ammonia produced by decomposition of cow urine caused brass cartridge cases to split open. Today, the army requires stress relief of all brass cartridge cases.

REFERENCES

Roger Pierce, *Sheet Metal Forming*, Adam Hilger, 1991.

Edward M. Meilnik, *Metalworking Science and Engineering*, McGraw Hill, 1991.

Z. Marciniak, J. L. Duncan and S. J. Hu, *Mechanics of Sheet Metal Forming*, Butterworth Heinemann, 2002.

PROBLEMS

15.1. Calculate the height-to-diameter ratio for drawing ratios of 1.8, 2.0, 2.25, and 2.5. Assume a constant thickness.

15.2. Calculate the slope of *LDR* vs. $\bar{R}$ for $\bar{R} = \ln h = 0.75$ according to

a) Whiteley's equation, 15.10.

b) Whiteley's analysis (equation 15.10) using equation 15.11 for β.

15.3. Show that the frictional force acting on the die lip is approximately equal to $\pi \eta F_d/2$.

15.4. A typical aluminum beverage can is 5.25 in. high and 2.437 in. in diameter with a wall thickness of 0.0005 in. and a bottom thickness of 0.016 in. The starting material is aluminum alloy 3004-H19 that has already been cold rolled over 80%.

a) What diameter blank is required?

b) Is a redrawing step necessary (assume that a safe drawing ratio is 1.8)? If so, how many redrawing steps are necessary?

c) How many ironing stages are required? (Assume a deformation efficiency of 50%.)

d) If this operation were carried out with one continuous stroke of a telescoping punch and if each operation were completed before the start of the next one, how long would the punch stroke be?

15.5. In the analysis of deep drawing the work in bending and unbending at the die lip was neglected.

a) Derive an expression for the energy expended in bending and unbending as a function of the strain ratio, R, the yield strength, X, the thickness, t, and the die-lip radius. Assume planar isotropy.

b) What fraction of the total drawing load might come from this source for a sheet with $R = 1$, $t = 1.0$ mm, $r = 0.8$ mm, $d_0 = 50$ mm, and $d_1 = 25$ in.?

15.6. Significant increases in the limiting drawing ratio can be achieved using pressurized water on the punch side of the blank as shown in Figure 15.28. Limiting drawing ratios of three have been achieved for material that would with conventional tooling have an *LDR* of 2.1.

a) Explain how the pressurized water can increase the drawability.

b) Estimate the level of pressure required to achieve $LDR = 3$ with $\eta = 0.75$ and $Y = 250$ MPa.

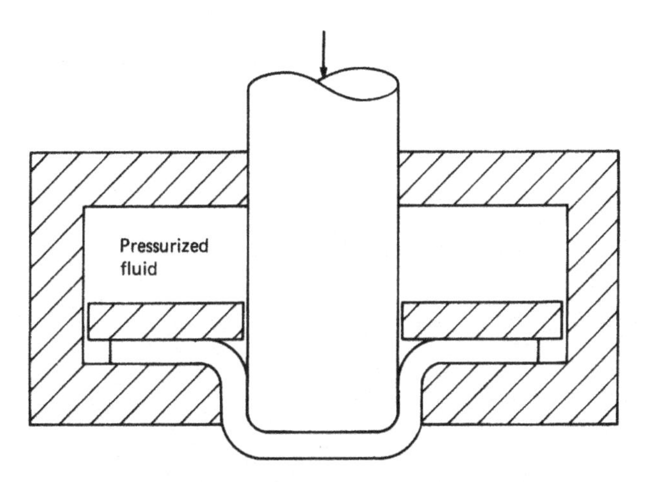

Figure 15.28. Cup drawing with pressurized water on the blank.

15.7. Using the procedure in Section 15.7 estimate $\Delta h/h = 4[2h_{45} - (h_0 + h_{90})]/[2h_{45} + (h_0 + h_{90})]$ for a drawing ratio of 1.8 for materials having the following strain ratios:

a) $R_0 = 1.8$, $R_{90} = 2.0$, $R_{45} = 1.4$
b) $R_0 = 1.2$, $R_{90} = 1.4$, $R_{45} = 1.6$
c) $R_0 = 0.6$, $R_{90} = 0.4$, $R_{45} = 0.8$
d) $R_0 = 0.8$, $R_{90} = 0.8$, $R_{45} = 0.5$
For each material, calculate $2\Delta R/\bar{R}$ and $\Delta h/h$ with Figure 15.14.

15.8. During the drawing of flat-bottom cups, the thickness, t_w, at the top of the wall is usually greater than t_0 at the bottom of the wall as shown in Figure 15.29. Derive an expression for t/t_0 in terms of the blank diameter, d_0, and the cup diameter, d_1. Assume isotropy and neglect the blank holder pressure. Find the ratio of t_w/t_0 for a drawing ratio of 1.8. (Consider the stress state at the outside of the flange.)

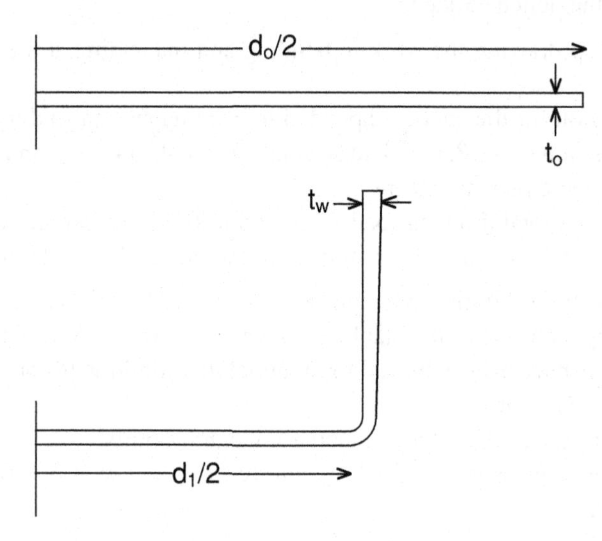

Figure 15.29. Wall thickness variation.

16 Forming Limit Diagrams

16.1 LOCALIZED NECKING

Many sheet forming operations involve biaxial stretching in the plane of the sheet. Failures occur by the formation of a sharp local neck. Localized necking should not be confused with diffuse necking, which precedes failure in round tensile specimens. Diffuse necking of sheet specimens involves contraction in both the lateral and width directions. In sheet tensile specimens, local necking occurs after diffuse necking. During local necking the specimen thins without further width contraction. Figure 16.1 illustrates local necking. At first the specimen elongates uniformly. At maximum load, a diffuse neck forms by contraction of both the width and thickness when $\varepsilon_1 = n$ (Figure 16.1a). Finally a local neck develops (Figure 16.1b).

In a tension test the strain in the width direction cannot localize easily, but eventually a condition is reached where a sharp local neck can form at some characteristic angle, θ, to the tensile axis. Typically the width of the neck is roughly equal to the thickness so very little elongation occurs after local necking. The strain parallel to the neck, $d\varepsilon_{2'}$, must be zero, but

$$d\varepsilon_{2'} = d\varepsilon_1 \cos^2\theta + d\varepsilon_2 \sin^2\theta = 0. \tag{16.1}$$

For an isotropic material under uniaxial tension in the 1-direction, $d\varepsilon_2 = d\varepsilon_3 = -d\varepsilon_1/2$.

Substituting into equation 16.1, $\cos^2\theta - \sin^2\theta/2 = 0$, or

$$\tan\theta = \sqrt{2}, \quad \theta = 54.74^\circ \tag{16.2}$$

If the metal is anisotropic,

$$d\varepsilon_2 = -R/(R+1)d\varepsilon_1. \tag{16.3}$$

In this case

$$\tan\theta = \sqrt{(R+1)/R}. \tag{16.4}$$

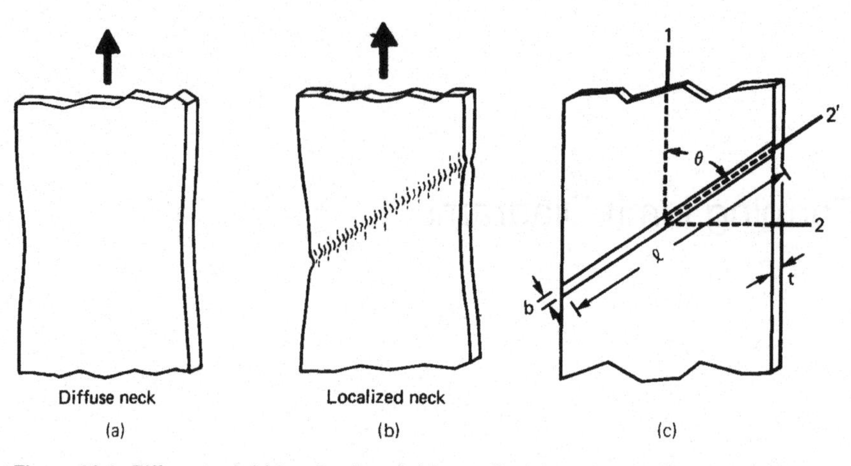

Figure 16.1. Diffuse neck (a) localized neck (b) coordinate system used in analysis (c).

The cross-sectional area of the neck, A', itself is $A' = \ell t$. Because ℓ is constant, $dA'/A' = dt/t = d\varepsilon_3$, and the area perpendicular to the 1-axis is $A = A' \sin\theta$. However, θ is also constant so a local neck can form only if the load can fall under the constraint $d\varepsilon_{2'} = 0$. Since $F = \sigma_1 A$,

$$dF = 0 = \sigma_1 dA + A d\sigma_1, \tag{16.5}$$

or $d\sigma_1/\sigma_1 = -dA/A = -d\varepsilon_3$. Since $d\varepsilon_3 = -d\varepsilon_1/2$,

$$d\sigma_1/\sigma_1 = d\varepsilon_1/2. \tag{16.6}$$

If $\sigma_1 = K\varepsilon_1^n$, $d\sigma_1 = nK\varepsilon_1^{n-1}d\varepsilon_1$. Therefore the critical strain, ε^*, for localized necking in uniaxial tension is

$$\varepsilon_1^* = 2n. \tag{16.7}$$

In sheet forming, the stress state is rarely uniaxial tension but the same principles can be used to develop the conditions for localized necking under a general stress state of biaxial tension. Assume that the strain ratio $\rho = \varepsilon_2/\varepsilon_1$ remains constant during loading. (This is equivalent to assuming that the stress ratio $\alpha = \sigma_2/\sigma_1$ remains constant.) Substitution of $\varepsilon_2/\varepsilon_1 = \rho$ into equation 16.1 gives $\varepsilon_1 \cos^2\theta + \rho\varepsilon_2 \sin^2\theta = 0$, or

$$\tan\theta = 1/\sqrt{-\rho}. \tag{16.8}$$

The angle, θ, can have a real value only if ρ is negative (i.e., ε_2 is negative). If ρ is positive, there is no angle at which a local neck can form.

The critical strain for localized necking is also influenced by ρ. For constant volume, $d\varepsilon_3 = -(1+\rho)d\varepsilon_1$. Substituting into $d\sigma_1/\sigma_1 = -d\varepsilon_3$,

$$d\sigma_1/\sigma_1 = (1+\rho)d\varepsilon_1. \tag{16.9}$$

With power-law hardening the condition for local necking is

$$\varepsilon_1^* = \frac{n}{1+\rho}. \tag{16.10}$$

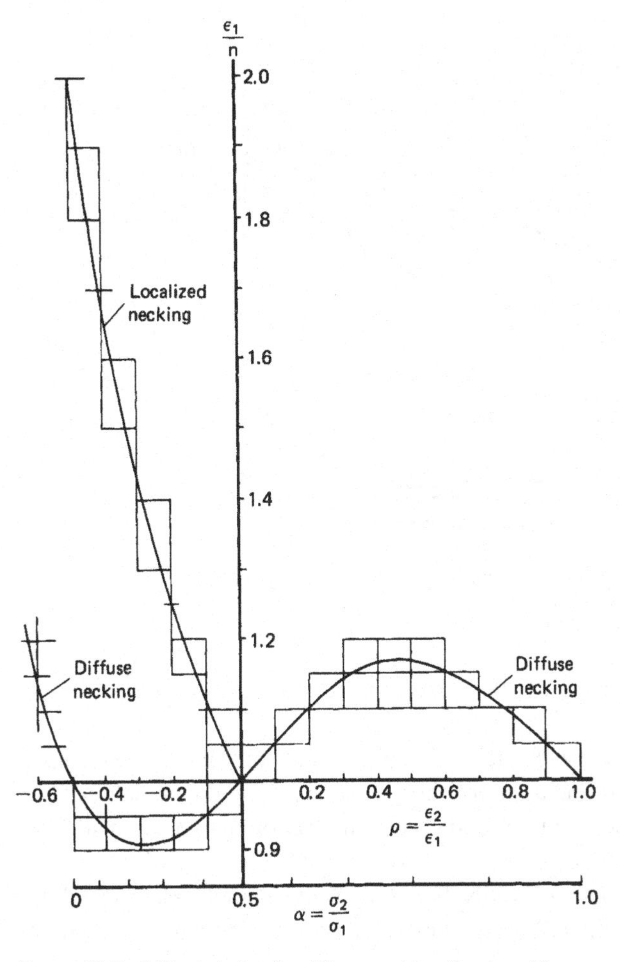

Figure 16.2. Critical strains for diffuse and localized necking according to equations 16.10 and 16.11. Loading under constant stress ratio and constant strain ratio are assumed. Note that under these conditions, no localized necking can occur if ε_2 is positive.

Equation 16.10 implies that the critical strain for localized necking, ε_1^*, decreases from $2n$ in uniaxial tension to n for plane-strain tension.

Swift* showed that diffuse necking can be expected when

$$\varepsilon_1^* = \frac{2n(1+\rho+\rho^2)}{(\rho+1)(2\rho^2-\rho+2)}. \tag{16.11}$$

The criteria for local necking (equation 16.10) and diffuse necking (equation 16.11) are plotted in Figure 16.2.

The previous analysis seems to imply that local necks cannot form if ε_2 is positive. However, this is true only if ρ and α remain constant during loading. If they do change during stretching, local necking can occur even with ε_2 being positive. What is critical

* H. W. Swift, *J. Mech. and Phys. of Solids*, v 1 (1952).

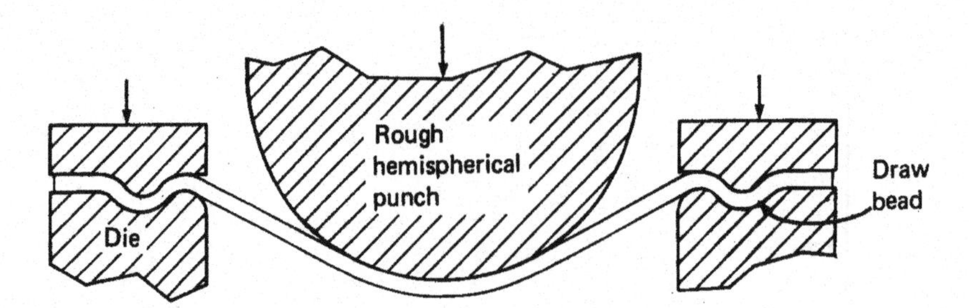

Figure 16.3. Sketch of a rough hemispherical punch.

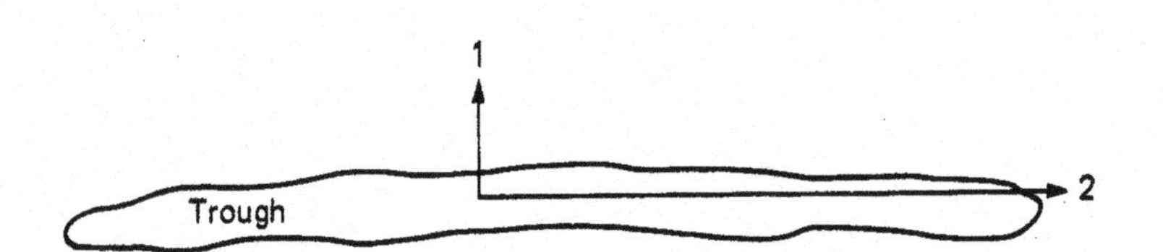

Figure 16.4. Sketch of a trough parallel to the 2-axis.

is that $\rho' = \mathrm{d}\varepsilon_2/\mathrm{d}\varepsilon_1$ become zero rather than that ratio, $\rho = \varepsilon_2/\varepsilon_1$, of total strains be zero.

Often, tool geometry causes a change of strain path. Consider a sheet being stretched over a hemispherical dome as shown in Figure 16.3. The flange is locked to prevent drawing. If friction between punch and sheet is high enough to prevent sliding, deformation ceases where the sheet contacts the punch. Elements in the region between the punch and die are free to expand biaxially. However as the element approached the punch the circumferential strain is constrained by neighboring material on the punch so $\mathrm{d}\varepsilon_2/\mathrm{d}\varepsilon_1 \to 0$.

It has been argued* that because of variations of sheet thickness, grain size, texture, or solute concentration, there can exist local troughs that are softer than the surroundings that lie perpendicular to the 1-axis (Figure 16.4). Although such a trough is not a true neck, it can develop into one. The strain, ε_1, in the trough will grow faster than ε_1 outside of it, but the strain, ε_2, in the trough can grow only at the same as outside the trough. Therefore, the local value of $\rho' = \mathrm{d}\varepsilon_2/\mathrm{d}\varepsilon_1$ decreases during stretching. Once ρ' reaches zero, a local neck can form. The trough consists of material that is either thinner or weaker than material outside of it.

As the strain rate, $\dot{\varepsilon}_1$, inside the groove accelerates, the ratio, $\dot{\varepsilon}_2/\dot{\varepsilon}_1$, approaches zero. Figure 16.5 shows how the strain paths inside and outside of the groove can diverge. The terminal strain outside the groove is the limit strain. Very shallow grooves are sufficient to cause such localization. How rapidly this happens depends on n and to a lesser extent on m.

* Z. Marciniak, *Archiwum Mechanikj Stosowanaj*, 4 (1965) and Z. Marciniak and K. Kuczynski, *Int. J. Mech. Sci.*, 9 (1967).

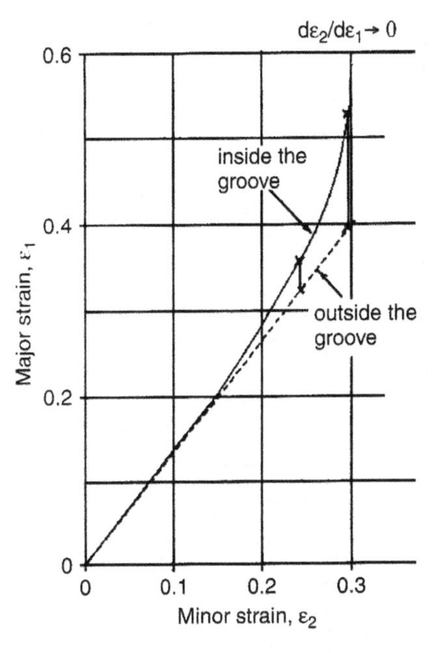

Figure 16.5. Strain paths inside and outside a pre-existing groove for a linear strain path imposed outside the groove. The forming limit corresponds to the strain outside the groove when $d\varepsilon_2/d\varepsilon_1 \rightarrow \infty$.

16.2 FORMING LIMIT DIAGRAMS

The strains, ε_1^*, at which local necks first form, have been experimentally observed for a wide range of sheet materials and loading paths. A widely used technique is to print or etch a grid of small circles of diameter, d_0, on the sheet before deformation. As the sheet is deformed, the circles become ellipses. The principal strains can be found by measuring the major and minor diameters after straining. By convention, the engineering strains e_1 and e_2 are reported. These values at the neck or fracture give the failure condition, while the strains away from the failure indicate *safe* conditions as shown in Figure 16.6. A plot of these measured strains for a material form its forming limit diagram (FLD), also called a Keeler-Goodwin diagram, because of early work by S. P. Keeler* and G. Goodwin.† Figure 16.7 shows a typical FLD for a low-carbon steel.

A plot of the combination of strains that lead to failure is called a *forming limit diagram* or FLD. Figure 16.7 is such a plot for a low-carbon steel. Combinations of ε_1 and ε_2 below the curve are safe and those above the curve lead to failure. Note that the lowest value of ε_1 corresponds to plane-strain, $\varepsilon_1 = 0$.

Figure 16.8 is a comparison of this FLD with equations 16.10 and 16.11 for local and diffuse necking. For negative values of e_2 the experimental and theoretical curves are parallel. The fact that the experimental curve is higher reflects the fact that a neck has to be developed before it can be detected.

* S. P. Keeler, SAE paper 680092, 1968.
† G. Goodwin, SAE paper 680093, 1968.

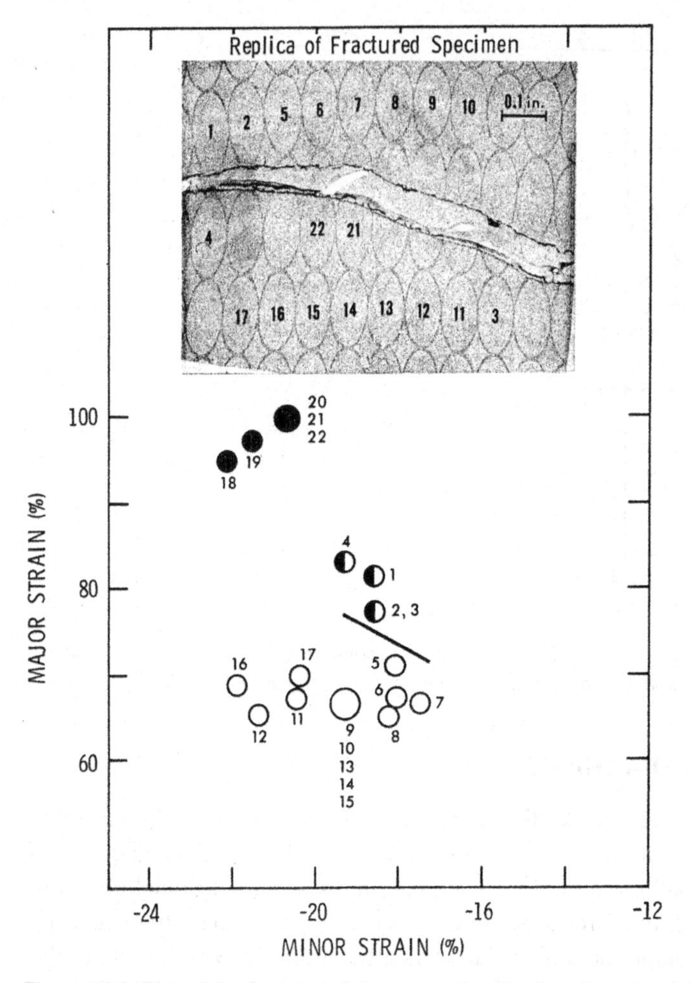

Figure 16.6. Distortion of printed circles near a localized neck and a plot of the strains in the circles. Solid points are for grid circles through which the failure occurred. Open points are for grid circles well removed from the failure and partially filled circles are for grid circles very near the failure. From S. S. Hecker, *Sheet Metal Ind.*, 52 (1975).

16.3 EXPERIMENTAL DETERMINATION OF FLDs

To determine forming limit diagrams experimentally, a grid of circles or squares is printed photographically on a sheet of metal, which is then dyed or lightly etched. Usually the circles are 0.100 in. in diameter but smaller ones may be used. Alternatively strains can be measured by speckle photography without contact on the specimen. A laser speckle pattern diffracts light. A doubly exposed negative is used to determine the shifts between exposures.

The specimens are stretched over a hemispherical dome (usually 4 inches in diameter) until a local neck is first observed (Figure 16.9). To achieve various strain paths the lubrication and specimen width are varied. Full width specimens deform in balance biaxial tension and very narrow strips are almost in uniaxial tension. Better lubrication moves the failure point toward the dome. Figure 16.10 is a photograph showing the tearing from a localized neck.

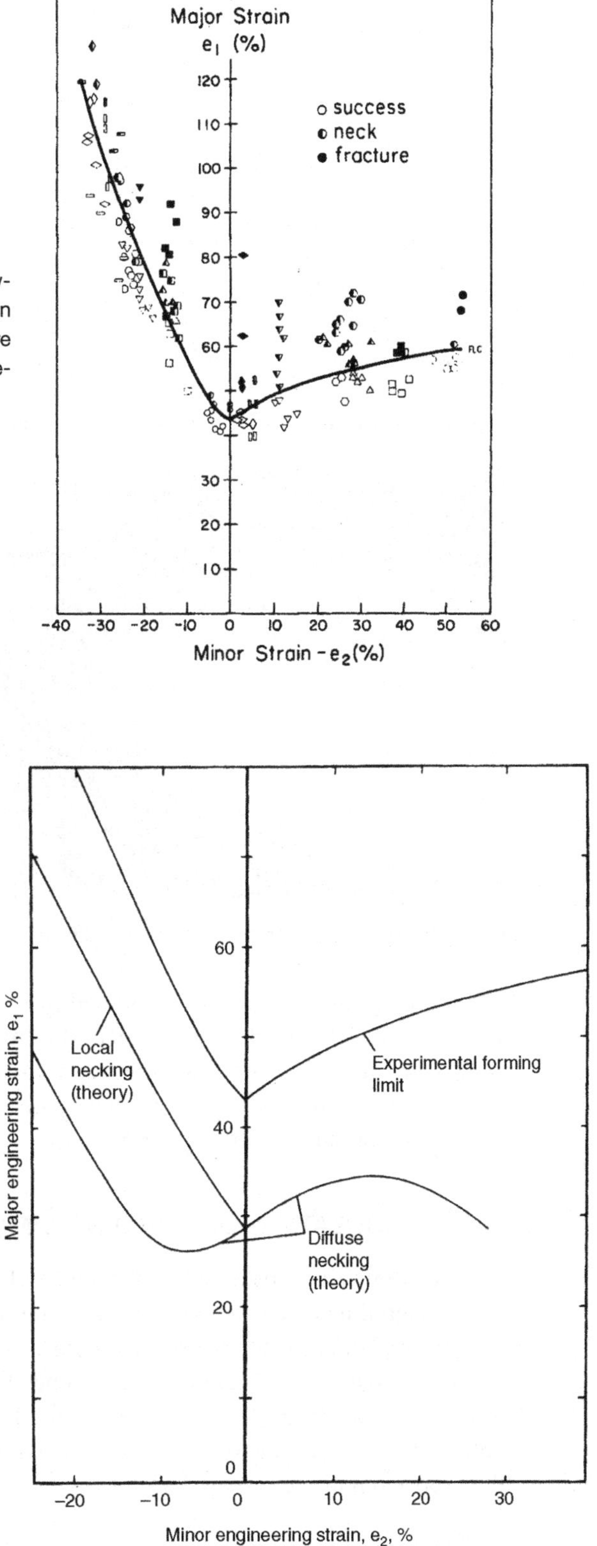

Figure 16.7. Forming limit diagram for a low-carbon steel determined from data like that in Figure 16.6. The strains below the curve are acceptable but those above the curve correspond to failure. From S. S. Hecker, *ibid.*

Figure 16.8. Comparison of experimental FLD in Figure 16.6 with predictions of equations 16.14 and 16.15.

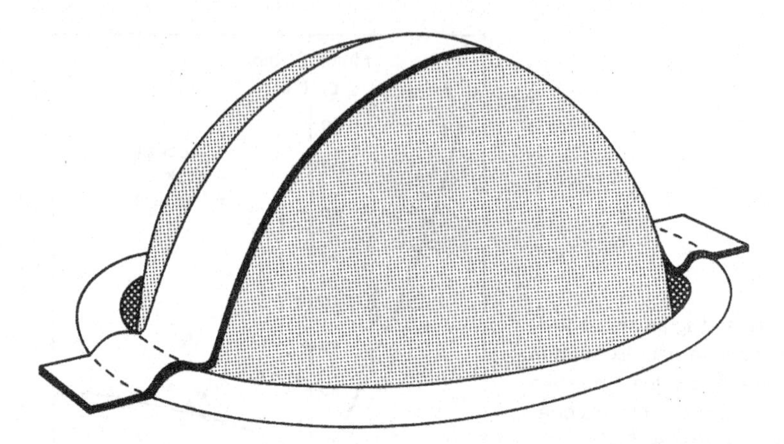

Figure 16.9. Schematic drawing of strip being stretched over a dome.

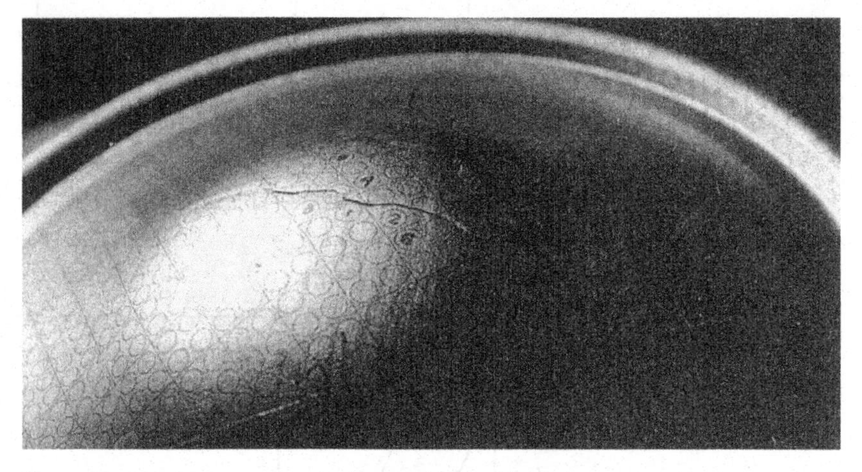

Figure 16.10. Failure near the top of a full width well-lubricated specimen.

Forming limit diagrams determined in different laboratories tend to differ somewhat. One cause is the determination of the "first observable neck" is subjective. There is also the question of how far from the center of a neck measurements can be called safe. Some authorities suggested that "safe readings" should be at least 1.5 circle diameters from the center of the neck.

16.4 CALCULATION OF FORMING LIMIT DIAGRAMS

Marciniak and Kuczynski* showed that the right-hand side of the forming limit diagram for a material may be calculated by assuming that there is a pre-existing defect which lies perpendicular to the major stress axis. For calculation purposes, this defect can be approximated as a region that is thinner than the rest of the sheet. Figure 16.11 illustrates this sort of defect. A ratio of the initial thicknesses inside and outside the defect, $f = t_{b0}/t_{a0}$ is assumed. Also it is assumed that the stress ratio, α_a, outside of the groove remains constant during loading.

* Z. Marciniak and K. Kuczynski, *Int. J. Mech. Sci*, 9 (1967).

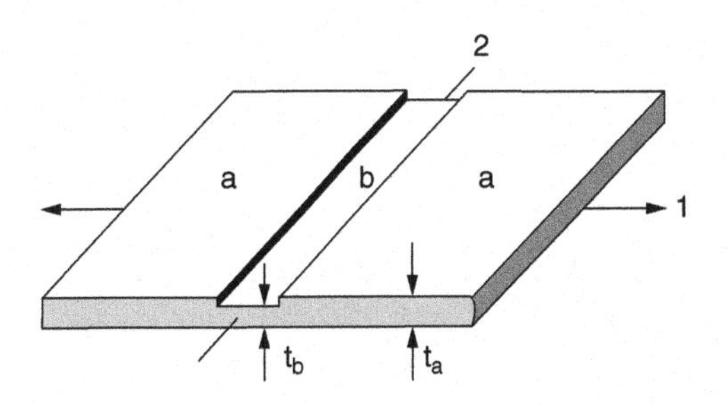

Figure 16.11. Schematic illustration of a pre-existing defect in a sheet.

The calculations are based on imposing strain increments $\Delta\varepsilon_{b1}$ on the material inside of the defect and finding the corresponding value of $\Delta\varepsilon_{a1}$. This involves an iterative procedure. First a value of $\Delta\varepsilon_{a1}$ must be guessed. (It will be somewhat less than $\Delta\varepsilon_{b1}$.) This value is used to calculate $\Delta\varepsilon_{a2} = \rho_a \Delta\varepsilon_{a1}$. Compatibility requires that $\Delta\varepsilon_{b2} = \Delta\varepsilon_{a2}$.

Then

$$\rho_b = \Delta\varepsilon_{b2}/\Delta\varepsilon_{b1} = \dot{\varepsilon}_{2b}/\dot{\varepsilon}_{1b}, \quad \rho_a = \Delta\varepsilon_{a2}/\Delta\varepsilon_{a1} = \dot{\varepsilon}_{2a}/\dot{\varepsilon}_{1a}. \tag{16.12}$$

Then α_a and α_b can be found from the associated flow rule. For Hill's criterion,

$$\alpha = [(R+1)\rho + R]/[(R+1) + R\rho]. \tag{16.13}$$

However, with the high-exponent yield criterion, the flow rule

$$\rho = [\alpha^{a-1} - R(1-\alpha)^{a-1}]/[1 + R(1-\alpha)^{a-1}] \tag{16.14}$$

must be solved by iteration.

Then $\beta = \varepsilon/\varepsilon_1$ can be found using the equation

$$\beta = \Delta\bar{\varepsilon}/\Delta\varepsilon_1 = \dot{\bar{\varepsilon}}/\dot{\varepsilon}_1 = (1 + \alpha\rho)/\varphi, \tag{16.15}$$

where φ_b and φ_a are given by

$$\varphi = \bar{\sigma}/\sigma_1 = \{[\alpha^a + 1 + (1-\alpha)^a]/(R+1)\}^{1/a}. \tag{16.16}$$

The thickness strain rate, $\dot{\varepsilon}_3 = \dot{t}/t$, is also given by

$$\dot{t}/t = -(1+\rho)\dot{\varepsilon} = -(1+\rho)\dot{\varepsilon}_2/\rho. \tag{16.17}$$

The effective stress is given by the effective stress-strain relation

$$\bar{\sigma} = K\dot{\bar{\varepsilon}}^m \bar{\varepsilon}^n. \tag{16.18}$$

In the calculations, the value of $\bar{\varepsilon}$ is incremented by $\Delta\bar{\varepsilon}$, so $\bar{\sigma}$ is calculated as

$$\bar{\sigma} = K\dot{\bar{\varepsilon}}^m (\bar{\varepsilon} + \Delta\bar{\varepsilon})^n. \tag{16.19}$$

Let F_1 be the force per length normal to the groove. The stress per unit perpendicular to the groove is

$$\sigma_1 = \varphi F_1/t, \tag{16.20}$$

$$\dot{\bar{\varepsilon}} = \beta\dot{\varepsilon}_1 = -\beta(\dot{t}/t)/(\rho+1). \tag{16.21}$$

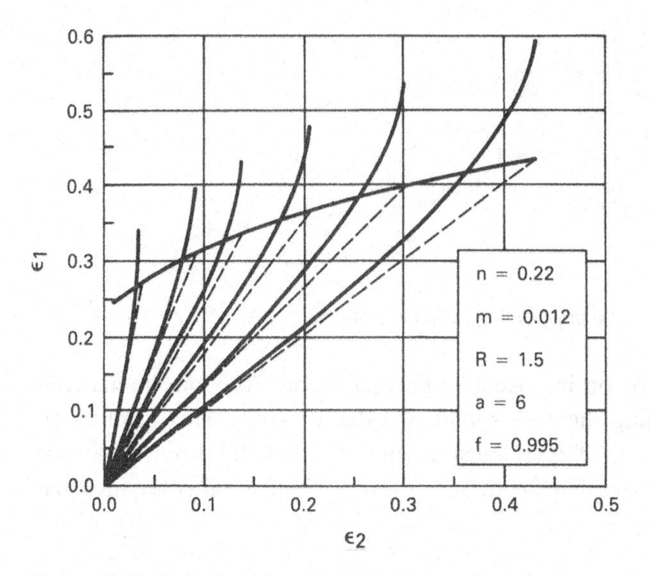

Figure 16.12. Calculated forming limit diagram for a hypothetical material using the high-exponent yield criterion. From A. Graf and W. F. Hosford, *Met. Trans A*, v. 21A (1990).

Combining equations 16.17, 16.18, 16.19, and 16.20 results in

$$F_1 = K(t/\varphi)(\bar{\varepsilon} + \Delta\bar{\varepsilon})^n(\beta\varepsilon_1/\rho)^m. \tag{16.22}$$

The values of F_1 and $\dot{\varepsilon}_2$ must be the same inside and outside of the groove, so

$$(t_a/\varphi_a)(\bar{\varepsilon}_a + \Delta\bar{\varepsilon}_a)^n(\beta_a\varepsilon_{2a}/\rho_a)^m = (t_b/\varphi_b)(\bar{\varepsilon}_b + \Delta\bar{\varepsilon}_b)^n(\beta_b\varepsilon_{2b}/\rho_b)^m. \tag{16.23}$$

Finally substituting $f = t_{b0}/t_{a0}$ and $t = t_0 \exp(\varepsilon_3)$,

$$(\bar{\varepsilon}_a + \Delta\bar{\varepsilon}_a)^n(\beta_a/\rho_a)^m/\varphi_a = f(\bar{\varepsilon}_b + \Delta\bar{\varepsilon}_b)^n(\beta_b/\rho_b)^m/\varphi_b. \tag{16.24}$$

For each strain increment, $\Delta\varepsilon_{1b}$, in the groove there is a corresponding strain increment, $\Delta\varepsilon_{1a}$, outside the groove. The procedure is to impose a strain increment, $\Delta\bar{\varepsilon}_b$, and then guess the resulting value of $\Delta\varepsilon_{1a}$ and use this value together with α_a to calculate β_b, φ_b, ρ_b and $\bar{\varepsilon}_b$. Then these are substituted into equation 16.24 to find $\Delta\bar{\varepsilon}_a$ and then calculate $\Delta\varepsilon_{1a}$ from the change in $\Delta\bar{\varepsilon}_a$. This value of $\Delta\varepsilon_{1a}$ is then compared to the assumed value. This process is repeated until the difference between the assumed and calculated values becomes negligible.

Additional strain increments, $\Delta\varepsilon_{b1}$, are imposed until the $\Delta\varepsilon_{a1} < 0.10\Delta\varepsilon_{b1}$ or some other criterion of approaching plane-strain is reached. The values of ε_{a2} and ε_{a1} at this point are taken as points on the FLD. Figure 16.12 shows an example of the calculated strain paths inside and outside of the defect using $\Delta\varepsilon_{a1} < 0.10\Delta\varepsilon_{b1}$, $\varepsilon_{3a} = 0.90\varepsilon_{3b}$ as the criterion for failure.

Calculated forming limit diagrams are very sensitive to the yield criterion used in the calculations. Figure 16.13 shows that the forming limit diagrams calculated with Hill's 1948 yield criterion are very dependent on the *R*-value. This is not in accord with experimental observations in which there is no appreciable dependence on *R*. Figure 16.14 shows that with the high exponent criterion, there is virtually no calculated dependence on the *R*-value.

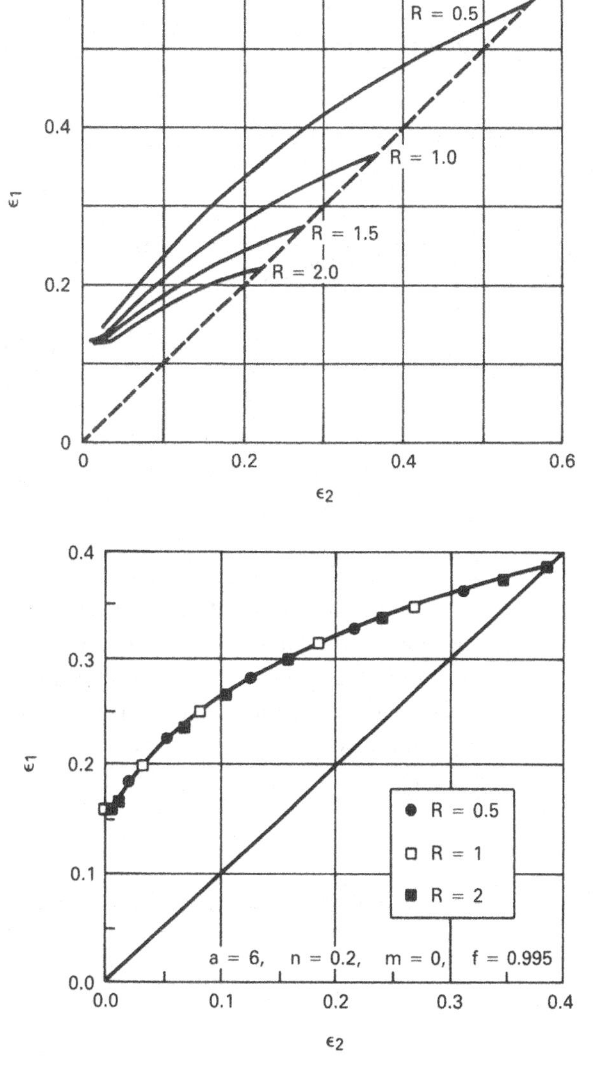

Figure 16.13. Forming limit diagrams calculated for several *R*-values using Hill's 1948 yield criterion. Values of $n = 0.20$, $m = 0$ and $f = 0.98$ were assumed. From Graf and Hosford, *ibid*.

Figure 16.14. Forming limit diagrams calculated for several *R*-values using $a = 6$ in the high-exponent yield criterion. Values of $n = 0.20$, $m = 0$, and $f = 0.995$ were assumed. From Graf and Hosford, *ibid*.

Hill showed that the left-hand side of FLD is easily calculated. Equation 16.10 for localized necking, $\varepsilon_1^* = n/(1+\rho)$, can be expressed as

$$\varepsilon_3^* = -n, \tag{16.25}$$

which corresponds to a simple condition of a critical thickness strain.

16.5 FACTORS AFFECTING FORMING LIMITS

The level of the FLD_0 at plane-strain, $\varepsilon_2 = 0$, depends primarily on the strain-hardening exponent, n. Theoretically, at plane-strain, $\varepsilon_1 = n$. An increased strain-rate exponent, m, raises the FLD_0 somewhat. Increasing values of n decrease the slope of the right-hand side of the FLD. These effects are illustrated in Figure 16.15.

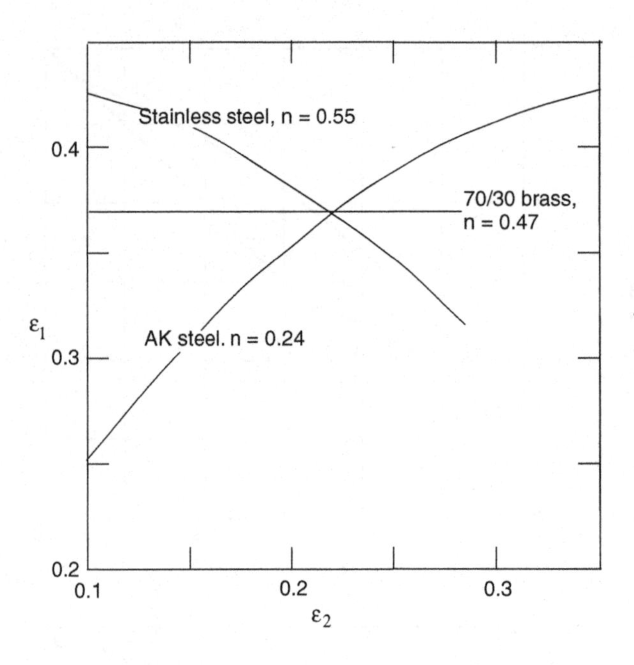

Figure 16.15. Increasing values of n raise the forming limit for plane-strain but lower the slope of the right-hand side of the diagram. Data from M. Azrin and W. A. Backofen, *Met. Trans*, v. 1 (1970).

Inhomogeneity and defects lower the FLD. These inhomogeneities may be local variations in sheet thickness, in composition, grain size or anything that affects yielding. Anisotropy has almost no effect on the shape or level of the FLD.

In biaxial tension, fracture may occur before local necking. Figures 16.16 and 16.17 show how fracture truncates the FLD caused by local necking.

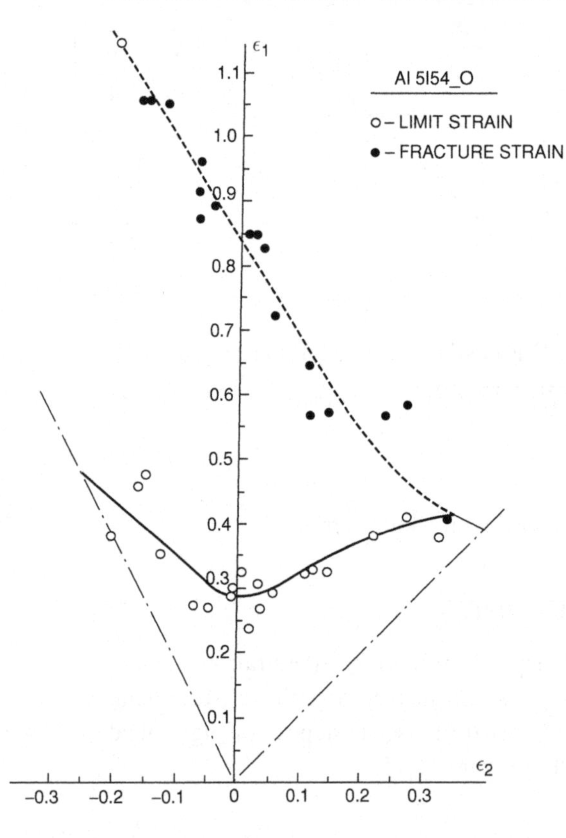

Figure 16.16. Forming limit diagram for aluminum alloy 5154-0. From G. H. LeRoy and J. D. Embury in *Formability: Analysis, Modeling and Experimentation*, A. K. Ghosh and H. L. Hecker eds. AIME, 1978.

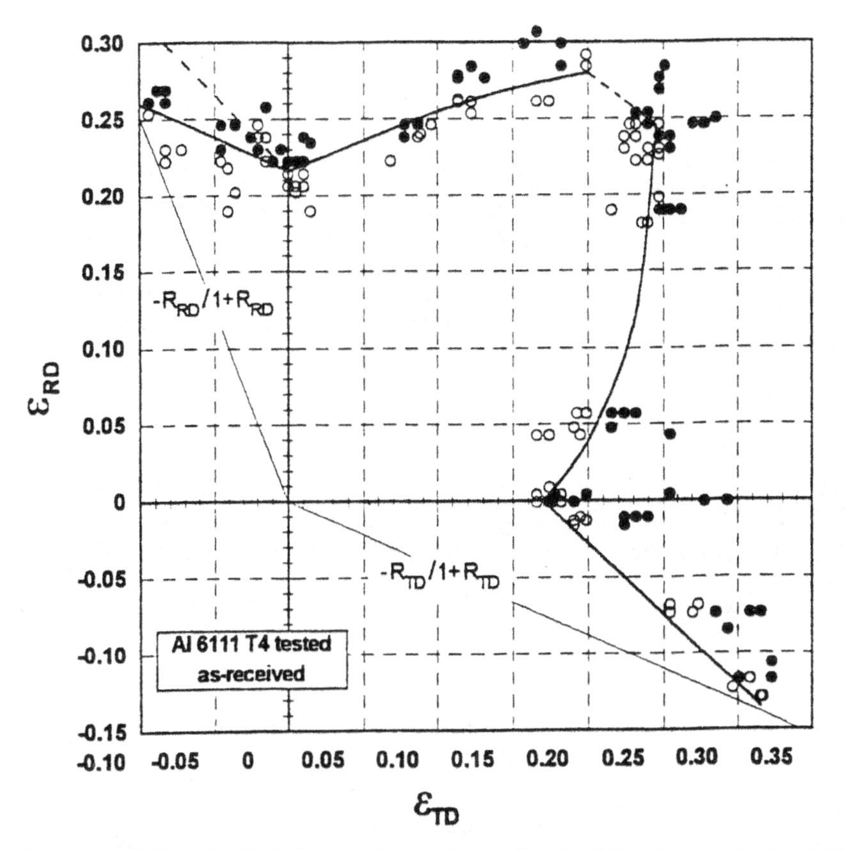

Figure 16.17. Forming limit diagram for aluminum alloy 6111-T4. From A. Graf and W. F. Hosford, *I. J. Mech. Sci.* v. 36 (1994).

It has been found experimentally that the level of the FLD increases with sheet thickness. Keeler and Brazier* have approximated this effect by

$$\begin{aligned} \mathrm{FLD}_0 &= n(1+0.72t) \quad \text{for} \quad n \le 0.2 \quad \text{and} \\ \mathrm{FLD}_0 &= (1+0.72t) \quad \text{for} \quad n \ge 0.2, \end{aligned} \tag{16.26}$$

where t is the sheet thickness in mm. This is illustrated in Figure 16.18.

Although the thickness effect is well established, the reason for it is still controversial. There is some evidence that the measured forming limits depend on the size of the grid used to measure strains. With thicker sheets, and a constant grid size, the measured strains at the middle of the neck are larger. If this is true, the effect would disappear if the ratio of grid size to thickness were held constant. Finally the level of the FLD may be affected by the method used to detect local necks.

Smith and Lee† have reported that the thickness effect in aluminum alloys is much less than that in steel (Figure 16.19). This can be attributed to the much sharper necks in aluminum alloys.

Forming limit diagrams have proved useful in diagnosing actual and potential problems in sheet forming. If previously gridded sheets are used in die tryouts, local

* S. P. Keeler and W. G. Brazier, in *Micro Alloying 75* (Union Carbide, 1977).

† P. E. Smith and D. Lee, *Proc. of the Intern. Body Engineering Conf.*, SAE Detroit, SAE pub. 331 (1998).

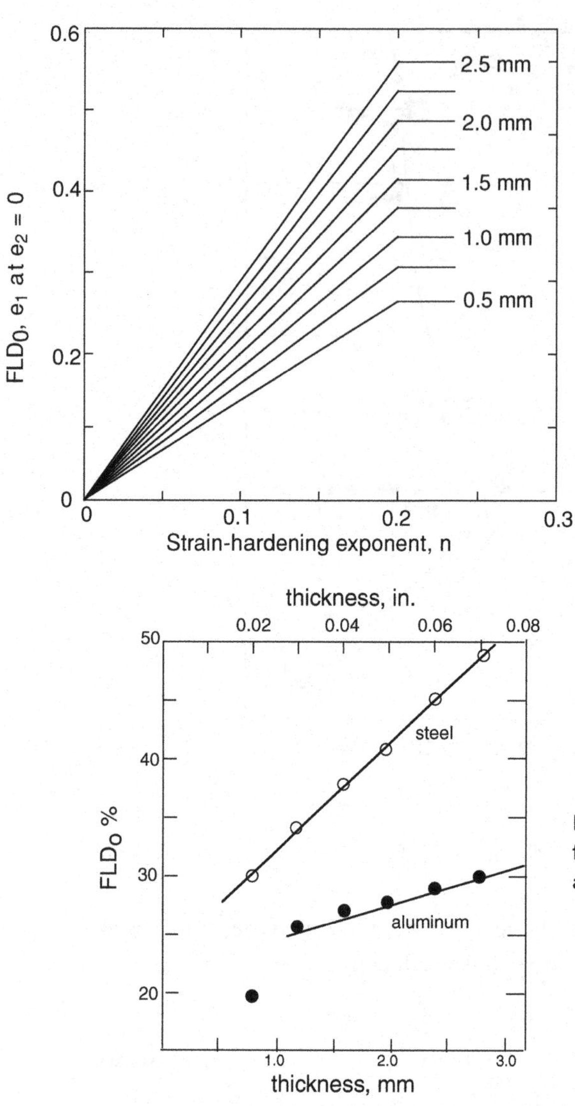

Figure 16.18. Effect of sheet thickness on FLDs according to equation 16.18.

Figure 16.19. Effect of thickness on measured forming limits in plane-strain. Adapted from Smith and Lee. *ibid.*

strains near failures or suspected trouble spots can be measured and compared with the forming limit diagram. This serves two purposes. The severity of strains at potential trouble can be assessed, If the strains are too near the limiting strains, failures are likely to occur in production because of variations in lubrication, tool alignment, tool wear, material properties, and thickness. The second reason is that the nature of the problem can be diagnosed. Because the lowest values of ε_1^* correspond to plane-strain, higher levels of ε_1^* can be achieved by either lowering ε_2 with less flange locking or raising ε_2, with better lubrication.

16.6 CHANGING STRAIN PATHS

Forming limit diagrams are usually determined with experiments that involve nearly linear strain paths (i.e., the ratio of principal stresses and principal strains is nearly

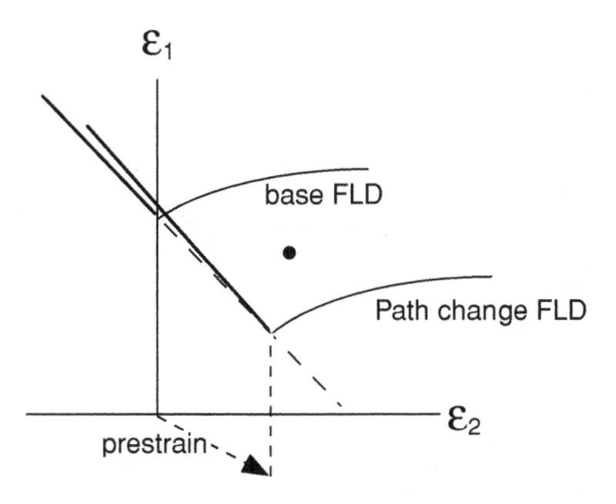

Figure 16.20. Tensile prestraining in 2-direction lowers the FLD along a line $\varepsilon_1 + \varepsilon_2 = C$. After the prestraining, strains corresponding to the black dot would cause failure. Data from Graf and Hosford, *Met. Trans.* v. 24A (1993).

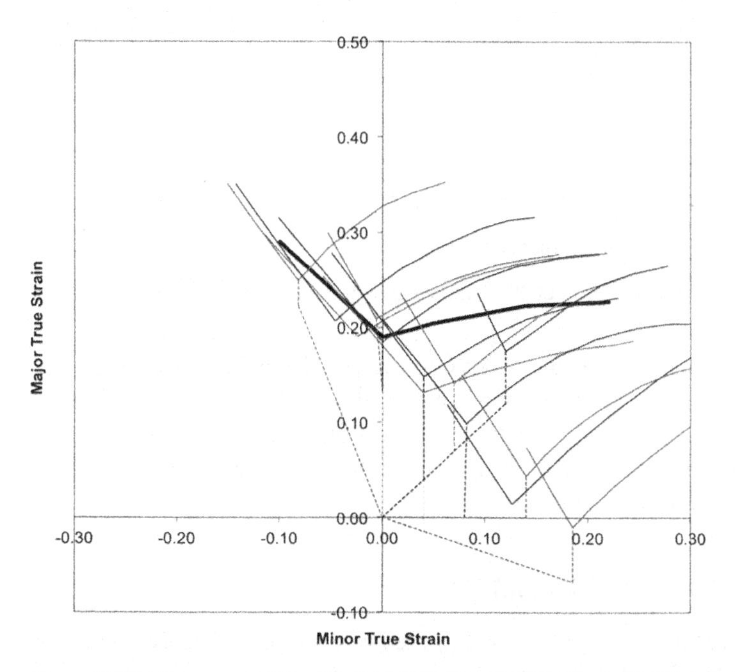

Figure 16.21. Summary of the effect of changing strain paths on forming limits of 2008 T4 aluminum. Data from Graf and Hosford, *ibid.* Graph from T. Stoughton, presentation at the NADDRG meeting at Oakland University, May 2010.

constant). However, in many real stampings, the stain path may vary. Experiments on aluminum alloys using two-stage strain paths have shown that changes in strain path have a very large effect on the forming limits.* Prestraining in tension along the 2-direction shifts the FLD downward and to higher values of ε_2, with the minimum following a line $\varepsilon_1 + \varepsilon_2 = C$ where C is a constant equal to the forming limit in plane-strain (Figure 16.20). Figure 16.21 summarizes the effects of prestraining on forming

* Alejandro Graf and W. F. Hosford, *Met. Trans.* v. 24A (1993).

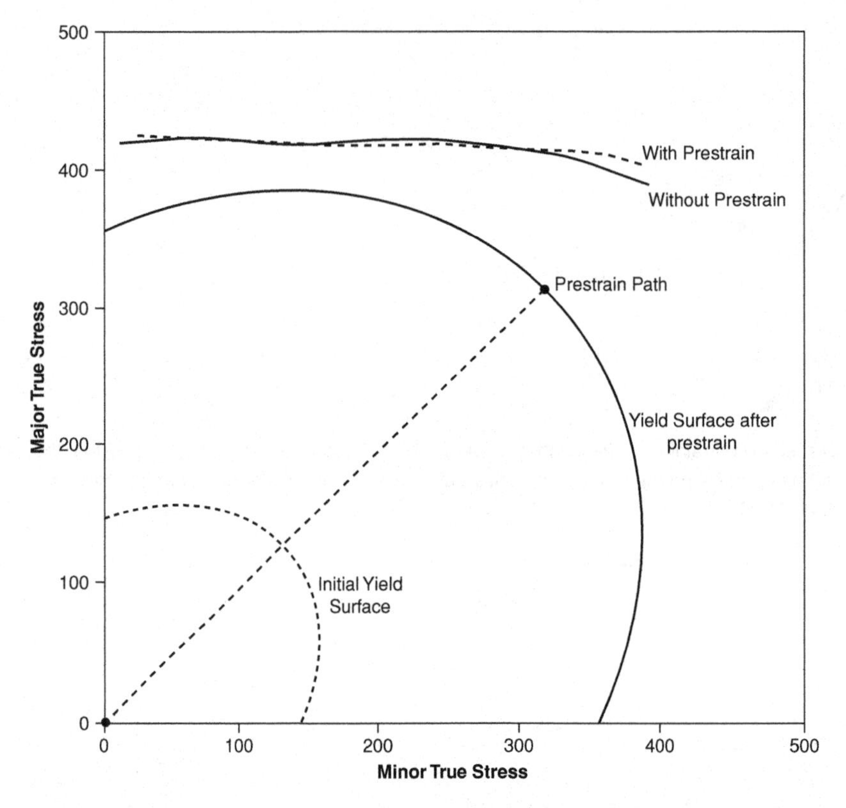

Figure 16.22. Stress-based forming limits of 2008-T4 before and after an equibiaxial prestrain of 0.07.

limits. Prestraining intension along the 1-direction raises the FLD curve along the line $\varepsilon_1 + \varepsilon_2 = C$. Prestraining in biaxial tension causes a shift of the FLD to higher values of ε_2. The net effect is that with changing strain paths, combinations of strains are possible well above the standard FLD without failure or failure may occur with combinations of strains well below the base FLD.

16.7 STRESS-BASED FORMING LIMITS

Stoughton* proposed that forming limits can be plotted in terms of the stresses at failure instead of the strains. He found that such plots are valid for changing strain paths as well as continuous straining. Figure 16.22 is such a plot for aluminum alloy 2008-T4. Critical forming stresses can be used in finite element codes to check for failure.

NOTE OF INTEREST

Stuart Keeler was born in Wassau, Wisconsin, in 1934. He received a BA from Ripon College and BS and ScD from MIT, where his thesis advisor was W. A. Backofen. His basic research, which led to the development of forming limit diagrams, has transformed sheet metal forming from an art to a science.

* T. B. Stoughton and J. W. Yoon, *Int. J. Mech. Sci.*, 47 (2005).

REFERENCES

Alejandro Graf and W. F. Hosford, *Met. Trans.*, 21A (1990).
Alejandro Graf and W. F. Hosford, *Met. Trans.*, 24A (1993).
A. Graf and W. Hosford, *Int. J. Mech. Sci.*, 36 (1994).
S. S. Hecker, A. K. Ghosh and H. L. Gegel, eds. *Formability Analysis, Modeling and Experimentation*, AIME 1978.
R. Hill, *J. Mech. Phys. Solids*, 1 (1952).
S. P. Keeler and W. A. Backofen, *Trans. ASM*, 56 (1963).
S. P. Keeler and W. G. Brazier, in *Microalloying 75*, Union Carbide, 1977.
S. L. Semiatin and J. J. Jonas, *Formability and Workability of Metals*, ASM, 1984.
R. H. Wagoner, K.S. Chan, and S. P. Keeler, eds. *Forming Limit Diagrams: Concepts, Methods and Applications*, TMS-AIME, 1989.

PROBLEMS

16.1. Derive an expression for the critical strain, ε_1^*, to produce a diffuse neck as a function of n and the stress ratio, α. Assume the von Mises criterion and loading under constant α. [Hint: Start with equation 16.11.]

16.2. In principle one can determine the R-value by measuring the angle, θ, of the neck in a strip tension test. How accurately would one have to measure θ to distinguish between two materials having R-values of 1.6 and 1.8?

16.3. It has been suggested that failures in sheets occur when the thickness strain reaches a critical value. Superimpose a plot of e_1 versus e_2 for this criterion on Figure 16.8. Adjust the constant so that the curves coincide for plane-strain, $e_2 = 0$. How good is this criterion?

16.4. Repeat Problem 16.3 for a criterion that predicts failure when the absolutely largest principal strain, $|e_i|_{max}$, reaches a critical value.

16.5. Ironing of a two-piece beverage can thins the walls from 0.015- to 0.005-in. thickness. This strain far exceeds the forming limits. Explain why failure doesn't occur.

16.6. The true minimum of a forming limit diagram determined by in-plane stretching (without bending) occurs at plane-strain, $\varepsilon_2 = 0$. However, forming limit diagrams are determined by stretching strips over a 4-in-diameter dome. Assume that the forming limit is reached when the center of the strip reaches the limit strain for in-plane stretching. If the limit strain is $\varepsilon_1 = 0.30$ for in-plane stretching, what values of ε_1 and ε_2 would be reported for stretching over a 4-in-diameter dome?

16.7. The forming limit diagram for a certain steel is shown in Figure 16.23.

a). Show the loading path of a specimen that fails at point N.

b). Plot as accurately as possible the locus of points corresponding to uniaxial tension, $\sigma_2 = 0$.

c). Describe qualitatively how the line in b) would be drawn differently for $R = 2$.

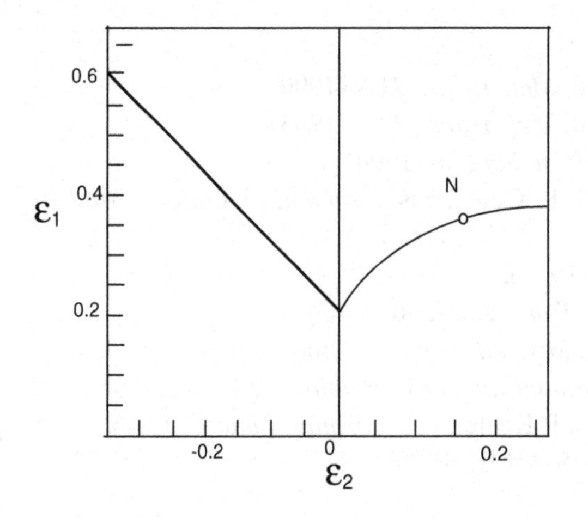

Figure 16.23. Schematic forming limit diagram for Problem 16.7.

16.8. Frequently the minimum on experimentally determined forming limit diagrams occurs at a slightly positive value of e_2 rather than at $e_2 = 0$. Explain why this might be.

16.9. The following points were taken from the forming limit diagram for aluminum alloy 2008. The strain-hardening exponent is 0.265, and $K = 538$ MPa. Plot the forming limits in stress space (i.e., values of σ_1 at failure as a function of σ_2). Assume isotropy.

ε_1	ε_2
0.44	−0.22
0.32	−0.10
0.22	0.0
0.225	0.10
0.23	0.23

17 Stamping

17.1 STAMPING

Operations called stamping, pressing and sometimes drawing involve clamping a sheet at it edges and forcing it into a die cavity with a punch as shown in Figure 17.1. The sheet is stretched rather than squeezed between the tools. Pressure on the draw beads controls how much additional material is drawn into the die cavity. In some cases there is a die, which reverses the movement of material after it is stretched over the punch.

17.2 DRAW BEADS

Draw beads (Figure 17.2) are used to create tension in the sheet being formed by preventing excessive drawing. As a sheet moves through a die bead it is bent three times and unbent three times. Each bend and each unbend there requires plastic work. Over each radius there is friction. Bending and unbending create resistance to movement of the sheet. If the resistance is sufficiently high, the sheet will be locked by the draw bead. The restraining force of the draw bead can be controlled by the height of the insert.

The restraining force has two components. One is caused by the work necessary to bend and unbend the sheet as it flows over the die bead and the other is the work to overcome friction. A crude estimate can be made of the restraining force per length resulting from the bending and unbending with the following simplifying assumptions:

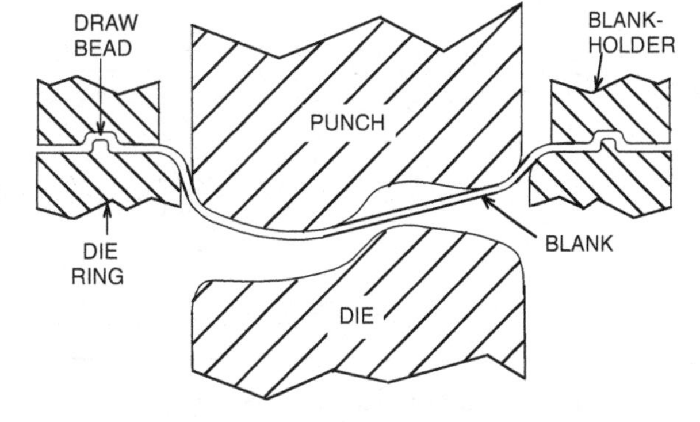

Figure 17.1. Schematic punch and die set with blank holder.

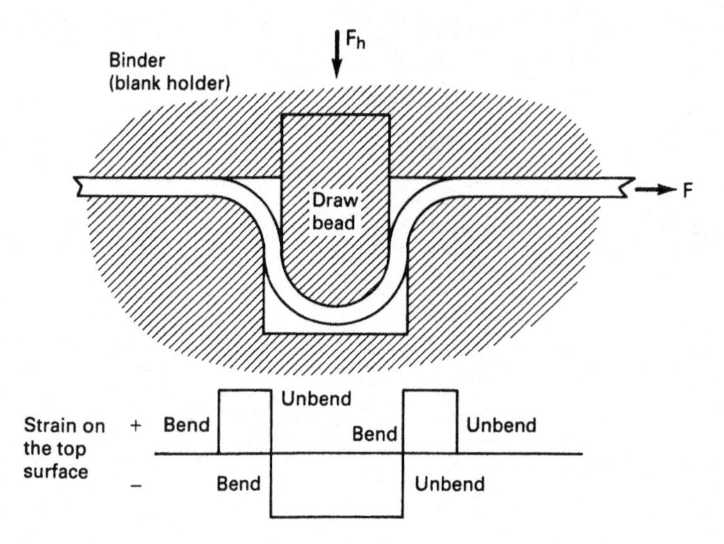

Figure 17.2. Sketch of a draw bead. The strains on the top surface are indicated. There is friction wherever the sheet contacts the tooling.

work hardening, elastic core, movement of the neutral plane, and the difference between engineering strain and true strain ($\varepsilon = e$) are neglected. The strain is given by $e = z/r$, where z is the distance from the neutral plane. In this case the work per volume for an element at z is $Y(z/r)$, where Y is the yield strength and r is the radius of the bend at the neutral axis. The work for all elements then is $\mathrm{d}W/\mathrm{d}L = 2\int_0^{t/2}(Y/r)z\mathrm{d}z = Yt^2/(4r)$. The restraining force caused by bending three times and unbending three times as the sheet moves through the draw bead is then

$$F_b = 6\mathrm{d}W/\mathrm{d}L = 1.5Yt^2/r. \quad (17.1)$$

This treatment neglects the Bauschinger effect, which lowers the restraining force somewhat. A more accurate determination of F_b can be obtained by pulling a strip through a fixture in which freely rotating cylinders simulate the draw beads (Figure 17.3).

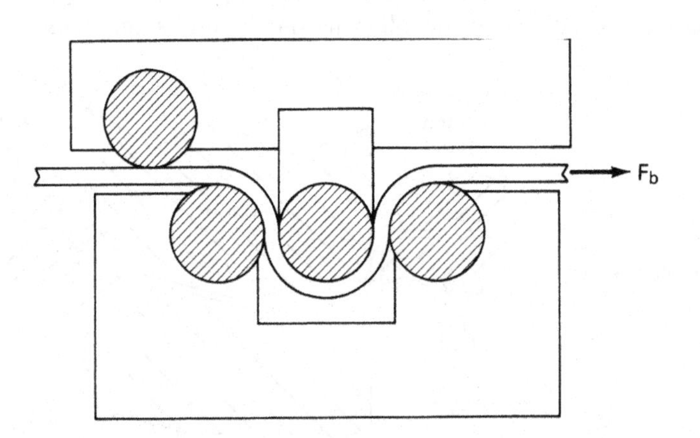

Figure 17.3. Determining the bending contribution to the draw bead restraining force.

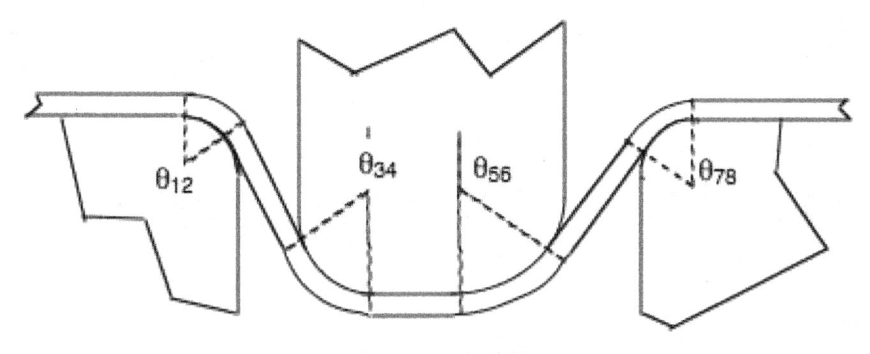

Figure 17.4. Draw beads with different radii.

Estimating the restraining force is more complicated if the bend radii are not all the same, as shown in Figure 17.4. In this case,

$$F_b = (Yt^2/4)(1/r_{12} + 1/r_{34} + 1/r_{56} + 1r_{78}). \quad (17.2)$$

The restraining force caused by friction is in addition to that from bending and unbending. For a single bend $F_1 = F_0(\exp(\mu\theta)$ so the restraining force from all of the bends is

$$F_f = \exp(\mu\theta_{12}) + \exp(\mu\theta_{34}) + \exp(\mu\theta_{56}) + \exp(\mu\theta_{78}). \quad (17.3)$$

The position of draw beads is important. They should be placed perpendicular to the direction of metal flow. If they are too close to the trim line, material being drawn over them may become part of the finished product and create a surface defect. If they are too far from the trim line their effectiveness in preventing drawing will be diminished.

17.3 STRAIN DISTRIBUTION

Forming limit diagrams do not completely describe the forming behavior of different metals. Two materials may have nearly the same forming limits ε_1^*, but differ substantially in formability. For example consider a symmetric part (Figure 17.5) with a line of length L_0 parallel to the 1-direction. The stress is not uniform along the line so ε_1 will vary from place to place. The total length of line will be given by

$$L = L_0 + \int_0^{L_0} \mathrm{d}x. \quad (17.4)$$

The height, h_{max}, of a hemispherical cup at failure depends on L/L_0 when $\varepsilon_{1\,max} = \varepsilon_1^*$. In other words, h_{max}, depends strongly on the distribution of ε_1 on x. The ratio of ε_1 in lightly loaded regions to that in heavily loaded regions depends primarily on the strain-hardening exponent n, but also to some extent on the strain-rate sensitivity, m. Increasing values of n and m distribute the strain and permit deeper parts to be formed. Figure 17.6 illustrates this.

When a sheet that is locked by draw beads is stretched over a hemispherical punch, the failure site depends on the friction between the sheet and the punch. With high friction, the failure usually occurs near the ring in unsupported material that is close

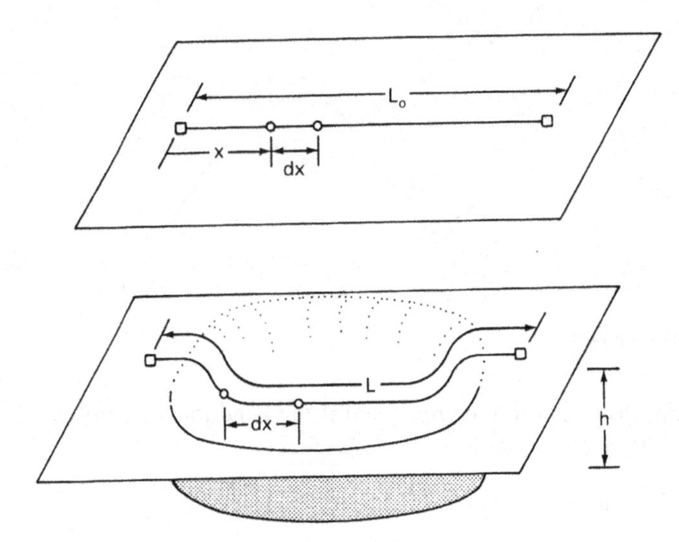

Figure 17.5. Increase of line length, L_0.

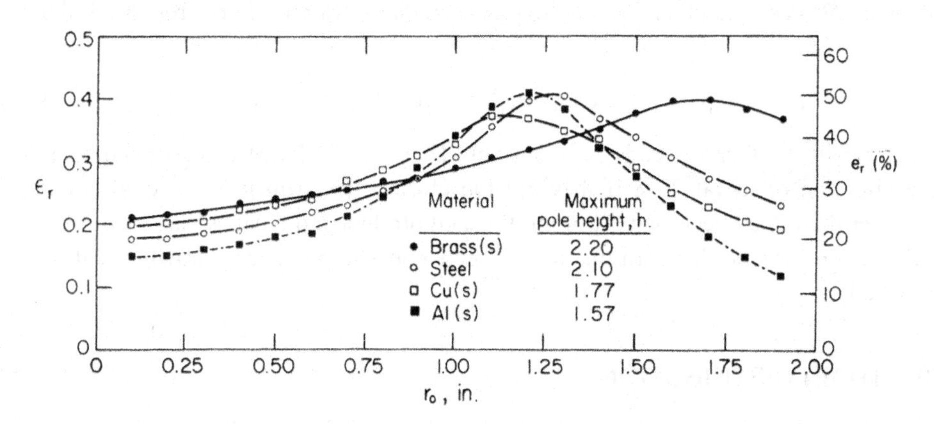

Figure 17.6. Distribution of strains during punch stretching of sheets of several metals. The punch was stopped at the same peak strain in each case. Materials with higher values of n had more widely distributed strains and formed deeper cups. From S. P. Keeler and W. A. Backofen, *Trans. Q. ASM,* v 56 (1963), pp. 25–48.

to being in contact with the punch. Just outside of the contact region, the deformation is in the plane-strain because of the hoop direction constraint from the neighboring material in contact with the punch. With lower friction, the failure site moves toward the dome and the depth of cup at failure increases.

17.4 LOOSE METAL AND WRINKLING

When flat sections are not stretched enough, they become loose or floppy. If a potential car buyer puts his hand on the roof of a car and finds that it isn't stiff, he will perceive poor quality. The same is true if the hood of a car vibrates at high speed. Loose metal can be avoided by ensuring that all parts of a stamping are deformed to some minimum effective strain (perhaps 4%).

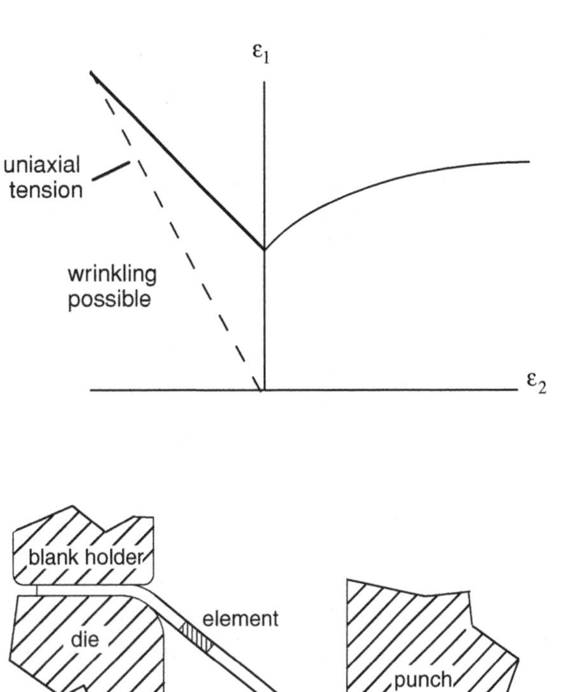

Figure 17.7. A forming limit diagram showing the path of a uniaxial tension test (dashed line). To the left of this line, the stress, σ_2, is compressive so wrinkling is possible.

Figure 17.8. In drawing of a conical cup, wall wrinkling may occur if there is not enough tensile stretching of the wall.

Wrinkling occurs when the stresses in the sheet are compressive. With thin sheets, wrinkling will occur under very low amounts of compressive stress. Figure 17.7 shows the region of wrinkling on a forming limit diagram.

Wrinkling is a potential problem in drawing of many parts. Consider the forming of a flat-bottom cup with conical walls as shown in Figure 17.8. As the punch descends, the shaded element is drawn closer to the punch so its circumference must shrink. Stretching of the wall in tension will reduce the circumference, but if the wall isn't stretched enough it will wrinkle. The amount of stretching can be increased by raising the blank-holder pressure. However, too much restraint of flange drawing in by excessive blank-holder pressure or draw beads may cause the wall to fail in tension. Therefore the blank-holder pressure must be controlled.*

There is a window of permissible levels of blank-holder force. Too little causes wrinkling; too much force results in wall failure. This is shown in Figure 17.9. Since more circumferential contraction occurs when material with high R-values is stretched, there is more contraction in the plane of the sheet so an increased R widens the left side of the window. With an increased value of n, more stretching can occur before failure. For any given material and required depth of draw, there is a window of permissible blank-holder force. Increasing the depth of draw beads has the same effect as increasing blank-holder force. It is possible to change the blank-holder force during

* J. Havranek, in *Sheet Metal Forming and Energy Conservation, Proc. 9th Bienn. Cong. Int. Deep Drawing Res. Group*, ASM Ann Arbor, MI (1976).

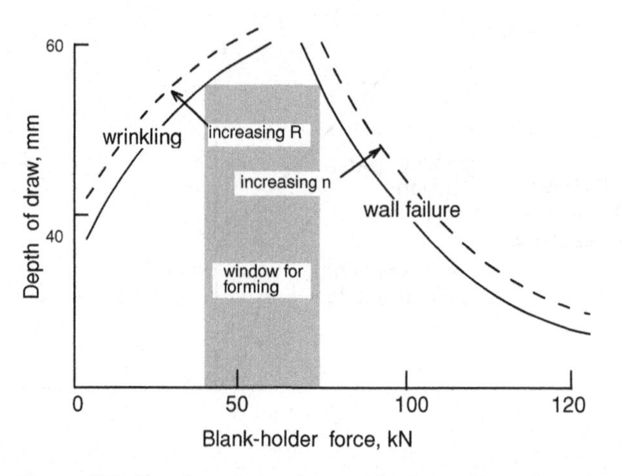

Figure 17.9. Forming window for a conical cup. For a draw depth of 55 mm, the blank-holder force must be between 40 and 75 kN.

the draw. Greater depths of draw are possible if the force is low in the early stages of the draw and increases in the later stages.*

17.5 SPRINGBACK

Control of springback is very important, especially in lightly curved sections like the bottom of the stamping in Figure 17.10. As indicated in Chapter 13, springback is minimized if tension is sufficient to cause tensile yielding across the whole cross section. Yet the wall force, F_2, must not cause a tensile stress that exceeds the tensile strength. The relative magnitudes of F_1 and F_2, can be calculated approximately as follows: a radial force balance on an element in the bend gives the normal force, $dN = Fd\theta$, so the frictional force on the element is $\mu\, dN = \mu F d\theta$, where μ is the coefficient of friction. Substituting $dF = \mu FN$ from a force balance in the circumferential direction, $dF = F\mu d\theta$.

Integrating,

$$\int_{F_1}^{F_2} dF/F = \mu \int_0^{\theta} d\theta \quad \text{so} \quad F_2 = F_1 \exp(\mu\theta). \tag{17.5}$$

Neglecting differences between the plane-strain and uniaxial tensile and yield strengths, $F_2 < (S_u)wt$ and $F_1(Y)wt$ where w is the dimension parallel to the bend, t is the sheet thickness, S_u is the tensile strength and Y is the yield strength. Therefore

$$Su/Y > \exp(\mu\theta). \tag{17.6}$$

* D. E. Hardt and R. C. Fenn, *Proc. ASM Symposium on Monitoring and Control of Manufacturing Processes* (1990).

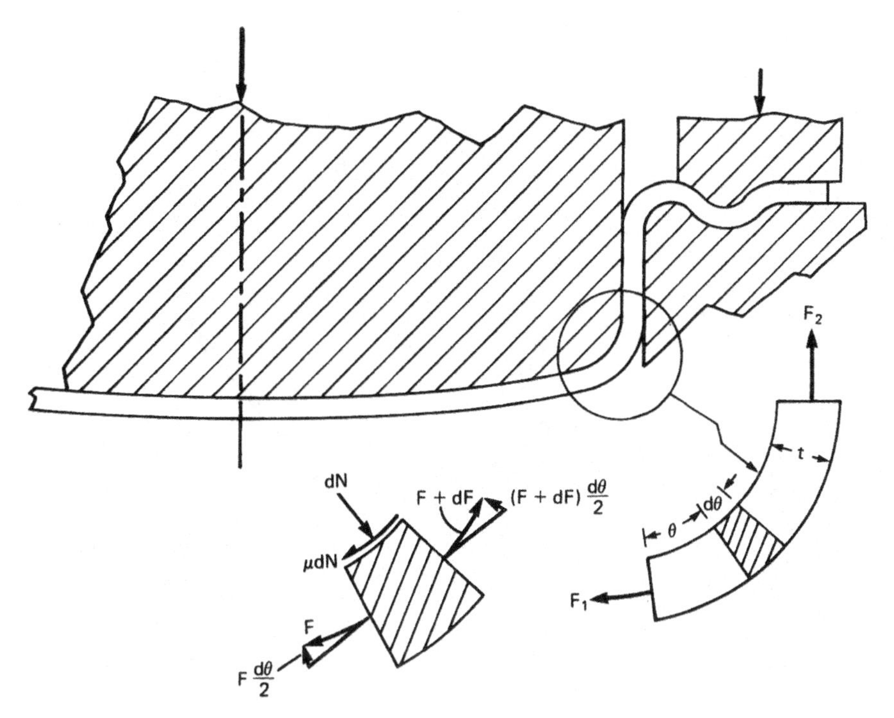

Figure 17.10. Sheet being bent over a punch.

Thus the tensile-to-yield strength ratio must exceed a value, which depends on both the coefficient of friction and the bend angle. For example if $\mu = 0.2$ and $\theta = \pi/2$, $S_u/Y > 1.37$.

In springback calculations it should be recognized that when a metal is stretched in tension and unloaded, the unloading is not linear. This is called the *Bauschinger effect.* Figure 13.5 shows typical behavior.

17.6 STRAIN SIGNATURES

The strain path varies from one place to another in a stamping. The strain path of a particular point is called its strain signature. Figures 17.11 and 17.12 show

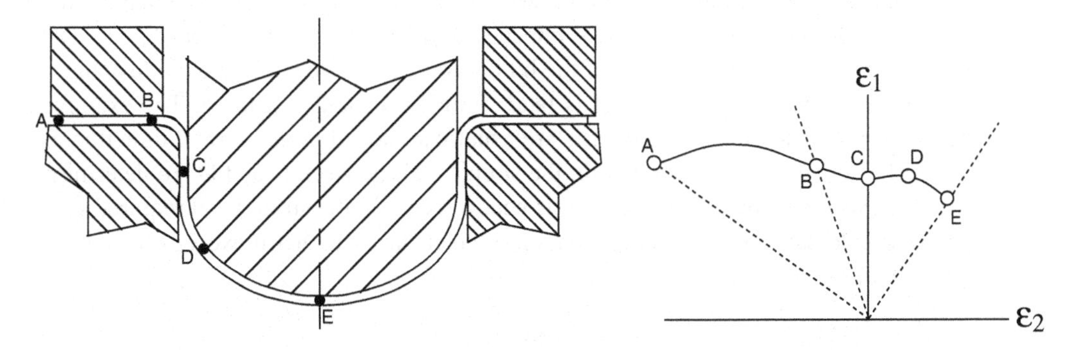

Figure 17.11. Strain signature in drawing of a hemispherical cup.

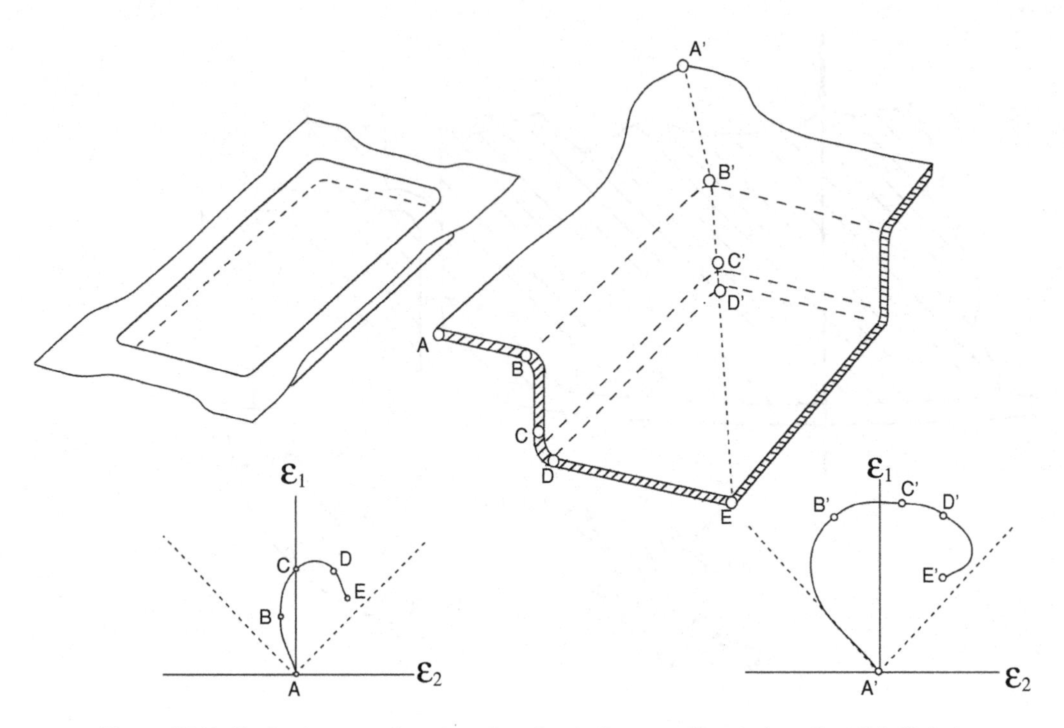

Figure 17.12. Strain signatures in a stamping of a shallow pan. The strain path at B is likely to cause wrinkling.

stampings of a hemispherical cup and a shallow pan with the strain signatures of several locations.

17.7 DIE DESIGN*

At the center of most sheet stampings a certain effective strain level is necessary to prevent "slack" or "loose" metal and to minimize springback. Otherwise, the panel may not retain the desired shape or it may feel floppy to the customer. A target strain at the center of the panel will also ensure enough work hardening to strengthen the panel. To achieve this target strain, there must be sufficient tension in the sheet at that point. Because of friction and bending, the tension at the binder must be much higher. Figure 17.13 illustrates a part being stamped. The final part is actually from A to C. The material between C and the edge of the binder must be stretched in the die to achieve the required stress at point A, but is later trimmed off and discarded. The example below is in plane-strain but real parts are usually stretched in the normal direction as well. The highest stress will occur in the stretch wall between C and D. To determine whether this stamping can be made, it is necessary to fix the strain level required at A and then calculate the stress in this region to determine whether it exceeds the tensile strength of the material.

* This section is a result of private discussions with Edmund Herman, Creative Concepts Company.

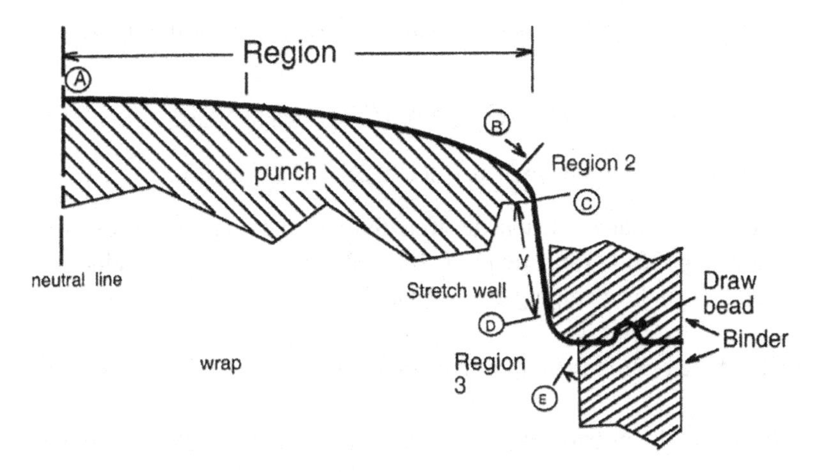

Figure 17.13. Schematic of a stamping.

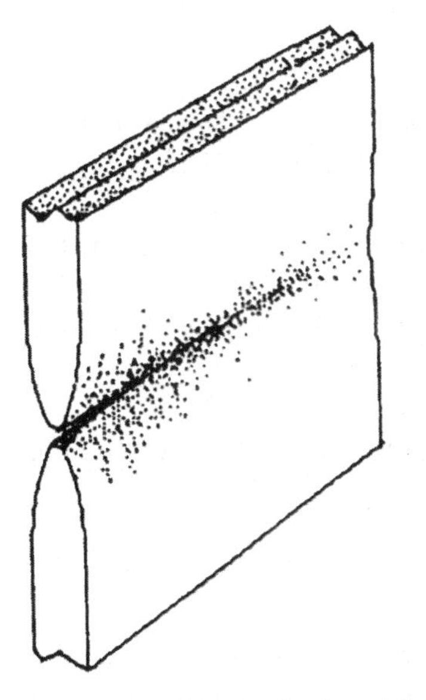

Figure 17.14. Sketch of a sheet failure by through-thickness necking.

EXAMPLE 17.1: Determine whether the stamping in Figure 17.13 can be made assuming a target strain at A of $\varepsilon_A = 0.035$. Other data are: the sheet thickness = 1 mm.; the stress strain relation is $\sigma = 520\varepsilon^{0.18}$ MPa; the radii are $r_1 = 8$ m, $r_2 = 0.10$ m, $r_3 = 0.025$ m; the angles are $\theta_1 = 0.25$, $\theta_2 = 1.05$, and $\theta_3 = 1.30$ radians and the friction coefficient is $\mu = 0.10$.

For simplicity, assume that this is a plane-strain stamping.

SOLUTION: The thickness at A, $t_A = t_0 \exp(-\varepsilon_A) = (1\text{ mm}) \exp(-0.035) = 0.9656\text{ mm}$. The tensile force per length at A, $F_A = 520 t_A \varepsilon_A^n = 520(.9656)(0.035)^{0.18} = 274.6\text{ kN/m}$.

The tensile force at B must be enough larger than this to overcome the friction in all the bends.

$$F_B = F_A \exp(\mu\theta) = 274.6 \exp[(0.10)(0.25)] = 281\,\text{kN/m}.$$

$$F_C = F_B \exp(\mu\theta) = 281 \exp[(0.10)(1.052)] = 312.7\,\text{kN/m}.$$

The maximum force corresponds to the tensile strength times the original thickness $= t_0 K n^n \exp(-n) = (0.001)(520)(0.18^{0.18}) \exp(-0.18) = 319.0\,\text{kN/m}$. The value of $F_C = F_D$ is less than this so the stamping can be made.

EXAMPLE 17.2: Find the required length of the stretch wall.

SOLUTION: Guessing that the horizontal component of the stretch wall is 0.04 m, the length of the wrap is $8 \times 0.25 + 0.1 \times 1.05 \times 0.25 + 0.04 = 2.145$.

The force per length at B, $F_B = F_A \exp(\mu\theta) = 274.6\,\text{kN/m}[(0.10)(0.25)] = 281.6$ but F_B also equals $t_B\sigma_B = t_0 \exp(-\varepsilon_B) K \varepsilon_B^{0.18}$ so $\varepsilon_B^{0.18} \exp(-\varepsilon_B) = F_A/(Kt_0) = 274.6/520 = 0.5281$.

Solving by trial and error, $\varepsilon_B = 0.0521$.

The force per length at C is 312.7 kN/m but F_C also equals $t_C\sigma_C = t_0 \exp(-\varepsilon_C) K \varepsilon_C^{0.18}$ so $\varepsilon_C^{0.18} \exp(-\varepsilon_C) = F_C/(Kt_0) = 312.7/(520) = 0.6013$.

Solving by trial and error, $\varepsilon_C = 0.108$.

Taking the average strain between A and B as $(.035 + .0521)/2 = 0.0436$, the increase in the length of line is $(8 \times 0.25)[\exp(0.0436) - 1] = 0.089$ m.

Taking the average strain between B and C as $(0.0521 + 0.108)/2 = 0.080$, the increase in the length of line is $(0.1 \times 1.105)[\exp(.08) - 1] = 0.0092$ m.

The change of length of the stretch wall, $.04[\exp(\varepsilon_C) - 1] = 0.00146$ m.

The total change of length is then $0.089 + 0.009 + 0.0015 = 0.01$ m or 10 mm. To this must be added any length of material that comes out of the binder.

17.8 TOUGHNESS AND SHEET TEARING

Figure 17.14 is a sketch of a plane-strain neck in a sheet. The volume of the plastic region is proportional to t^2L where t is the specimen thickness and L is the length of the crack. Since the area of the fracture is tL, the plastic work per area is proportional to the thickness, t, and the fracture toughness K_C is proportional to $\sqrt{t}$. Very thin sheets tear suddenly at surprisingly low stresses.

A fracture toughness can be associated with the process of localized necking and tearing. A simple way of measuring the toughness is to test a series of deeply edge-notched tensile specimens* like that shown in Figure 17.15. The deep notch confines the plastic deformation to a circular region of diameter. The work is composed of two terms. On is the plastic work in the circular region. The other is the work to propagate the local neck.

$$W = \alpha 5L^2 t_0 + GLt_0. \tag{17.7}$$

* B. Cotrerell and J. K. Redel, *Int. J. Fract.*, 13 (1977).

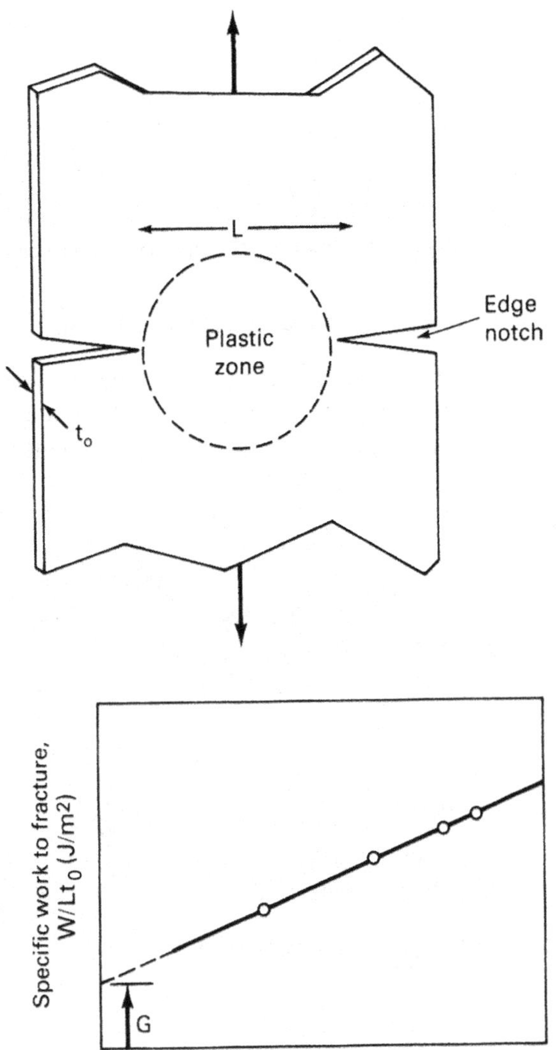

Figure 17.15. Plastic zone in a deeply-notched tensile specimen.

Figure 17.16. Schematic plot of the work per area. W/Lt_0 as a function of the distance between notches. G is the value at $L = 0$.

Here G is the fracture toughness or energy per area to cause the local neck. The constant, α, reflects the strain distribution in the circular area. G can be determined by plotting $W/(Lt_0)$ as a function of L as shown in Figure 17.16. The intercept at $L = 0$ is G.

At any instant during the necking of an ideally plastic material, plastic deformation is restricted to a region with a length, t, that is equal to the current thickness.* See Figure 17.17. In this case it can be shown that

$$G = t_0 K_{ps} \int_0^{\varepsilon_f} \varepsilon^n \exp(-2\varepsilon) d\varepsilon. \tag{17.8}$$

Here K_{ps} is the coefficient in the power-law expression for plane-strain.

* W. F. Hosford and A. G. Atkins, *J. Material Shaping Technol.*, 8 (1990).

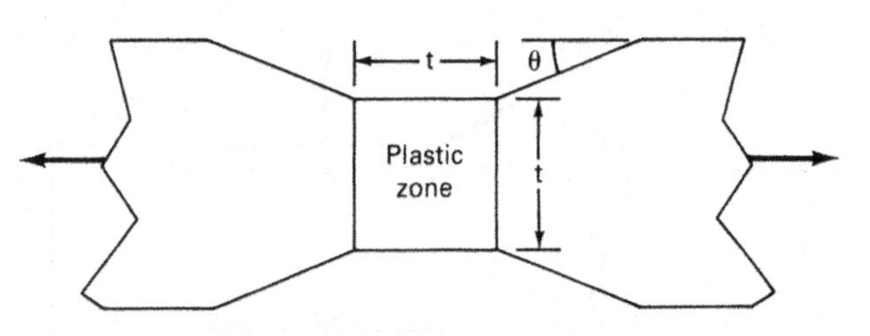

Figure 17.17. Shape of plastic zone during local necking of an ideally plastic material.

17.9 GENERAL OBSERVATIONS

The relative amount of stretching and drawing varies from one part to another. Where stretching predominates, formability depends mainly on n and m whereas if drawing predominates, $\bar{R}$ is most important. Figure 17.18 relates actual press performance to n and $\bar{R}$.

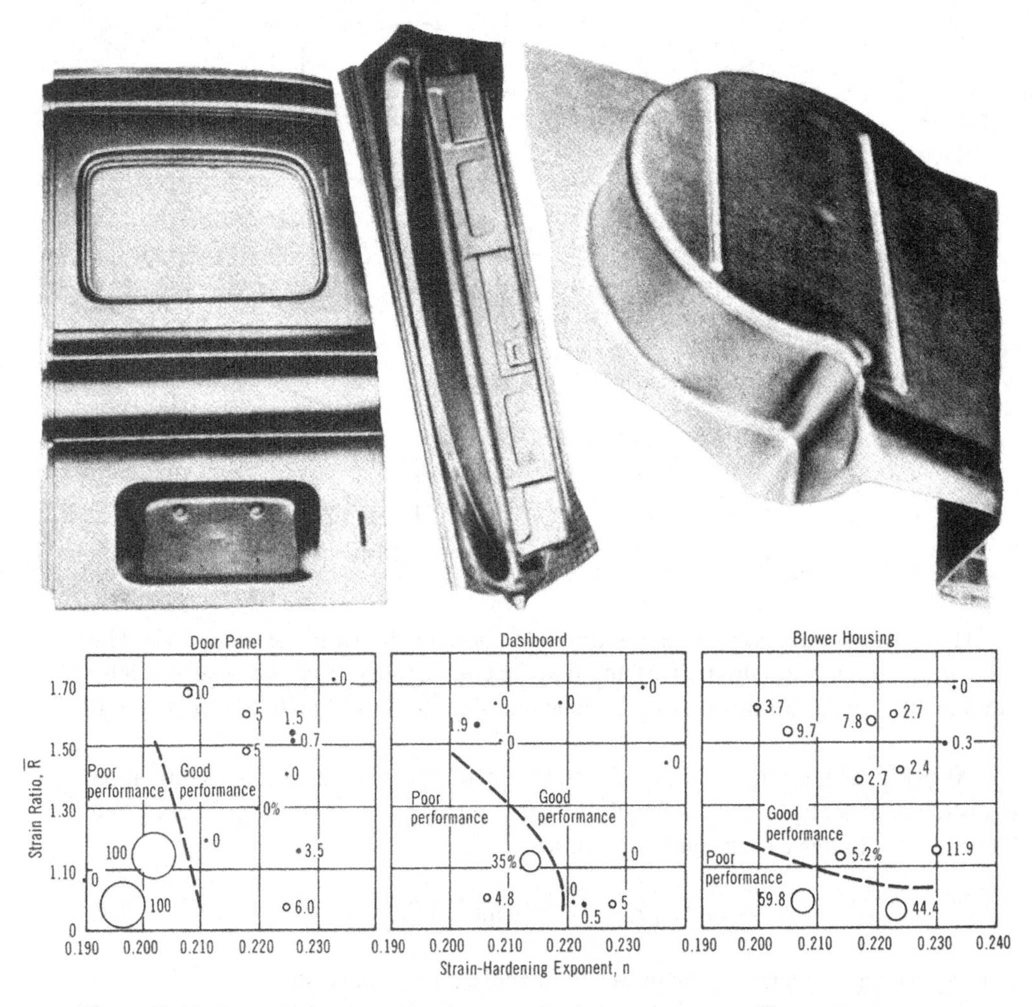

Figure 17.18. Dependence of press performance for three parts on n and $\bar{R}$.

NOTES OF INTEREST

Johann Bauschinger (1833–1893) was the director of the Polytechnical Institute of Munich. In this position, he installed the largest testing machine of the period and instrumented it with a mirror extensometer so that he could make very accurate strain measurements. He discovered that prior straining in tension lowered the compressive yield strength and vice versa. Bauschinger also did the original research on the yielding and strain aging of mild steel.

The aluminum pie dish is one of the very few stamped items with wrinkles that consumers accept willingly. The wrinkled sidewalls actually strengthen the pan.

REFERENCES

John H. Schey, *Metal Deformation Processes: Friction & Lubrication*, M. Decker, 1970.

Z. Marciniak, J. L. Duncan and S. J. Hu, *Mechanics of Sheet Metal Forming*, Butterworth and Heinemann, 2002.

E. M. Meilnik, *Metalworking Science and Technology*, McGraw-Hill, 1991.

PROBLEMS

17.1. Referring to Figure 17.6, determine the values of L/L_0 for brass and aluminum.

17.2. Plot the contour of $\bar{\varepsilon} = 0.04$ on a plot of ε_1 vs. ε_2 (i.e., on a forming limit diagram).

17.3. Show how the dashed line in Figure 17.6 would change for a material having an R value of 2.

17.4. A sheet of HSLA steel having a tensile strength of 450 MPa and a yield strength of 350 MPa is to be drawn over a 90° bend. What is the maximum permissible coefficient of friction?

17.5. Consider the deep drawing of a flat-bottom cylindrical cup. Sketch the strain path for an element initially on the flange halfway between the periphery and die lip.

17.6. Calculate the total drag force per length of a draw bead attributable to plastic bending if the bend angle entering and leaving the draw bead is 45° and the bend angle in the middle of the draw bead is 90°. Assume that the bend radii are 10 mm and the sheet thickness is 1 mm. Let the yield strength be Y and neglect strain hardening.

17.7. Calculate the drag force attributable to friction in Problem 17.6 assuming a friction coefficient of 0.10.

17.8. Carefully sketch the strain signature for the conical cup drawing in Figure 17.19.

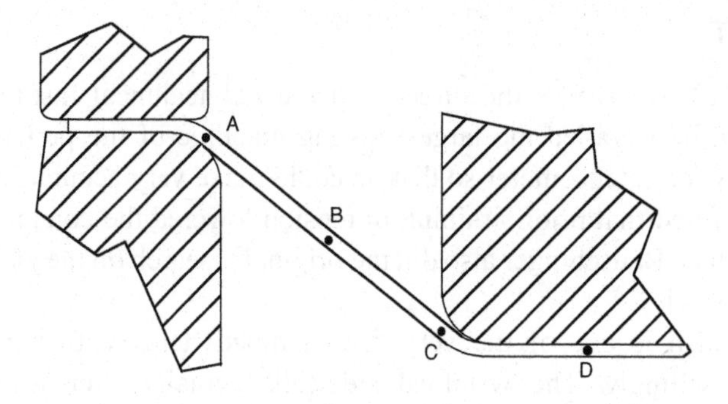

Figure 17.19. Drawing of a conical cup.

17.9. Why isn't the bending included in the drag in regions A-B and B-C in Example 17.1?

17.10. It has been noted that when a cup is drawn with a hemispherical punch, greater depths can be achieved before failure when the punch spins. Explain why this might be so.

18 Hydroforming

18.1 GENERAL

In the hydroforming of tubes, the tube walls are forced against a die by internal pressure. Hydroforming is used to produce such diverse products as plumbing fittings, bellows, and vehicle frames. For hydroforming, welded tubes are preferred over seamless tubes because their wall thickness is more uniform. In the manufacture of seamless tubes, any wandering of the mandrel causes the wall thickness to vary.

Usually the tubes are subjected to axial compression in addition to the internal pressure. Figure 18.1 shows the production of a plumbing T-fitting. In the case of vehicle frames, the tubes are often bent before hydroforming. Figure 18.2 shows the strain signatures for several locations on a hydroformed T-joint.

18.2 FREE EXPANSION OF TUBES

A simple balance of forces on the wall (Figure 18.3) results in $P = (t/r)\sigma_\theta$, where P is the internal pressure, r is the tube radius, t is the wall thickness, and σ_θ is the hoop

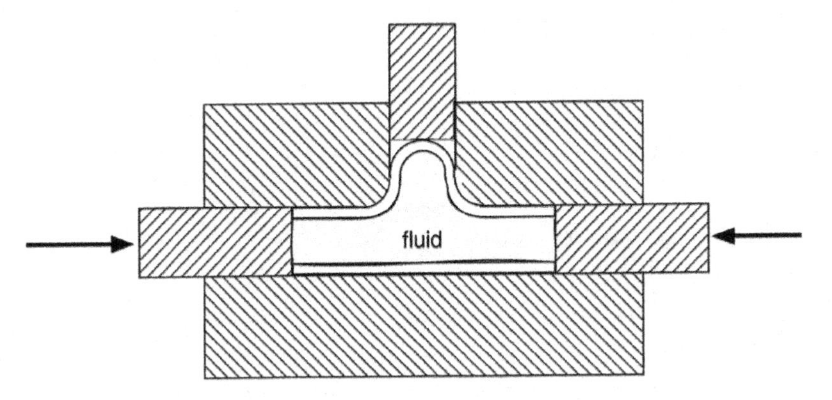

Figure 18.1. Production of a T-fitting by internal pressure and axial compression.

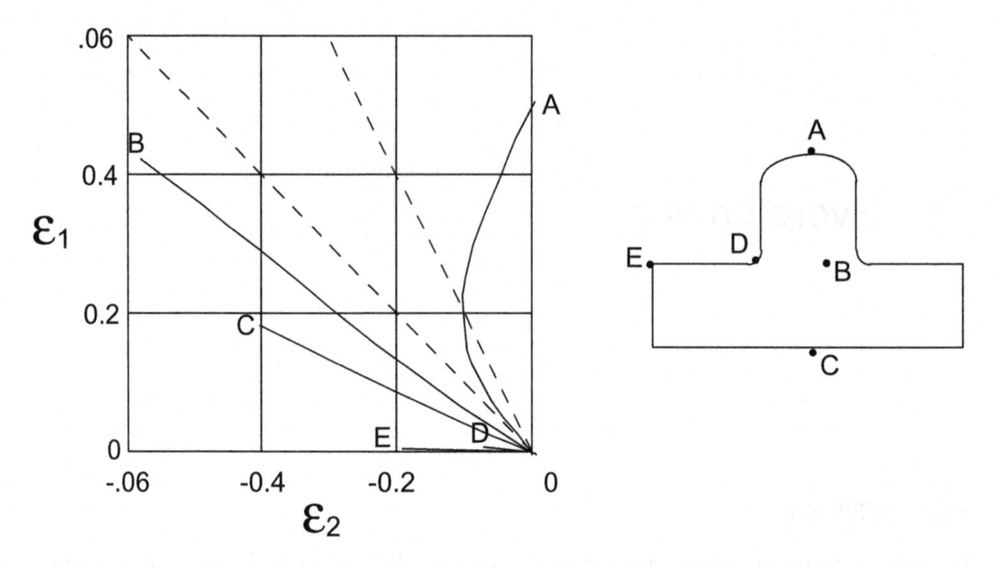

Figure 18.2. Strain signatures for several locations on a hydroformed T-joint. After T. Yoshida and Y. Kuriyama, *Sheet Metal Forming for the New Millennium*, IDDRG, 2000.

stress. If the length of the tube is fixed, this is a condition of plane-strain: $\sigma_z = \sigma_\theta/2$ and $\sigma_\theta = (2/\sqrt{3})\bar{\sigma}$. Therefore,

$$P = (2/\sqrt{3})(t/r)\bar{\sigma} = (2/\sqrt{3})(t/r)K\left[(2/\sqrt{3})\varepsilon_\theta\right]^n. \tag{18.1}$$

Substituting $\varepsilon_\theta = \ln(r/r_0)$ and $t = t_0 r_0/r$ into equation 18.1,

$$P = (2/\sqrt{3})(t_0 r_0/r^2)K\left[(2/\sqrt{3})\ln(r/r_0)\right]^n. \tag{18.2}$$

A plot of equation 18.2 (Figure 18.4) shows that, as the tube is expanded, the pressure first rises and then falls. The maximum pressure can be found by differentiating equation 18.1 and setting $\mathrm{d}P/\mathrm{d}r = 0$ as

$$\varepsilon_\theta = n/2. \tag{18.3}$$

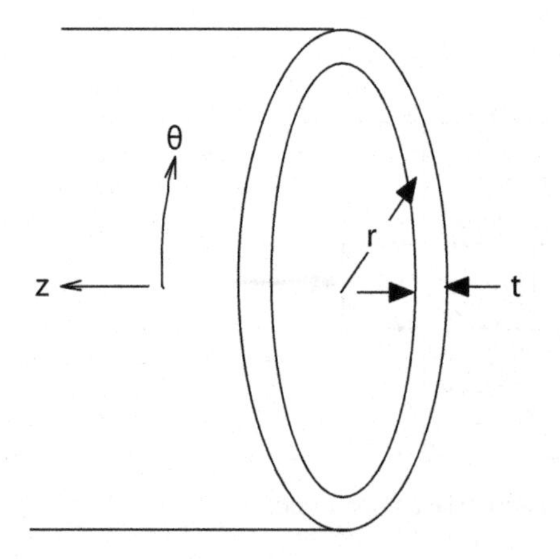

Figure 18.3. Force balance on a tube with internal pressure.

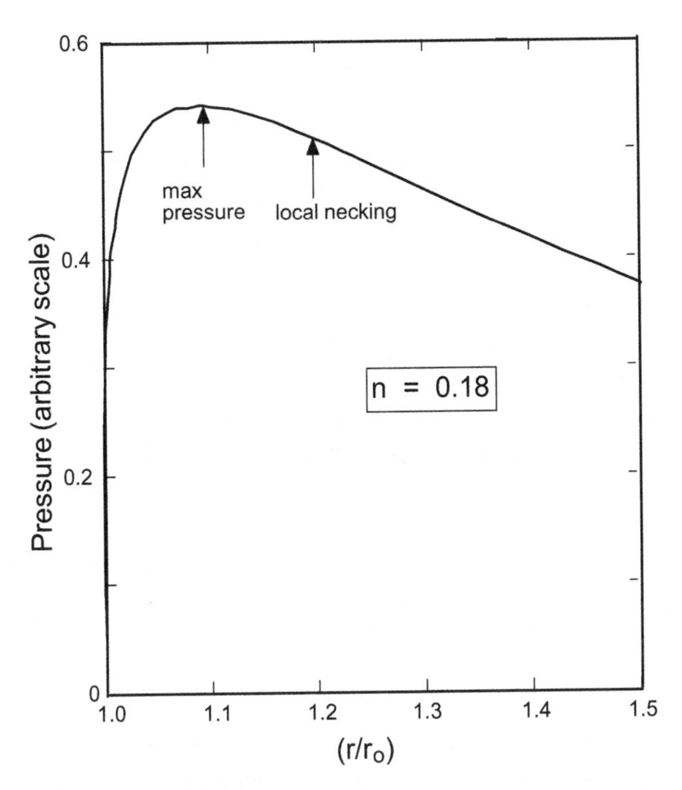

Figure 18.4. Variation of pressure with radius as a tube is expanded under plane-strain. Under internal pressure, the maximum pressure occurs when $\varepsilon_\theta = n/2$.

The maximum pressure, however, does not represent the condition for failure. Rather, failure occurs by plane-strain necking when $\varepsilon_\theta = n$.

The forming limit is increased if enough end pressure is applied to cause the tube to shorten. This shifts the stress state toward uniaxial tension on the left-hand side of the forming limit diagram.

The cross-sectional shape of a tube can be changed. Figure 18.5 illustrates the need for filling a tube with a liquid before changing its shape.

18.3 HYDROFORMING INTO SQUARE CROSS SECTIONS

Circular tubes are often formed into square or rectangular cross sections to increase their moment of inertia and hence stiffness. As a circular tube is expanded into a square tube, its radius decreases (Figure 18.6). Friction between the tube and die plays an important role in controlling the strain distribution and limiting radius. Two extreme cases will be considered.

With frictionless conditions, the stress and strain are the same everywhere. If there is no movement along the tube axis,

$$\varepsilon = \ln[4/\pi + (1 - 4/\pi)(r/r_0)]. \tag{18.4}$$

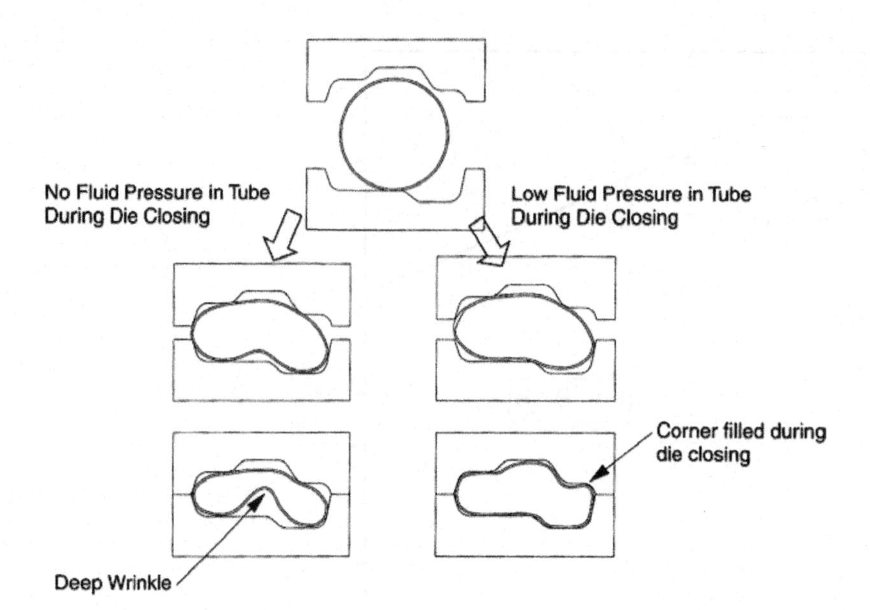

Figure 18.5. Changing the cross-sectional shape of a tube. From Harjinder Singh, *Fundamentals of Hydroforming*, SME, 2003.

With sticking friction, only the portion of the tube that is not in contact with the die wall can deform. In this case, the strain in the unsupported region is

$$\varepsilon = (1 - 4/\pi)\ln(r/r_0). \tag{18.5}$$

Figure 18.7 shows how the strain depends on how sharp a radius is formed for these extreme cases. Sliding friction will fall between these extremes.

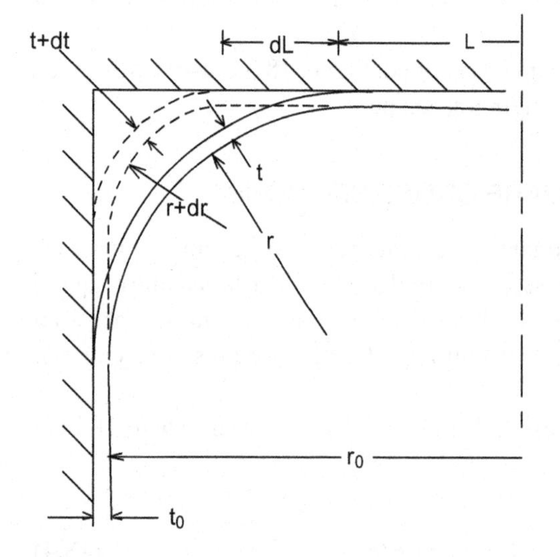

Figure 18.6. Incremental expansion of a circular tube into a square.

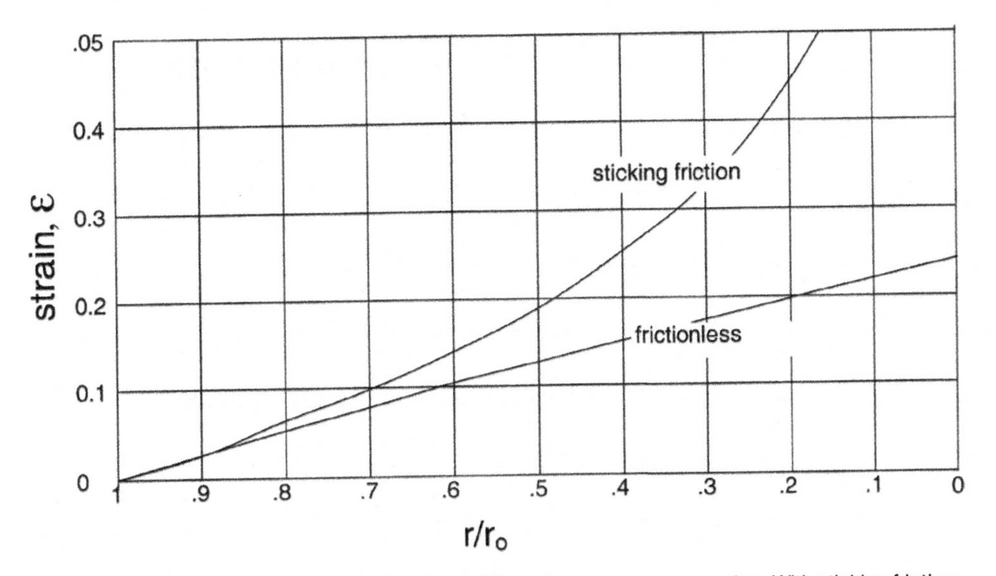

Figure 18.7. Strain in unsupported region during filling of a square cross section. With sticking friction, the strains become much higher than with frictionless conditions.

The pressure is given by $P = \sigma t/r$. Substituting $t = t_0 \exp(-\varepsilon)$,

$$P = \sigma t_0 \exp(-\varepsilon)/r. \tag{18.6}$$

For power-law hardening,

$$P = K\varepsilon^n t_0 \exp(-\varepsilon)/r. \tag{18.7}$$

In the *low-pressure* process, the tube is initially filled before being forced into a die cavity that has about the same perimeter as the tube. The low pressure avoids stretching of the walls so smaller radii can be achieved. The internal pressure keeps the walls from collapsing.

End feeding increases the amount of expansion possible.

Tubes are classified as *thin wall* if $d/t < 10$, *medium wall* if $10 < d/t < 50$, and *heavy wall* if $d/t > 50$.

The advantage of square tubes is that they have greater stiffness than circular tubes of the same wall thickness and mass. Table 18.1 illustrates this.

Hydroforming is also used to form bellows. Figure 18.8 illustrates how a bellows can be formed.

18.4 BENT SECTIONS

Many parts are made from tubes that are bent before hydroforming. Mandrels are used in the bending to keep the tube from collapsing. End feeding by friction or pressure block during bending can induce a net compression during bending. End feeding moves the neutral axis toward the inside of the bend, raises the forming limits during subsequent hydroforming, and moves the failure site from the outside of the bend toward the inside. These effects are understandable in terms of how the forming

Table 18.1. Comparison of the stiffness of square and circular tubes

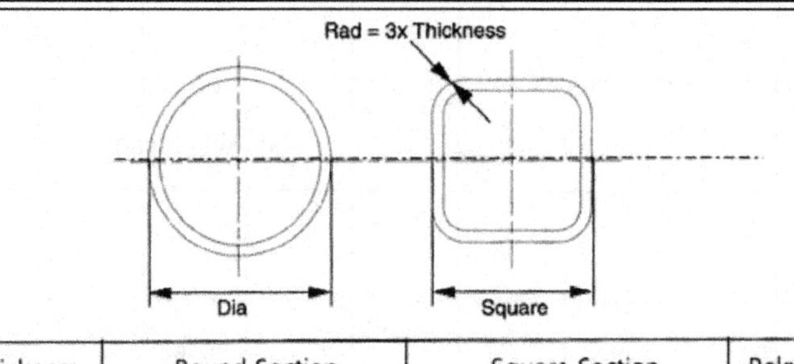

Thickness	Round Section		Square Section		Relative
mm	Dia (mm)	Modulus	Side (mm)	Modulus	Mass
3.00	63.5	1.00	53.7	1.06	1.00
1.60	83.5	1.00	67.6	1.06	0.72
2.24	83.5	1.37	68.5	1.45	1.00

From Harjinder Singh, *Fundamentals of Hydroforming*, SME, 2003, p. 41.

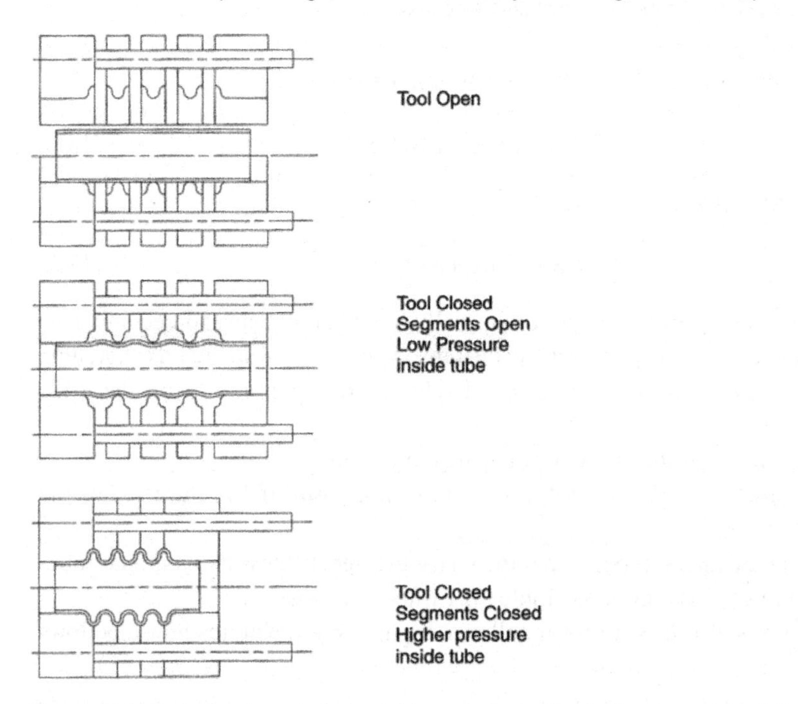

Figure 18.8. Hydroforming of a bellows. From Harjinder Singh, *Fundamentals of Hydroforming*, SME, 2003, p. 28.

limits are affected by strain-path changes (Section 16.6). Worswick* has found good agreement with stress-based forming limits (Section 16.7).

Bending and cross-sectional shape change can be combined in forming tubular frame sections as illustrated by Figure 18.9.

* M. J. Worswick, private communication.

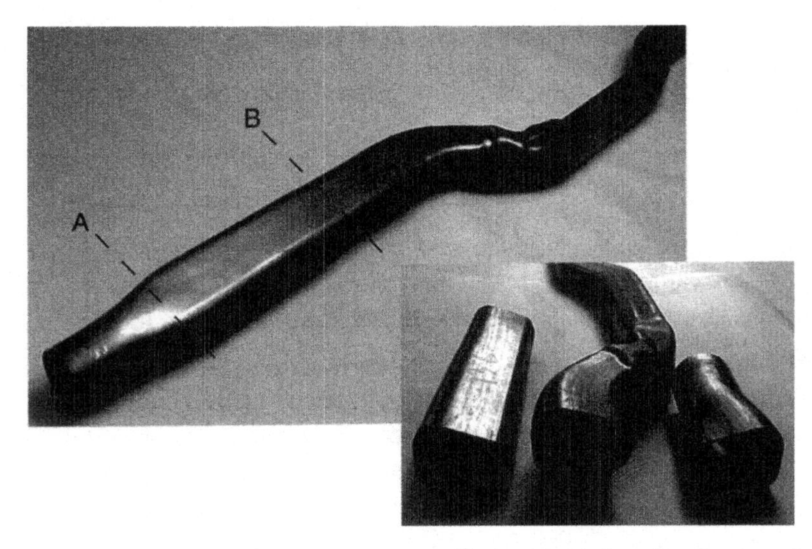

Figure 18.9. Long hydroformed component with an expanded section. Courtesy Copperweld Automotive, Copied from Harjinder Singh, *Fundamentals of Hydroforming*, SME, 2003, p. 57.

NOTE OF INTEREST

A 1903 patent describes using the pressure of molten lead to expand tubes. In 1917, a patent was issued for forming bent sections of wind musical instruments by hydroforming. It wasn't until the 1980s, however, that large frame members were produced by hydroforming.

REFERENCES

H. Singh, *Fundamentals of Hydroforming*, SME, 2003.

Z. Marciniak, J. L. Duncan, and S. J. Hu, *Mechanics of Sheet Metal Forming*, Butterworth-Heinemann, 2002.

Metals Handbook, 9th ed., 14, *Forging and Forming*, ASM.

PROBLEMS

18.1. Calculate the pressure to expand a tube from a radius r to r_0 if the ends are allowed to move freely. Assume $\sigma = K\varepsilon^n$. Also find the strain at maximum pressure.

18.2. A tube with an external radius of 25 mm and 1.0-mm thickness is to be expanded by internal pressure into a square 50 mm on a side. Assume that $\sigma = 650\varepsilon^{0.20}$ MPa. Determine the corner radius if the maximum pressure available is 80 MPa, the ends are fixed, and the die is frictionless.

18.3. Calculate the percent increase of the moment of inertia when a circular tube of 1-inch outside diameter and 0.9-inch inside diameter is hydroformed into a square cross section. Assume the wall thickness and tube length are unchanged.

18.4. A thin-wall tube with a radius, *R*, and a wall thickness, *t*, is pressurized in a square die. The effective stress–strain curve is $\bar{\sigma} = K(\varepsilon_0 + \bar{\varepsilon})^n$. Assume the length remains constant. Find the relation between the corner radius and the pressure if:

a) There is no friction.

b) Sticking friction prevails.

18.5. A steel tube of 50-mm diameter and 1-mm wall thickness is expanded into a square cross section. Assume that the effective stress–strain relation is $\bar{\sigma} = 650\bar{\varepsilon}^{0.20}$. Determine the minimum corner radius assuming frictionless conditions.

18.6. Water is used as the pressurized fluid in hydroforming. Why isn't air used?

19 Other Sheet Forming Operations

There are a number of sheet forming operations that cannot be classified as drawing, stamping or hydroforming. Among these are roll forming, spinning, making foldable shapes, incremental sheet forming, shearing, flanging and hemming.

19.1 ROLL FORMING

Roll forming transforms a flat sheet into another shape. As the new shape is formed over some distance, its width generally decreases as illustrated in Figures 19.1 and 19.2.

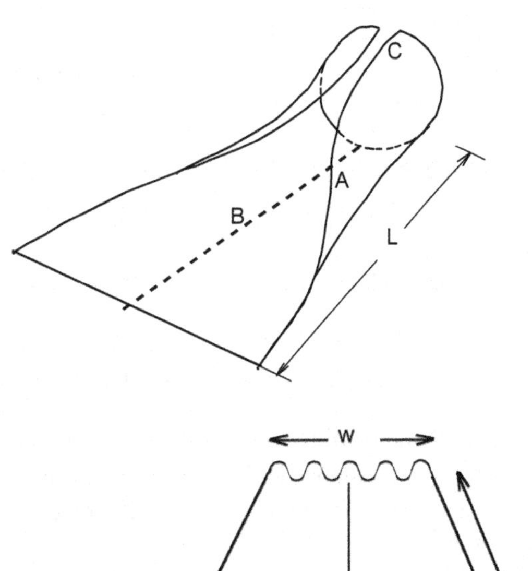

Figure 19.1. Roll forming of a tube. The length of the edge must first increase and then decrease during roll forming.

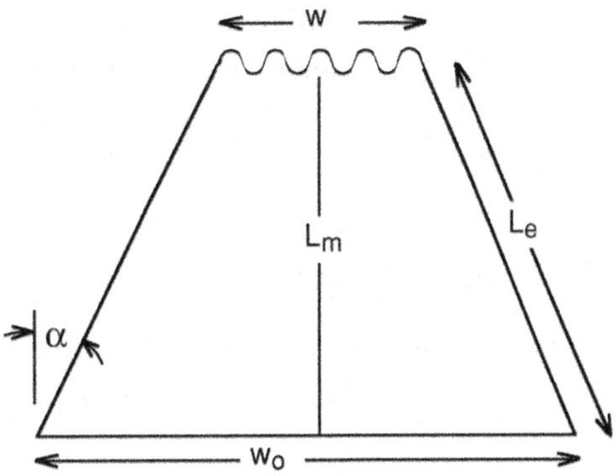

Figure 19.2. During corrugation of a sheet, the length of the edge elongates.

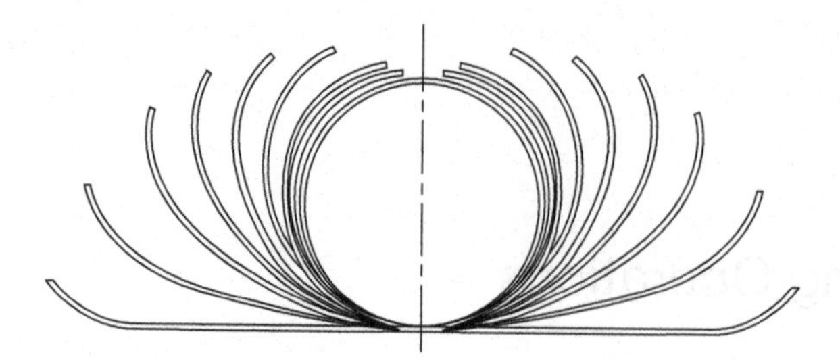

Figure 19.3. Progressive shapes in roll forming of a tube from a flat sheet. From H. Singh, *Fundamentals of Hydroforming*, SME, 2003.

During the roll forming process the edges are elongated. If the lead-in length, L, is insufficient, the edges will elongate plastically during the lead-in with the result that in the final region they will be too long after they become straight. As a result they will wrinkle.

During this transformation, the length of the edge must increase. The length must be sufficiently long so that the stretching is elastic; the edge will be longer than the centerline after the forming is complete, causing the formed sheet to warp.* The strain along the edge is

$$\varepsilon = \ln(L_e/L_m), \tag{19.1}$$

but

$$(L_e/L_m) = 1/\cos\alpha, \tag{19.2}$$

where

$$\alpha = \arctan(W_0 - W)/2L_m. \tag{19.3}$$

EXAMPLE 19.1: How long must L_m be if a sheet 6 feet wide contracts to 4 feet wide? Assume that yielding will occur when the strain reaches 0.002.

SOLUTION: $L_e/L_m = \exp(0.002) = 1.002.$

$\alpha = \arccos(1/1.002) = 3.6°.\ (W_0 - W)/2L_m = \tan\alpha = \tan 3.6° = 0.63.$
$L_m = (W_0 - W)/[2(.063)] = 15.9\,\text{ft}.$

Roll forming is used to make welded tubes. Figure 19.3 shows the progressive change of shape. Roll-formed tubes are used for hydroforming because their wall thicknesses are much more uniform than the walls of tubes made by extrusion or drawing.

19.2 SPINNING

Spinning is a sheet forming process that is suitable for axially symmetric parts. A tool forces the sheet metal disc to conform to a mandrel as shown in Figure 19.4.

* The author is indebted to J. L. Duncan for this concept.

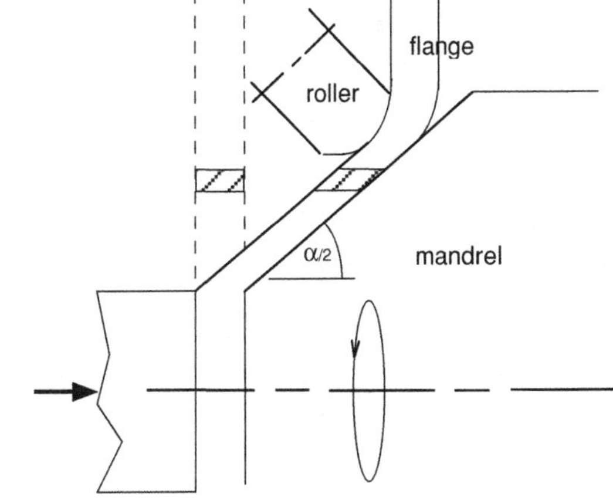

Figure 19.4. Sketch of a cup spinning operation. With pure shear, there is no deformation in the flange.

The tool usually consists of a small wheel. The process should be controlled so that all of the deformation is pure shear under the tool. In this case the reduction, r, is given by

$$r = (1 - \sin(\alpha/2), \tag{19.4}$$

and the shear strain, γ, by

$$\gamma = \cot(\alpha/2). \tag{19.5}$$

If the tool causes less thinning than would be produced by shear, the unsupported flange will wrinkle as shown in Figure 19.5. Even though spinning is slow, it is suitable for low production items because the tooling costs are low.

Spinning may also be used to reduce the wall thickness of tubes and cups shown in Figure 19.6.

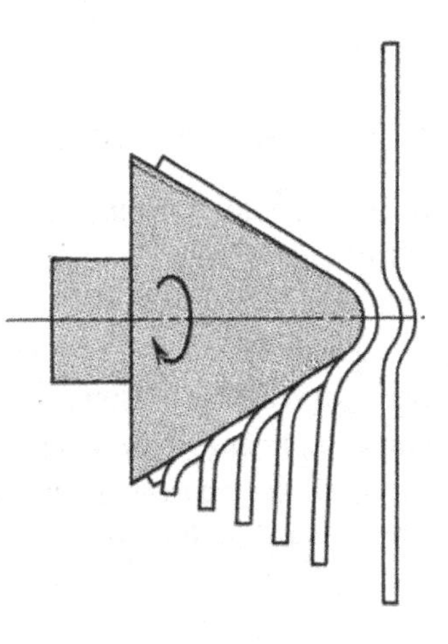

Figure 19.5. Spinning with no reduction of thickness requires the flange to shrink. This tends to cause wrinkling in the unsupported flange. Figure from Serope Kalpakjian, *Mechanical Processing of Metals*, Van Nostrand, 1967, p. 215.

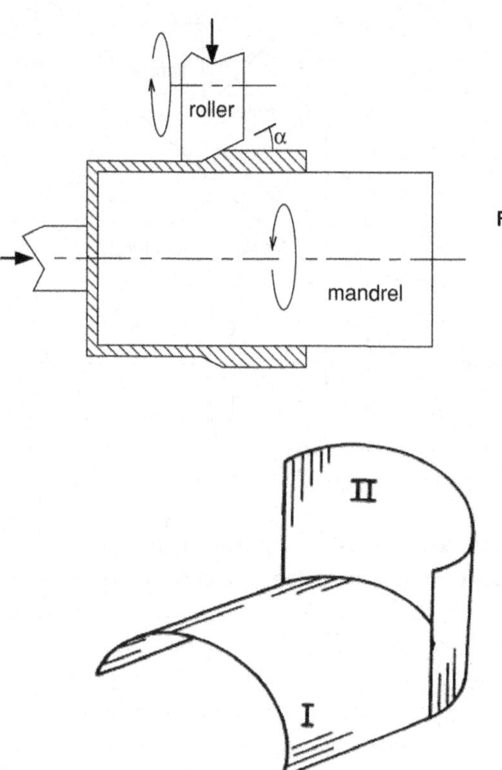

Figure 19.6. Reduction of wall thickness by spinning.

Figure 19.7. An example of a foldable shape.

19.3 FOLDABLE SHAPES

Some shapes that can be formed by bending without any stretching or drawing can be made from sheets with low ductility. Cones and cylinders are examples. Duncan and Duncan* have shown that these simple surfaces can be combined to form complex shapes. Figure 19.7 is an example of how cylindrical surfaces can be joined to form a complex shape. The boundary between the two cylinders does not need to be a generator of either surface. Even a martensitic steel with very low ductility can be formed into a developable shape used for reinforcing auto doors.

19.4 INCREMENTAL SHEET FORMING

In incremental forming, at any one time, only a small portion of a sheet is being deformed. A small punch moves along a prescribed path as shown in Figures 19.8 and 19.9.

There are several reasons that the achievable strain is far in excess of the forming limit curves. One reason is because of the high normal force on the tool. Another is each element undergoes bending and reverse bending so the strain path is not monotonic.

Very deep parts can be made by this process. An example is shown in Figure 19.10. The process is slow so it is not very attractive for mass production.

* J. P. Duncan and J. L. Duncan, Folded Developables, *Proc. R. Soc. Lond.* A 383 (1982), p. 191.

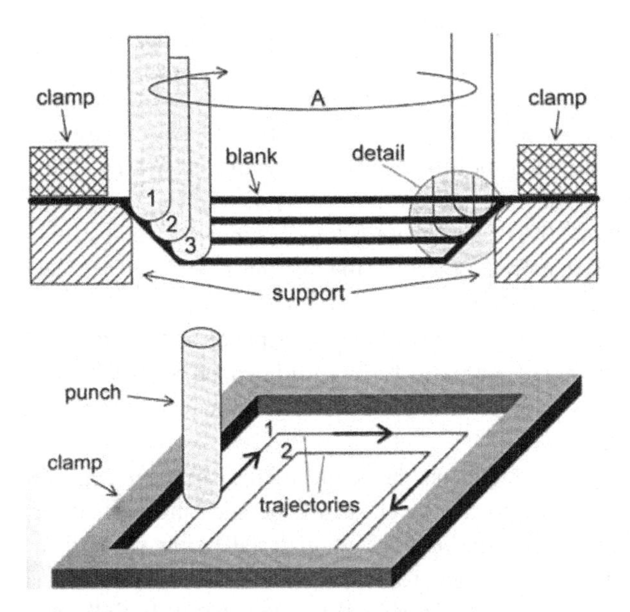

Figure 19.8. System used in incremental sheet forming. From W. C. Emmens, D. H. van der Weijde and A. H. van der Boogaard, in *Material Properties for More Effective Numerical Analysis*, B. S. Levy, D. K. Matlock and C. J. Van Tyne, Proceedings of the IDDGR, 2000.

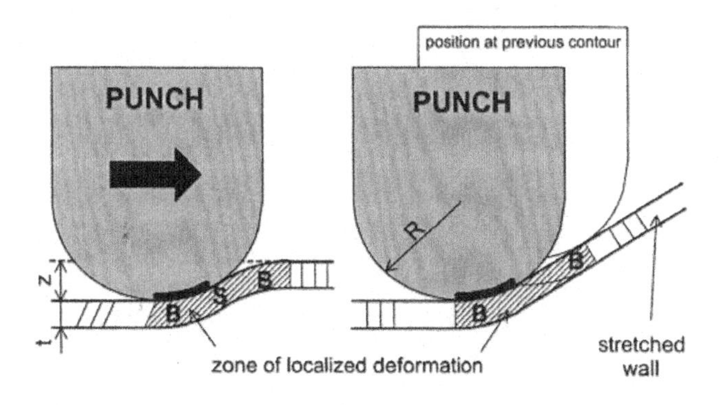

Figure 19.9. Tool traveling over sheet surface. W. C. Emmens, D. H. van der Weijde and A. H. van der Boogaard, *ibid.*

Figure 19.10. Pyramidal part formed by incremental sheet forming. W. C. Emmens, D. H. van der Weijde and A. H. van der Boogaard, *ibid.*

19.5 SHEARING

Sheets are sheared in preparing blanks for stamping and in trimming finished stampings. Punching of holes is also a form of shearing. The major concerns in shearing are the nature of the sheared edge, tool wear, and tool forces. The shape of the sheared edge, shown in Figure 19.11, depends on the clearance and the sharpness of the tools. The heights of burrs are usually less than 10% of the sheet thickness. For low-carbon steels, clearance between punch and die is usually 5 to 12% of the thickness. Tight clearances result in cleaner holes but with lower tool life.

Formability during subsequent hole expansion or flanging depends on the nature of the flanged edges. The presence of burrs decreases forming limits (see Section 19.6). The sharpness of the tools decreases with life as wear occurs. The wear rate is greater in higher strength sheets.

The forces required for shearing depend on the angle of the shearing edge to the sheet. If the cutting edge is parallel to the sheet, all of the shearing will occur simultaneously. By increasing the angle, the amount of material being sheared at any instant decreases. The shearing force is increased with smaller clearances, duller tools, and higher strength sheets. The force also varies during the stroke. Initially the shearing occurs through the entire thickness. As the tool penetrates, the area being sheared decreases.

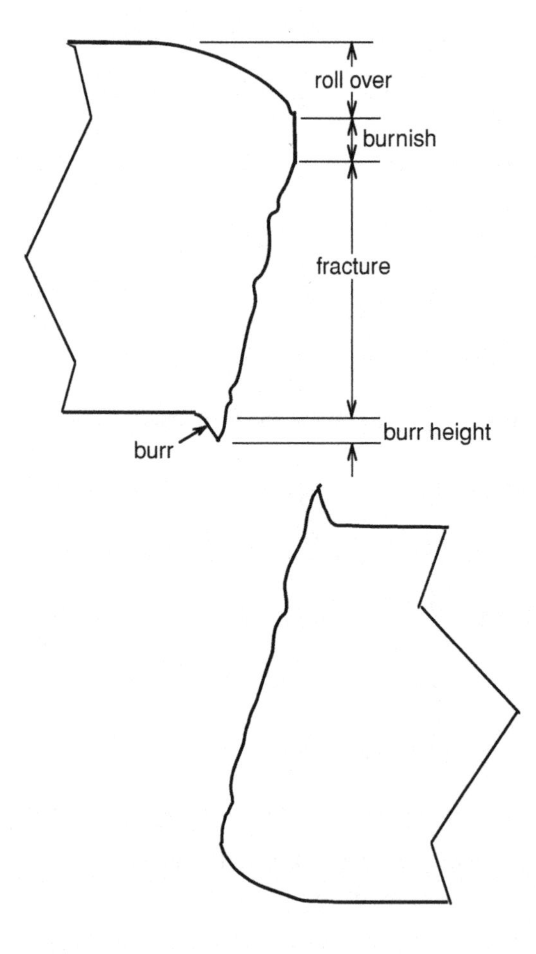

Figure 19.11. Sheared edge showing several regions: Roll over is part of original surface; burnish is where tools smeared the shear portion of the fracture; the fracture surface itself; and the burr.

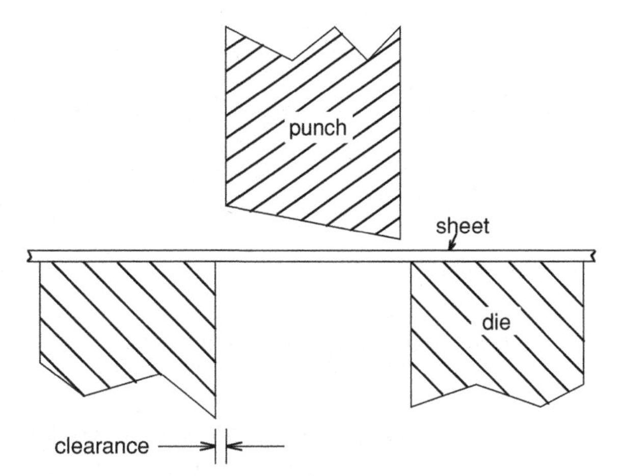

Figure 19.12. Tools for hole punching. Note that the cutting edge of the punch is on an angle so the shearing does not occur simultaneously over the entire surface.

To avoid excessive shearing forces, the tools for shearing and hole punching are made so that as the tools descend, the shearing progresses from one location to another (Figure 19.12.)

19.6 FLANGING, HOLE EXPANSION, AND BEADING

Often edges of sheets are flanged. If the flange is concave, cracking may occur at sheared edges where the edges are elongated in tension. This occurs when a hole is expanded or a concave edge is flared as shown in Figure 19.13. The tensile strain in such a case is

$$\varepsilon_1 = \ln(d_0/d), \tag{19.6}$$

where the diameter before and after flanging are d_0 and d. The tendency is aggravated if shearing has left a burred edge. On the other hand, if the flange is convex, the contraction of the flanged edge may lead to wrinkling.

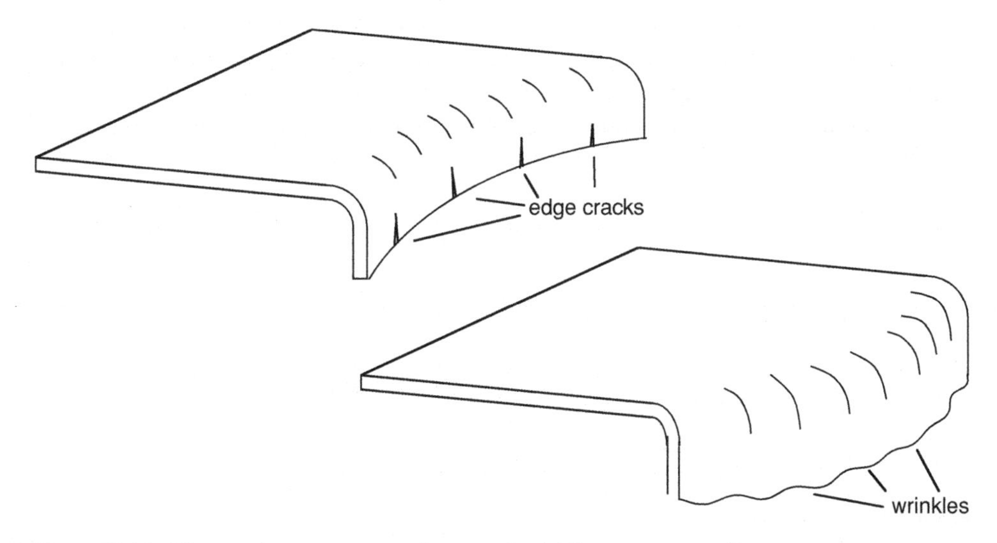

Figure 19.13. Edge cracks on a concave flange and wrinkling on a convex flange.

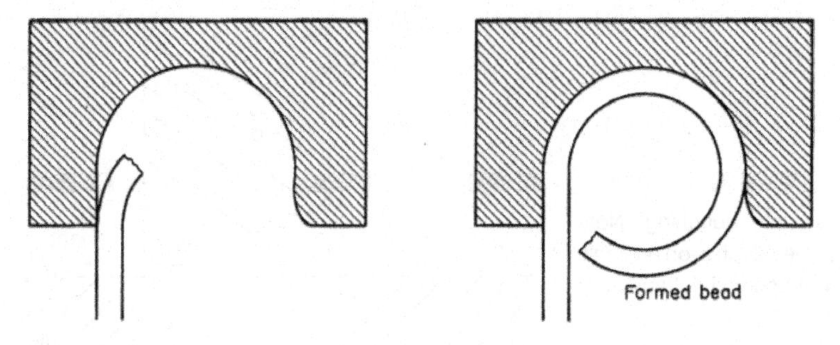

Figure 19.14. Tooling for forming a bead.

Expansion of pierced holes is a common forming operation and problems associated with this operation are not unusual. Measured failure strains in hole expansion of high strength steels are usually considerably lower than those predicted by forming limit curves. The behavior of steels with finished holes is much better than those with sheared edges. Microstructure, chemical composition, and tensile strength affect failures during hole expansion. A high ratio of yield strength to tensile strength is desirable. A high silicon and low-carbon contents improve hole expansionability.

The strains associated with cracking of edges in either flanging or hole expansion are usually lower than those predicted by the forming limit diagrams. In hole expansion the strains around the edge are not uniform so some regions have higher strains than would be calculated from the increase of diameter.

Often a bead has to be formed on a sheared edge. Figure 19.14 shows how this can be accomplished.

19.7 HEMMING

Hemming is a bending operation that bends and folds an edge of sheet. In the automotive industry hemming is used to join inner and outer panels such as hoods, doors, and tailgates. An outer panel is folded over the inner one as shown in Figure 19.15. The accuracy of hemming is critical to the final appearance of car. There are two methods of hemming, one uses traditional hydraulically operated stamping presses.The other, roller hemming, uses a robot controlled roller to progressively bend the outer panel over the inner panel.

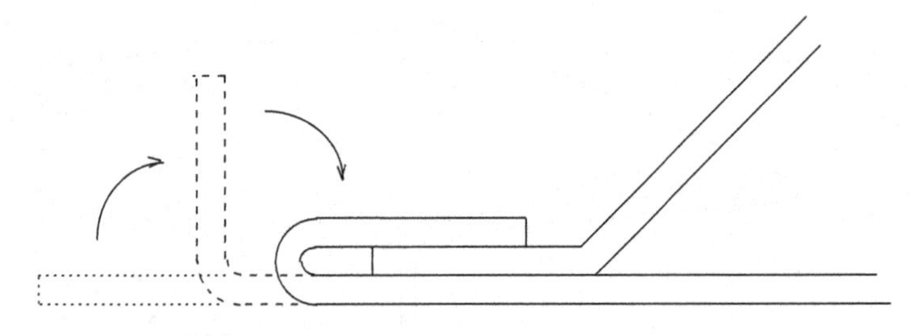

Figure 19.15. Hemming to form an edge.

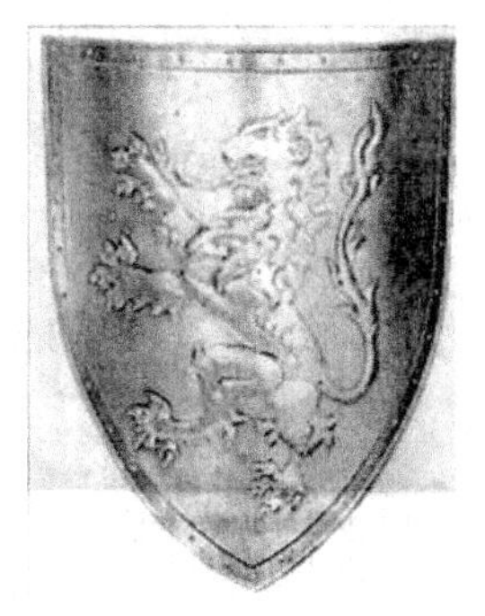

Figure 19.16. Repoussé on a steel shield. From Wikopedia repoussé and chasing.

NOTE OF INTEREST

In a very old process called *repoussé*, sheet metal is shaped by hammering from the reverse side so as to raise an image. Figure 19.16 is an example.

REFERENCES

Serope Kalpakjian, *Mechanical Processing of Materials*, Van Nostrand, 1967.
Z. Marciniak, J. L. Duncan, and S. J. Hu, *Mechanics of Sheet Metal Forming*, Butterworth-Heinemann, 2002.
Metals Handbook, 9th ed., v. 14, Forging and Forming, ASM.
H. Singh, *Fundamentals of Hydroforming*, SME, 2003.

PROBLEMS

19.1. Consider roll forming of a circular tube having a radius of 12 in. as illustrated in Figure 19.1. What is the minimum ratio of the lead in length, L, to the initial sheet width, W, if yielding of the edge is to be prevented? Assume a yield strength of 40 ksi and a modulus of 30×10^6 psi.

19.2. Consider the roll forming of a corrugated sheet. Assume that the corrugations can be approximated as a series of semicircles as illustrated in Figure 19.2. What is the minimum ratio of the lead in length, L, to the initial sheet width, W, if yielding of the edge is to be prevented? Let Y be 50,000 psi and $E = 30 \times 10^6$ psi.

19.3. A cup with 45° walls is to be spun from sheet. What should the thickness of the sheet be if the cup walls are to be 1-mm thick?

19.4. Calculate the percent increase of the moment of inertia when a circular tube of 1-in. outside diameter and 0.9-in. inside diameter is hydroformed into a 1-in. square cross section. Assume the wall thickness and tube length are unchanged.

19.5. Water is used as the pressurized fluid in hydroforming. Why isn't air used?

19.6. The very deep parts made by the incremental forming described in Figures 19.8 and 19.9 produce strains much greater than the forming limit curves predict. Explain how this is possible.

20 Formability Tests

20.1 CUPPING TESTS

The Swift cup test is the determination of the limiting drawing ratio for flat-bottom cups. In the Erichsen and Olsen tests, cups are formed by stretching over a hemispherical tool. The flanges are very large so little drawing occurs. The results depend on stretchability rather than drawability. The Olsen test is used in America and the Erichsen in Europe. Figure 20.1 shows the set up.

The Fukui conical cup test involves both stretching and drawing over a ball. The opening is much larger than the ball so a conical cup is developed. The flanges are allowed to draw in. Figure 20.2 shows the set up. A failed Fukui cup is shown in Figure 20.3.

Figure 20.4 shows comparison of the relative amounts of stretch and draw in these tests.

20.2 LDH TEST

The cupping tests discussed above are losing favor because of irreproducibility. Hecker* attributed this to "insufficient size of the penetrator, inability to prevent inadvertent draw

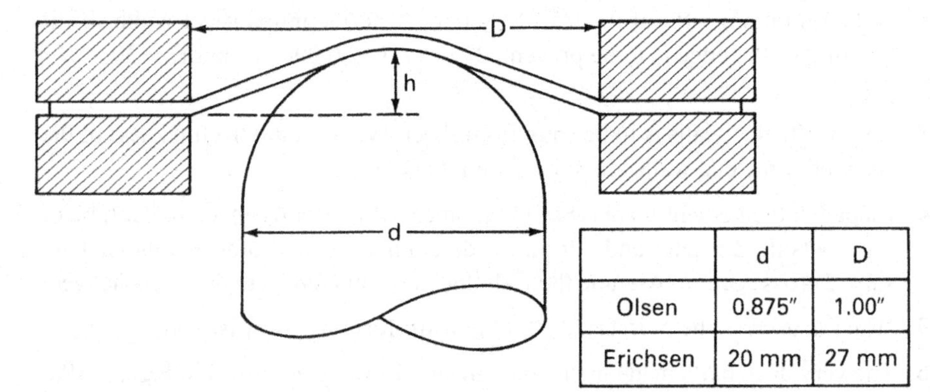

	d	D
Olsen	0.875″	1.00″
Erichsen	20 mm	27 mm

Figure 20.1. Olsen and Erichsen tests.

* S. S. Hecker, *Met. Engr Q.* v. 14 (1974).

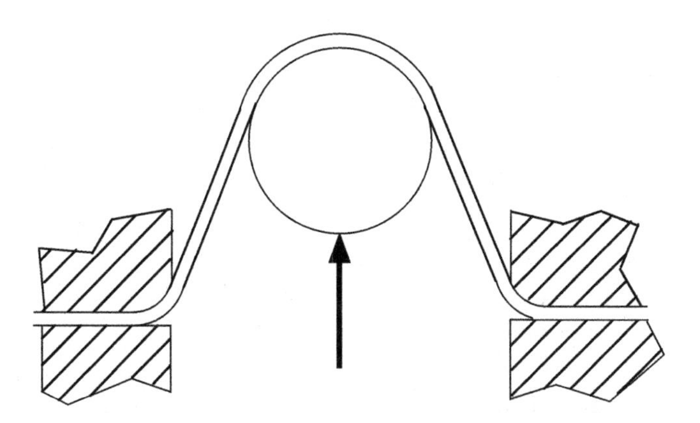

Figure 20.2. Fukui test.

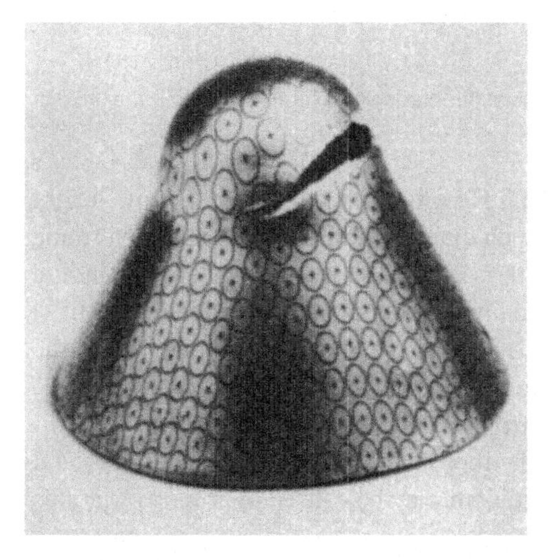

Figure 20.3. Failed Fukui cup. Courtesy of *Institut de Researches de la Siderurgie Francaise.*

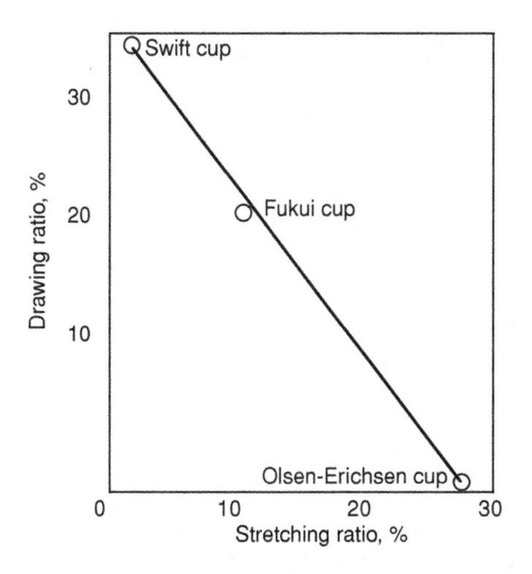

Figure 20.4. Relative amounts of stretching and drawing in these cupping tests.

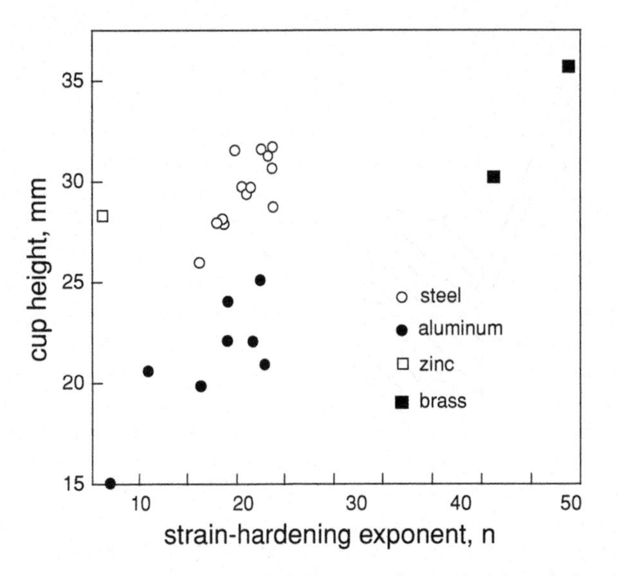

Figure 20.5. Failure of cup height to correlate with the strain-hardening exponent, which is the uniform strain before necking. Data from S. S. Hecker, *ibid*.

in of the flange and inconsistent lubrication." He proposed the limiting dome height (LDH) test which uses the same tooling (4 inch diameter punch) as used to determine forming limit diagrams. The specimen width is adjusted to achieve plane-strain and the flange is clamped to prevent draw-in. The limiting dome height is greatest depth of cup formed with the flanges clamped. The LDH test results correlate better with the total elongation than with the uniform elongation as shown in Figures 20.5 and 20.6. The total elongation includes the post-uniform elongation.

A major problem with the LDH test is the reproducibility within a laboratory and between different laboratories. Part of the problem may be caused by minor variations in details of the clamping.

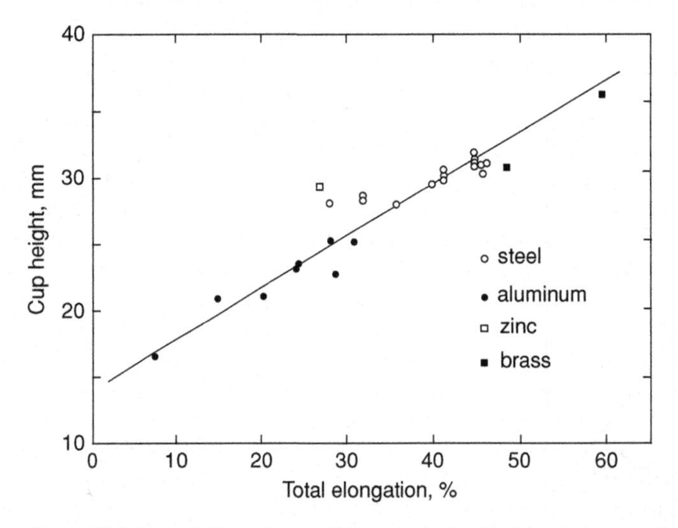

Figure 20.6. Correlation of the LDH cup height with the total elongation in a tension test. Data from S. S. Hecker, *ibid*.

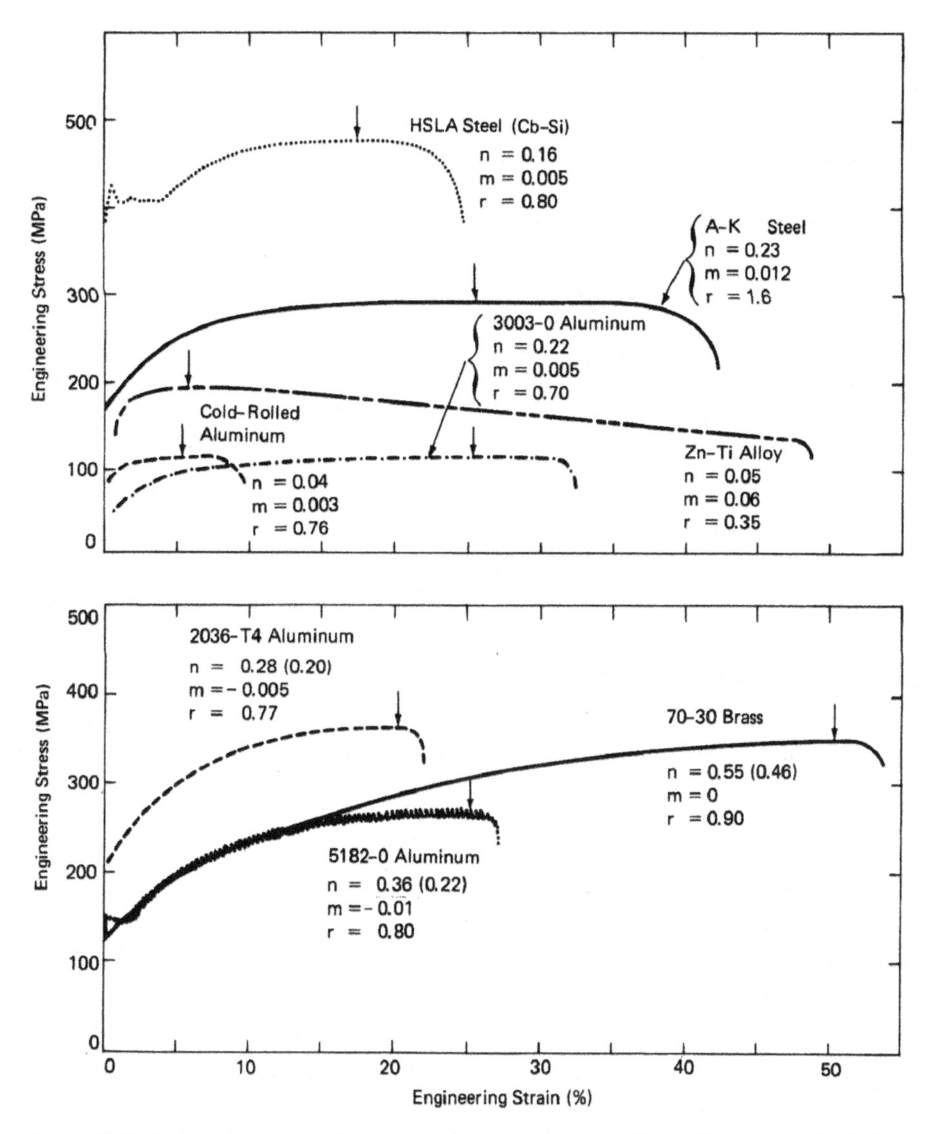

Figure 20.7. Engineering stress-strain curves for several metals. The uniform elongation is indicated by an arrow. Note that the post-uniform elongation is greater for materials with higher values of the strain-rate sensitivity, *m*. From A. K. Ghosh, *J. of Eng. Matls. and Tech, Trans ASME Series H*, v 99, (1977).

20.3 POST-UNIFORM ELONGATION

Figure 20.7 shows the engineering stress-strain curves for various sheet metals. It can be seen that the elongation that occurs after necking correlates well with the strain-rate sensitivity. This is plotted in Figure 20.8.

20.4 OSU FORMABILITY TEST

Wagoner and coworkers have proposed a more reliable test that involves cylindrical punches instead of spherical punches. Three different proposed punch

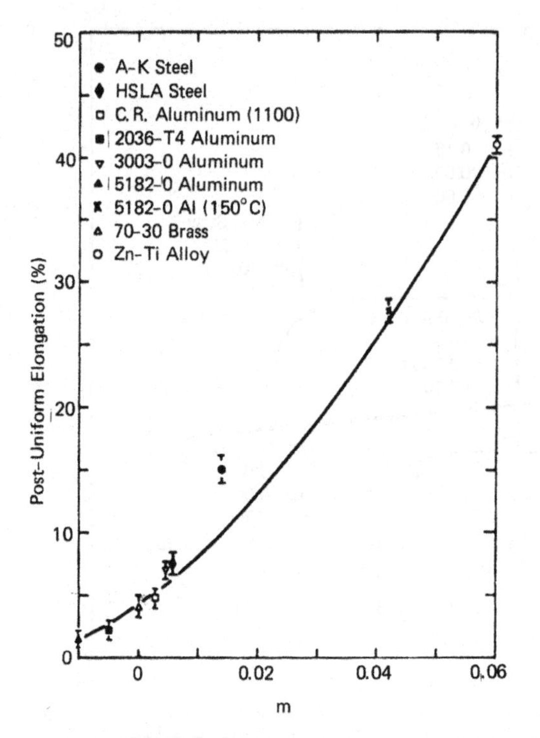

Figure 20.8. Post-uniform elongation as a function of the strain-rate sensitivity, m. From A. K. Ghosh, *ibid.*

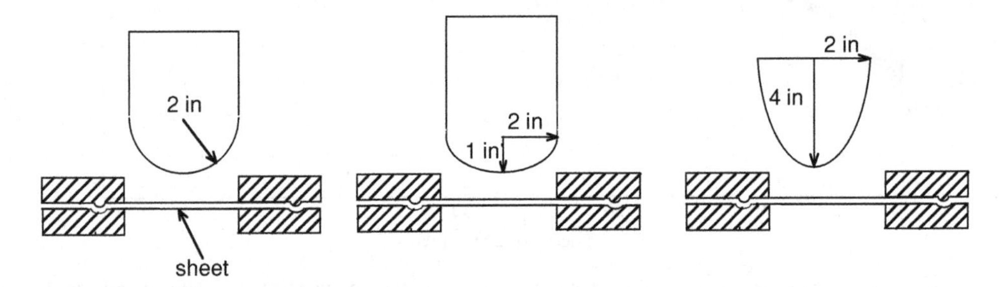

Figure 20.9. Tooling for OSU formability test. All three indenters are cylinders.

geometries are illustrated in Figure 20.9. The height of the draw at failure is measured.

20.5 HOLE EXPANSION

In the hole expansion test, a punch forces a hole to expand as shown in Figure 20.10. This increases the radius. The engineering strain at which failure occurs is $e = \Delta r/r_0$. The results depend strongly on the height of burrs made when the hole was punched. This dependence is illustrated schematically in Figure 20.11. Failures usually occur

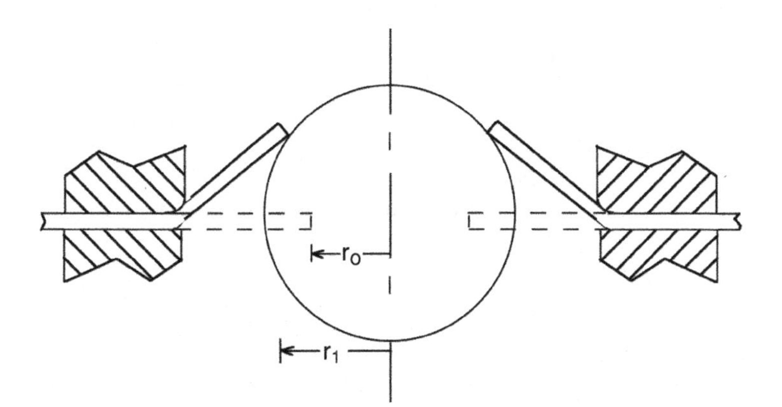

Figure 20.10. Hole expansion test.

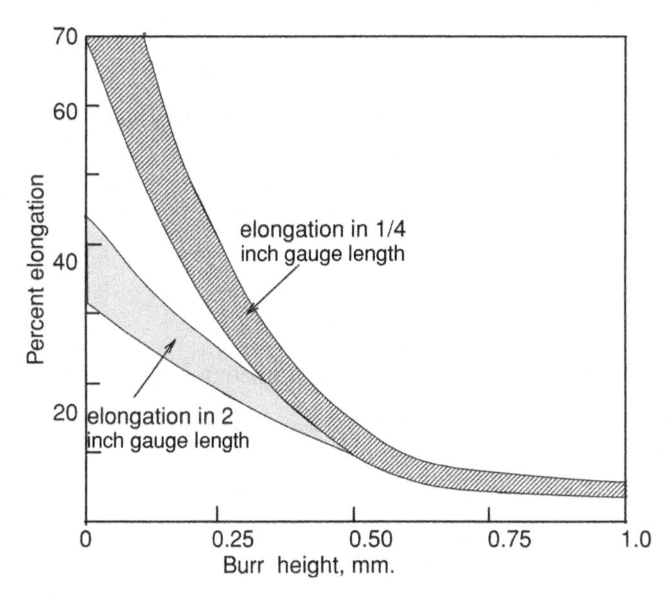

Figure 20.11. The amount a hole can be expanded decreases significantly with burr height.

with cracks running parallel to the rolling direction because ductility in the transverse direction is lower than in the longitudinal direction.

20.6 HYDRAULIC BULGE TEST

Much higher strains are possible in a hydraulic bulge test than in a tension test, so the effective stress-strain relations can be evaluated at higher strains. Figure 20.12 shows the set up for a bulge test. The sheet is placed over a circular hole, clamped, and bulged outward by the oil pressure, P.

Consider a force balance on a circular element of radius, ρ, near the pole (Figure 20.13). The radius, r, of this element is $r = \rho\Delta\theta$. The vertical component of the force acting on the circumference of this element is $2\pi r t \sigma \Delta\theta = 2\pi t \rho \sigma \Delta\theta^2$.

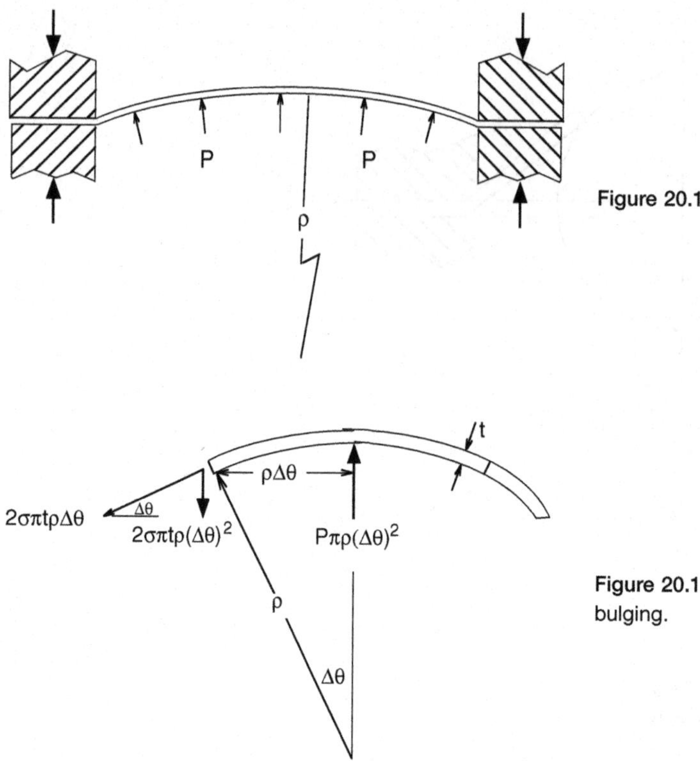

Figure 20.12. Hydraulic bulge test.

Figure 20.13. Force balance in hydraulic bulging.

This is balanced by the force of the oil, $\pi r^2 P = \pi(\rho\Delta\theta)^2 P$. Equating, $2\pi t\rho\sigma\Delta\theta^2 = \pi(\rho\Delta\theta)^2 P$ or

$$\sigma = P\rho/(2t). \tag{20.1}$$

The radial strain, ε_r, can be used to find the thickness,

$$t = t_0 \exp(-2\varepsilon_r). \tag{20.2}$$

To obtain the stress-strain relation, there must be simultaneous measurement of ε_r, ρ and P.

20.7 DUNCAN FRICTION TEST*

This consists of stretching a strip between two fixed cylinders as indicated in Figure 20.14. The strain in sections A and B are measured. From these the stresses, σ_A and σ_B, and the thicknesses, t_A and t_B, can be deduced from $\sigma = K\varepsilon^n$ and $t = t_0 \exp\varepsilon$. Now F_A and F_B can be determined as $F = \sigma t w$ where w is the strip width. Finally the friction coefficient can be found by solving $F_A/F_B = \exp(\mu\pi/2)$ for μ.

$$\mu = (2/\pi)\ln[(\varepsilon_A/\varepsilon_B)^n(\exp\varepsilon_A/\exp\varepsilon_B)]. \tag{20.3}$$

* J. L. Duncan, B. S. Shabel and J G. Filho, *SAE paper 780391*, 1978.

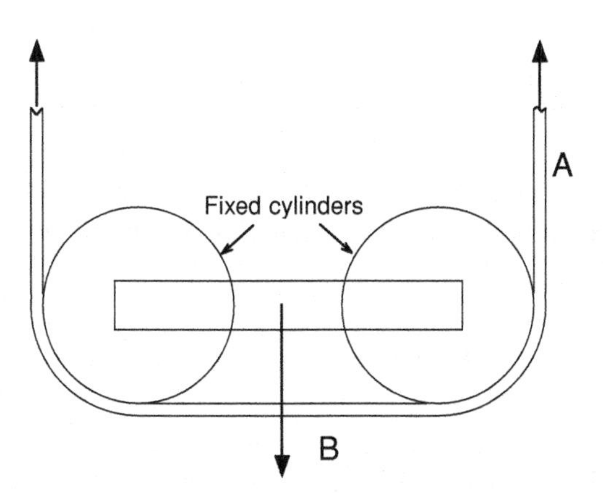

Figure 20.14. The Duncan friction test.

REFERENCES

E. M. Mielnik, *Metalworking Science and Engineering*, McGraw-Hill, 1991.
M. P. Miles, J. L. Siles, R. H. Wagoner, and K. Narasimhan, *Met. Trans.*, 24A (1993).

PROBLEMS

20.1. A strip was tested using the friction test in Figure 20.14. Strains of $\varepsilon_A = 0.180$ and $\varepsilon_B = 0.05$ were measured. The tensile stress-strain curve is approximated by $\sigma = 800\varepsilon^{0.20}$ MPa. Calculate the coefficient of friction between the strip and the cylinders.

20.2. Consider a bulge test on a sheet, which is clamped at the periphery of a circular hole of radius, R (Figure 20.15). The bulge height, $h = r/2$. Assume for simplicity that the shape of the bulged surface is spherical and that the radial strain is the same everywhere. (Neither of these assumptions is strictly correct.)

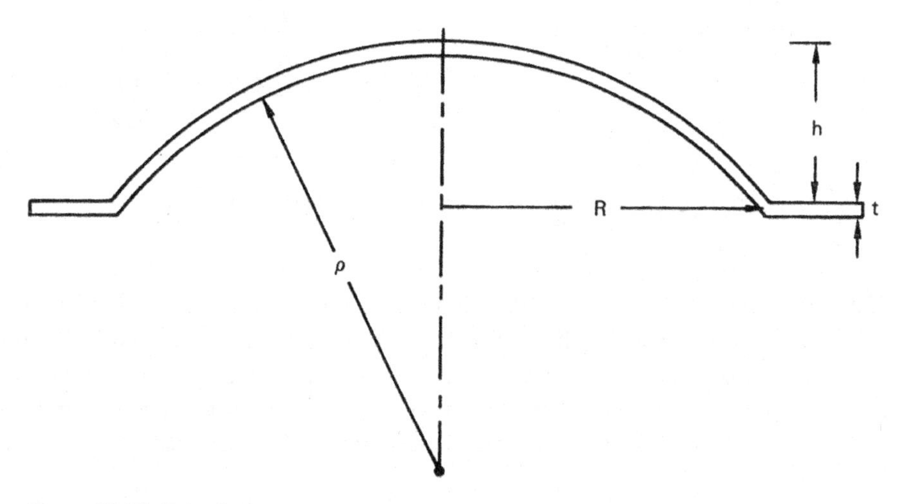

Figure 20.15. Bulge test.

a) Calculate the radial strain, ε_r.

b) Calculate the circumferential strain, ε_c, as a function of the initial radial position, r_0.

20.3. Assume that the bulged surface in a hydraulic bulge test is a section of a sphere. Show that with this assumption the radius of curvature, ρ, is related to the bulge height, h, and the die radius, R, by

$$\rho = \frac{R^2 + h^2}{2h}.$$

20.4. If it is further assumed that the thickness of the bulged surface is the same everywhere (not a particularly good assumption), the thickness can be related to the bulge height, h, and the die radius, R. Derive this relation.

20.5. During a laboratory investigation of springback, strips were bent between three cylinders and unloaded as illustrated in Figure 20.16. It was noted that, on unloading, the force dropped abruptly from F_1 to F_2 before noticeable unbending occurred. It was postulated that the drop was associated with the reversal of the direction of friction between the strips and the cylinders. Assuming that this is correct, derive an expression for the coefficient of friction in terms of F_1, F_2, and the bend angle α.

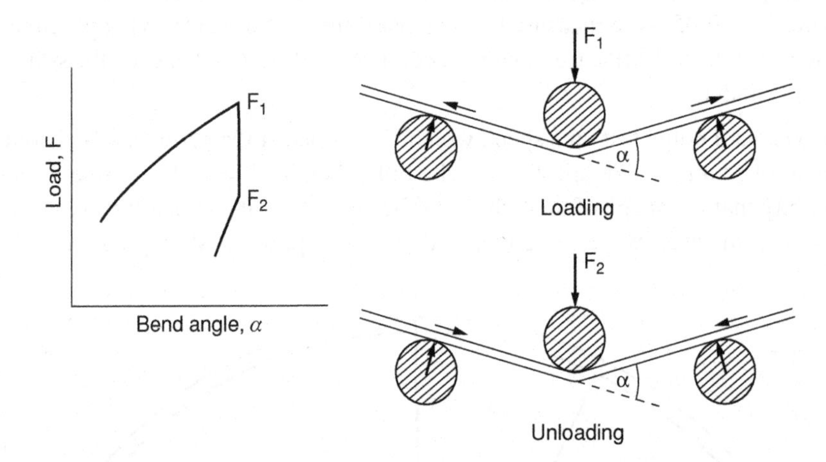

Figure 20.16. Change of direction of friction on unloading from bending.

20.6. The apparatus in Figure 20.17 is used to induce balanced biaxial stretching. Draw beads lock the sheet. The amount of stretching is limited by failure of the walls in plane-strain tension. Estimate the maximum biaxial strain that can be achieved in a low-carbon steel with a strain-hardening exponent of $n = 0.22$. Neglect any strain-rate dependence and assume that the friction coefficient is 0.12.

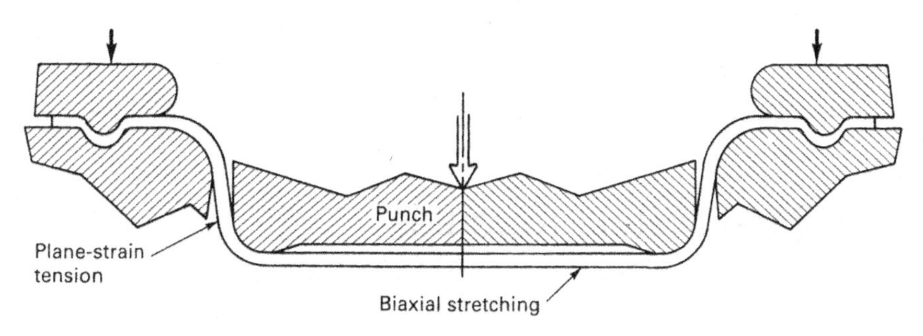

Figure 20.17. Apparatus for biaxial stretching.

20.7. Describe the state of stress at the edge of a hole during hole expansion.

21 Sheet Metal Properties

21.1 INTRODUCTION

Properties of sheet metals vary from one class of materials to another. Table 21.1 gives typical ranges of n, m, and $\bar{R}$ within several classes of materials. It should be noted that while these values are typical, higher or lower values may be encountered.

For both fcc and bcc metals, the highest values of $\bar{R}$ correspond to textures with {111} planes oriented parallel to the sheet. Grains with {100} planes oriented parallel to the sheet tend to have very low $\bar{R}$ values. The recrystallization textures of bcc metals after cold rolling tend to have strong {111} textural components parallel to the sheet and the $\bar{R}$-values depend mostly on the amount of the weaker {100} component. It has been shown* that pure rotationally symmetric {111} textures in bcc metals have a maximum $\bar{R}$-value of about 3. In contrast, cold-rolled and recrystallized fcc metals (aluminum, copper, and austenitic stainless steels) usually have very little {111} textural component parallel to the surface and consequently have $\bar{R}$-values less than 1.

Table 21.1. Typical Sheet Metal Properties†

Metal	n	$\bar{R}$	m
Low-carbon steel	0.20–0.25	1.4–2.0	0.015
Interstitial-free steel	0.30	1.8–2.5	0.015
HSLA steels	0.10–0.18	0.9–1.2	0.005–0.01
Ferritic stainless steel	0.16–0.23	1.0–1.2	0.010–0.015
Austenitic stainless steel	0.40–0.55	0.9–1.0	0.010–0.015
Copper	0.35–0.45	0.6–0.9	0.005
Brass (70–30)	0.40–0.60	0.8–0.9	0.0–0.005
Aluminum alloys	0.20–0.30	0.6–0.8	−0.005 to +0.005
Zinc alloys	0.05–0.15	0.4–0.6	0.05–0.08
α-titanium	0.05	3.0–5.0	0.01–0.02

† Although these values are typical, there is a great deal of variation from one lot to another, depending on the composition and the rolling and annealing practices. In general as the strength levels are increased by cold work, precipitation or grain-size refinement, the levels of n and m decrease.

* J. O'Brien, R. Logan and W. Hosford, in *Proceedings of the Sixth International Symposium on Plasticity and Its Current Applications*. A. S. Kahn ed. 1997.

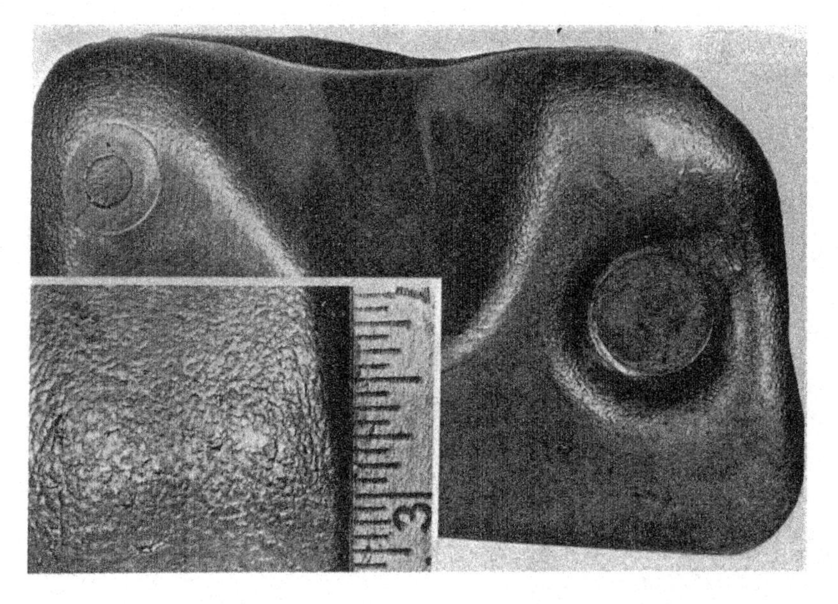

Figure 21.1. Example of orange peel. Courtesy of American Iron and Steel Institute.

21.2 SURFACE APPEARANCE

Surface appearance is of great importance for most parts formed from sheet metal. Several types of defects may occur during forming. *Orange peel* (Figure 21.1) is a surface roughening on the scale of the grain size. It occurs because of different orientations of neighboring grains on the surface. During elongations some grains contract more in the direction normal to the surface and others contract more in a direction in the surface. Orange peel is observed only on free surfaces. The effect can be reduced by using a material with a finer grain size.

A related phenomenon is *roping* or *ridging* of ferritic stainless steels and some aluminum alloys. In this case, whole clusters of grains have two different orientations. These clusters are elongated in the direction of prior working. Figure 21.2 is an example of roping in an automobile hubcap.

Another defect is the formation of *stretcher strains,* which are incomplete Lüders bands. These are very apparent where the overall strain is small. Figure 21.3 shows stretcher strains in a 1008 steel. They occur in low-carbon steels having a yield point and in some nonferrous materials that have a negative strain-rate sensitivity.

21.3 STRAIN AGING

Low-carbon steels (%C $\approx$ 0.06 or less) are usually finished by cold rolling and annealing except in heavy gauges ($\gtrsim$2 mm). They are marketed after annealing at 600°C to 700°C. Historically they were produced by casting into ingots and were classified as either *rimmed steel* or *aluminum-killed steel.* A rimmed steel was one that was not deoxidized before ingot casting. During the freezing, dissolved oxygen and carbon react to form CO. Violent evolution of CO bubbles threw sparks into the air. The bubbles stirred the molten metal breaking up boundary layers. This allowed segregation of carbon to

Figure 21.2. Ridging in a 430 stainless steel. From *Making, Shaping and Treating Steels, 9th ed.* United States Steel Corp. 1971.

the center of the ingot producing a very pure iron surface. In contrast killed steels are deoxidized with aluminum so the violent reaction is "killed." The solidification is quiet and so a boundary layer forms and prevents surface-to-center segregation.

Today continuous casting has almost completely replaced ingot casting. As a result, almost all steels are killed. There has been a trend to casting thinner sections, which require less rolling. This saves money but does not refine the grain structure as much.

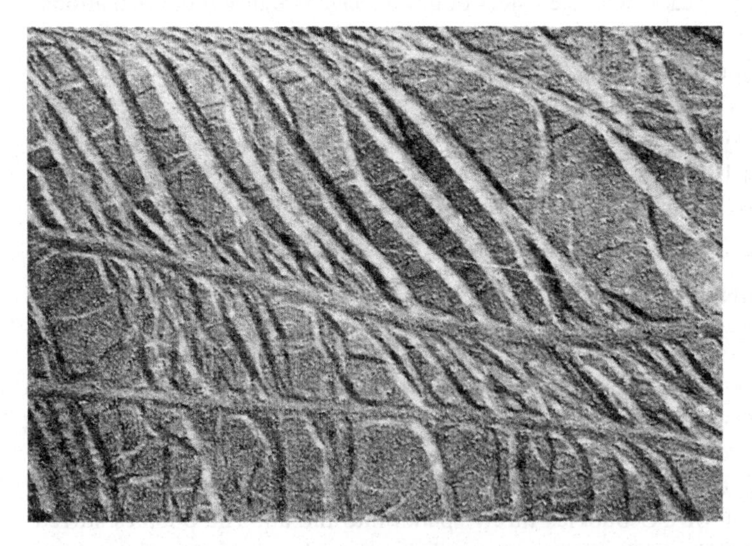

Figure 21.3. Stretcher strains in a 1008 steel sheet. From *Metals Handbook, 8th Ed. vol 7, ASM*, 1972.

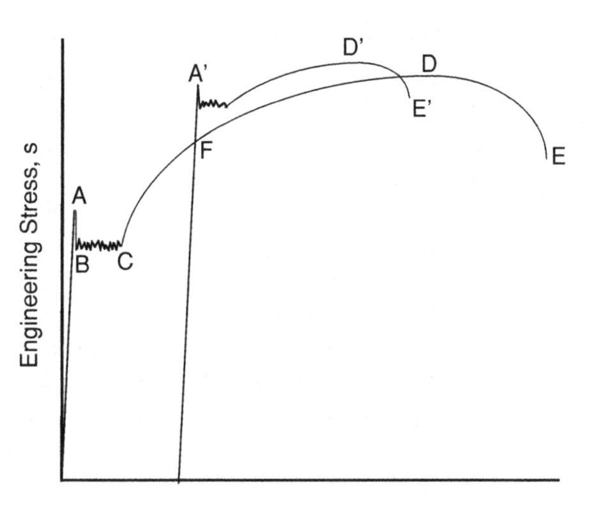

Figure 21.4. Engineering stress-strain curve for a low-carbon steel.

Figure 21.4 shows a tensile stress-strain curve of an annealed low-carbon steel. Loading is elastic until yielding occurs (point A). Then the load suddenly drops to a lower yield stress (point B). Continued elongation occurs by propagation of the yielded region at this lower stress until the entire specimen has yielded (point C). During the extension at the lower stress, there is a sharp boundary or *Lüders band* between the yielded and unyielded regions as shown in Figure 21.5. Behind this front, all of the material has suffered the same strain. The Lüders strain or yield point elongation is typically from 1 to 3%. Only after the Lüders band has traversed the entire specimen does strain hardening occur. Finally the specimen necks at point F.

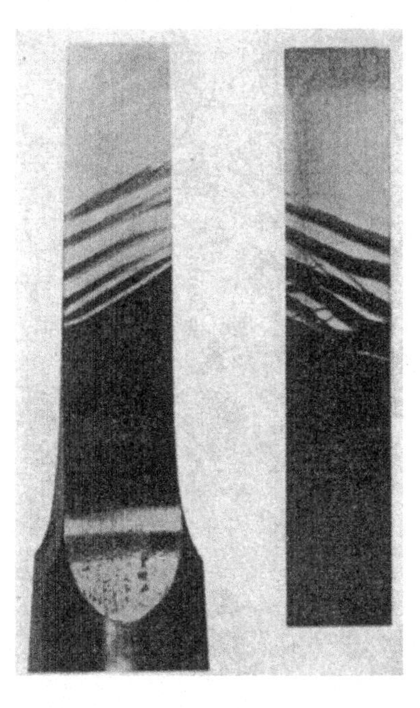

Figure 21.5. Tensile specimen of a low-carbon steel during extension. Deformation occurs by movement of a Lüder's band through the specimen. From F. Körber, *J. Inst. Metals* v. 48, 1932.

If the specimen in Figure 21.4 were unloaded at some point D after the lower yield region and immediately reloaded, the stress-strain curve would follow the original curve. However, if the steel were allowed to strain age between unloading and reloading, a new yield point, A′, would develop. The tensile strength would be raised from E to E′ and the elongation would be decreased. From the standpoint of forming, the yield point effect and strain aging are undesirable because of the reduction of n and the surface appearance of stretcher strains.

During strain aging, interstitially dissolved nitrogen and carbon segregate to dislocation lowering their energy. A higher stress is required to move the dislocations away from these interstitials than to continue their motion after they have broken free.

Aluminum-killed steels are much more resistant to strain aging than rimmed steels, but strain aging will occur at the temperatures of paint baking. At this point the strain aging is beneficial because the parts have already been formed and the strain aging adds strength.

21.4 ROLLER LEVELING AND TEMPER ROLLING

In commercial practice, the initial yield point is eliminated by either *roller leveling or temper rolling*. Both of which impart very small strains (typically $<0.5\%$) to the sheet.

In roller leveling the sheet is bent back and forth by a series of small diameter rolls as shown in Figure 21.6. This causes local regions to yield without much change in thickness. Roller leveling is often used to remove the yield point in low-carbon steels, as well as to straighten steel plates or strip after final rolling, heat treatment, or cooling operations.

Temper rolling (also called pinch passing or skin rolling) is also used to remove the yield point in steels. The reduction is very light (typically ½%). This is effective even though not all of the material deforms, as shown in Figure 21.7.

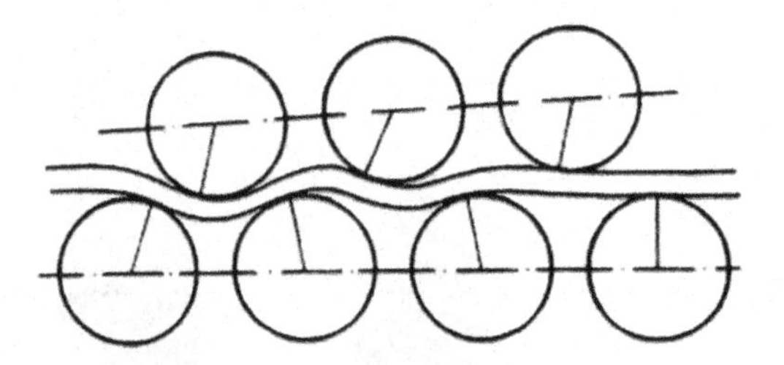

Figure 21.6. Schematic of roller leveling.

Figure 21.7. Deformed regions in roller-leveled material. From D. J. Blickwede, *Metals Progress*, July 1969.

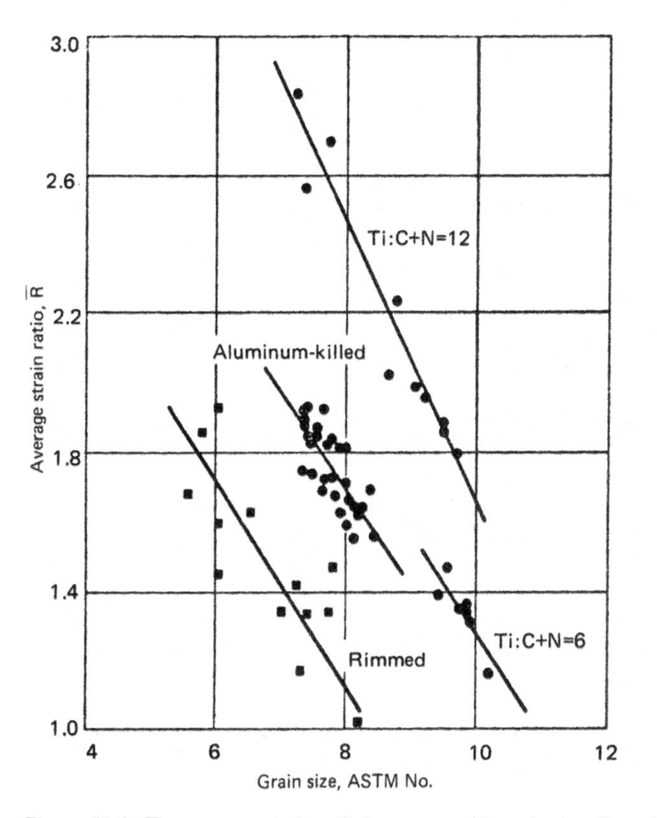

Figure 21.8. The average strain ratio increases with grain size. From D. J. Blickwede, *Metals Progress*, April 1969.

21.5 PROPERTIES OF STEELS

For killed steels, $\bar{R}$ is usually between 1.4 and 2.0. Much higher values are common in interstitial-free steels as shown in Figure 21.8. The level of $\bar{R}$ increases with grain size but grain sizes larger than ASTM 7 are usually avoided because of excessive orange peel. The $\bar{R}$-values for hot-rolled steels are usually about 1.0.

Properties of sheet steels vary from the interstitial-free steels (IF) with tensile strengths of 145 MPa and tensile elongations of 50% to martensitic steels with tensile strengths of 1400 MPa and tensile elongations of 5%. Table 21.1 lists typical properties of various grades of sheet steels. In general, the strain-hardening exponent and tensile elongation decrease with increased yield strength as shown in Figure 21.9 and Figure 21.10.

21.6 GRADES OF LOW-CARBON STEEL

There are a number of grades of low-carbon steels. Among these are:

Interstitial-free (IF) *steels*: These are steels from which carbon and nitrogen have been reduced to extremely low levels (less than 0.005%). After vacuum degassing, titanium is added to react with any carbon or nitrogen in solution. Titanium reacts

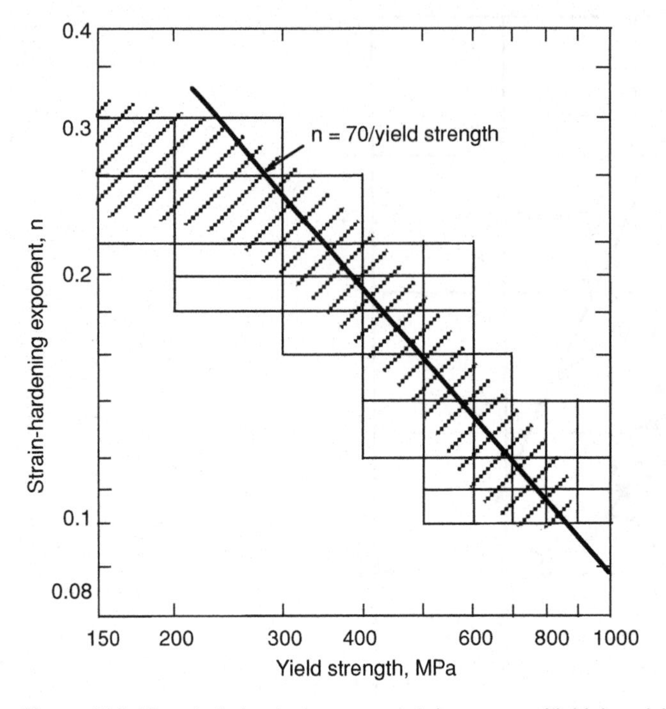

Figure 21.9. The strain-hardening exponent decreases with higher yield strengths. Adapted from S. P. Keeler and W. G. Brazier in *Micro Alloying 75*, Union Carbide (1977).

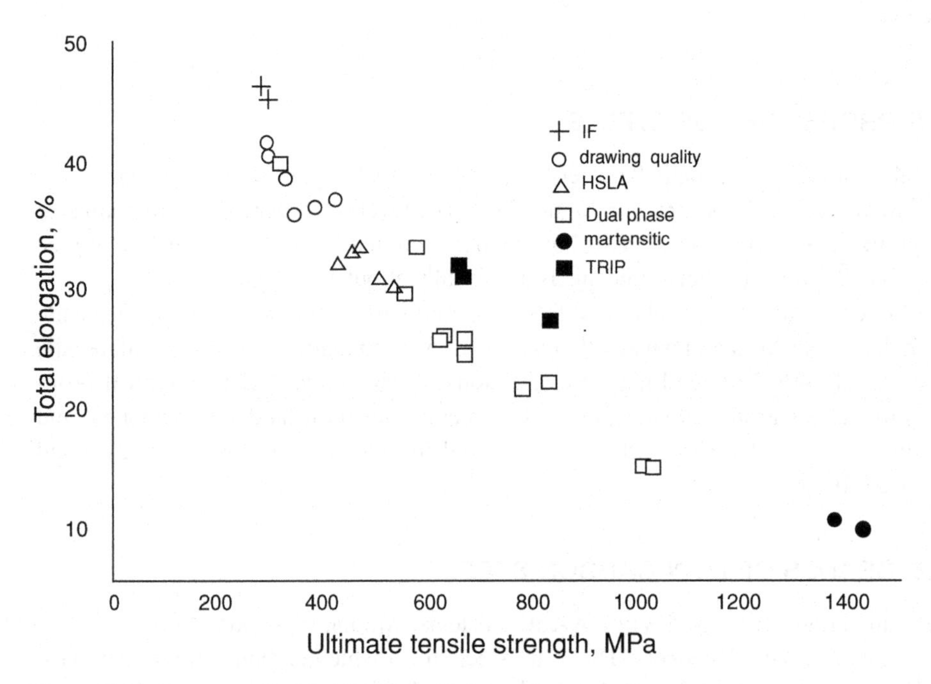

Figure 21.10. Decrease of total elongation with increasing strength.

preferentially with sulfur so the stochiometric amount of titanium that must be added to eliminate carbon and nitrogen is

$$\%\text{Ti} = (48/14)(\%\text{N}) + (48/32)(\%\text{S}) + (48/12)(\%\text{C}). \tag{21.1}$$

A typical composition is 0.002% C, 0.0025%N, 0.025%Ti, 0.15%Mn, 0.01%Si, 0.01%P, 0.04%Al, and 0.016%Nb. Niobium may be used instead of titanium but it is more expensive. IF steels typically have very low strengths. The advantage of these steels is that they are very formable. Control of crystallographic texture is also fundamental in producing exceptional deep drawability. Typically $\bar{R} = 2.0$. (R-values of traditional aluminum killed steels rarely exceed 1.8). Formability is also enhanced by high n-values ($\approx$2.5).

Fine grain sizes and higher strengths can be achieved by alloying with Nb. High-strength IF steels are solution-hardened with small amounts of Mn, Si, and P. The tensile strength is increased 4 MPa by 0.1% Mn, 10 MPa by 0.1%Si and 100 MPa by 0.1%P in solution. The presence of titanium reduces the phosphorus in solution by forming FeTiP. The increased strength comes at the expense of a somewhat reduced formability. A composition of 0.003%C, 0.003%N, 0.35%Mn, 0.05%P, 0.03% Al, 0.035%Nb, 0.2%Ti and 0.001% B has the following properties: YS = 220 MPa, TS = 390 MPa, elongation = 37%, $\bar{R} = 1.9$, and $n = 0.21$.

AK steels: Dissolved carbon and oxygen in steels having carbon contents 0.05 to 0.10% will react to form CO on freezing. This causes a violent rimming action as the CO bubbles are emitted, causing tiny drops of iron to burn in the air. Aluminum is added in the ladle to react with oxygen removing it in the form of Al_2O_3, which rises to the surface and is scraped off. This "kills" the rimming action. Aluminum also ties up dissolved nitrogen as AlN. Because of the removal of nitrogen, strain aging only occurs at elevated temperatures. Typically an aluminum-killed steel will have an n-value of about 0.22 and an R-value of 1.8.

Bake-hardenable steels: Steels having enough carbon and/or nitrogen in solution to strain-age at the temperatures for paint baking are termed bake-hardenable. The low-yield strength without a yield point prior to forming and the high strength caused by strain-aging are useful for producing dent resistance or down-sizing the thickness for weight reduction. Bake-hardenable steels are used in hoods, quarterdeck panels, roofs, doors, and fenders. With a low-carbon level they have good weldability.

HSLA steels (high-strength low-alloy): HSLA steels are much stronger than plain-carbon steels. They are used in cars, trucks, cranes, bridges, and other structures where stresses may be high. A typical HSLA steel may contain 0.15% C, 1.65% Mn and low levels (under 0.035%) of P and S. They may also contain small amounts of Cu, Ni, Nb, N, V, Cr, Mo, and Si. The term "micro-alloying" is often used because of small amounts of alloying elements. As little as 0.10% niobium and vanadium can have profound effects on the mechanical properties of a 0.1% C, 1.3% Mn steel. The Mn provides a good deal of solid solution strengthening. The other elements form a fine dispersion of precipitated carbides in an almost pure ferrite matrix. Rapid cooling produces a fine grain size which also contributes to the strength. Yield strengths are typically between 250 and 590 MPa (35,000 to 85,000 psi). The ductility and n-values are lower than in plain-carbon steels. HSLA steels are also more rust-resistant than most carbon steels, due to their lack of pearlite. An alloy with small amounts of Cu, called Cor-ten, forms a rust that is quite adherent and is used architecturally.

Dual-phase (DP) steels: These have been heat-treated to obtain 5 to 15% martensite in a ferrite matrix. They replace many HSLA grades. Uses include front and rear rails, bumpers, and panels designed for energy absorption.

TRIP steels (transformation-induced plasticity): The microstructure of TRIP steels consists mainly of ferrite but there is also martensite, bainite, and retained austenite. The various levels of these phases give TRIP steels their unique balance of properties.

During forming, the retained austenite transforms to martensite. This results in a high rate of work hardening that persists to higher strains, in contrast to that of DP which decreases at high strains. This causes enhanced formability. The carbon content controls the strain level of retained austenite-to-martensite transformation. With low-carbon levels, transformation starts at the beginning of forming, leading to excellent formability and strain distribution at the strength levels produced. With high-carbon levels, retained austenite is more stable and persists into the final part. The transformation occurs at strain levels beyond those produced during stamping and forming. Transformation to martensite occurs during subsequent deformation, such as a crash event, and provides greater crash energy absorption. Spot welding of TRIP steels is made more difficult by the alloying elements.

Complex-phase (CP) steels: These have a very fine microstructure of ferrite with martensite and bainite. They are further strengthened by precipitation of niobium, titanium, or vanadium carbonitrides. They are used for bumpers and B-pillar reinforcements because of their ability to absorb energy.

Martensitic grades: The microstructures of martensitic grades are completely martensite. Tensile strengths vary between 900 and 1,500 MPa (130 and 220 ksi). These grades can be made directly at the steel mill by quenching after annealing or by heat treating after forming. Mill-produced material has a very low ductility so it is typically roll-formed.

The carbon content controls the strength level. The tensile strength in MPa is approximately

$$\mathrm{TS} = 900 + 2800 \times \%\mathrm{C}. \tag{21.2}$$

Manganese, silicon, chromium, molybdenum, boron, vanadium, and nickel are used in various combinations to increase hardenability. Typical applications for martensitic steels usually are those requiring high strength and good fatigue resistance, with relatively simple cross sections, including door intrusion beams, bumper reinforcement beams, side sill reinforcements, and belt line reinforcements.

Typical properties of low-carbon sheet steels are listed in Table 21.2.

Until the recent past, automobile bodies were made also entirely from aluminum-killed low-carbon steels. The current emphasis on increased fuel economy of lighter vehicles has led to greatly increased usage of thinner gauges permitted by higher strength steels; low-carbon steels are rapidly becoming a minor part of auto bodies. The term Advance High-Strength Steels (AHSS) has been applied to steels with yield strengths greater than 200 MPa.

Table 21.2. Properties of some grades of low-carbon sheet steels

Steel	*YS* MPa	*TS* MPa	*El* %	*n*	*R*	*m*
IF	150	300	45	0.28	2+	0.015
IF w/P	220	390	37	0.21	1.9	0.015
AKDQ	180	350	32–40	0.20 to 0.22	1.4 to 2	0.015
BH210/340	210	340	34–39	0.18	1.8	
BH260/370	260	370	29–34	0.13	1.6	
DP280/600	280	600	30–34	0.21	1.0	
DP300/500	300	500	30–34	0.16	1.0	
DP350/600	350	600	24–30	0.14	1.0	
DP400/700	400	700	19–25	0.14	1.0	
DP500/800	500	800	14–20	0.14	1.0	
DP700/1000	700	1000	12–17	0.09	1.0	
HSLA350/450	350	450	23–27	0.14	1.1	0.005–0.01
TRIP450/800	450	800	26–32	0.24	0.9	
Mart950/1200	950	1200	5–7	0.07	0.9	
Mart1250/15201250	1520	4–6	0.065	0.9		

The lower ductility of higher strength steel (Figure 21.10) results in lower formability. The formability problem is aggravated by the lowering of the forming limit curves by the thinner gauges (Figure 16.18). There is much current research aiming to increase the ductility of higher strength steels. One approach is to incorporate more austenite into the microstructures. Increased austenite levels together with more martensite and still finer grain sizes also raise the strength level.

Considerable heating can occur during the high strain rates of forming of high-strength steels. Typical strain rates in automotive stampings are 10/s so little heat is transferred from the deforming steel. According to equation 5.24, with a flow stress of 720 MPa and a strain of 0.50, the expected temperature rise is about 100°C. This temperature rise can lower the flow stress appreciably as shown in Figure 21.11 and

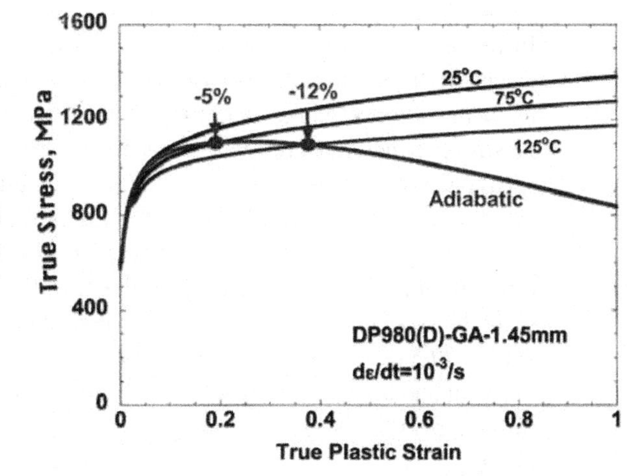

Figure 21.11. Adiabatic heating of a high-strength steel during forming can lead to early necking. From R. H. Wagoner, NADDRG presentation, May 4, 2010, Oakland University.

Figure 21.12. Shear failure of a high-strength steel during bending over a sharp radius. Courtesy of James Fekete.

lead to necking at very much reduced strains. There is also heating from friction. The rise in temperature tends to localize the deformation. However, in forming of most sheet metals (aluminum, copper, and lower strength steels) the localization caused by heating is a minor effect.

Adiabatic heating while bending of high-strength steels at high strain rates over sharp radii can lead to shear failures. Such a failure is illustrated in Figure 21.12.

21.7 TAILOR-WELDED BLANKS

In recent years parts are being stamped from blanks made by welding two or more sheets of different thickness or different base materials. The purpose is to save weight by using thinner gauge material where its strength is sufficient and using thicker or stronger material only where necessary. Some difficulties are encountered during forming. Offset blank-holder surfaces are required to assure adequate hold-down. Welding hardens the weld zone, which may reduce the formability and cause cross-weld failures if the direction of major strain is parallel to the weld. The best blank orientation is with the weld perpendicular to the major strain axis. In this case failure is likely to occur by splitting of the thinner material parallel to the weld. This problem can be alleviated to some extent by decreasing the hold-down pressure on the thicker material and allowing more of the thicker metal to flow into the die. The frequency of failures parallel to the weld are reduced by decreasing the movement of the weld in the die.

21.8 SPECIAL SHEET STEELS

Sandwich sheets with low-carbon steel on the outside around a polymer are used for sound dampening. For example, their use as the firewall between the engine and passenger compartment lowers engine noise.

Patterns can be impressed on the surface of sheets rolled with laser-textured rolls. It has been claimed that this permits better lubrication and better surface appearance after painting

21.9 SURFACE TREATMENT

Steel mills often sell prelubricated sheets or sheets coated with a polymer coating. Often steel is given a phosphate coating to help lubricants.

Steels are frequently galvanized (plated with zinc) for corrosion protection. Zinc is anodic to iron so it galvanically protects the underlying metal. Steels may be galvanized either by hot dipping or electroplating. In the more common hot dipping process, the thickness of the coating is controlled by wiping the sheet as it emerges from a molten zinc bath. Figure 21.13 shows the surface of a hot-dip galvanized sheet. In electroplating, the plating current and time control the thickness of electroplating. Usually, the thickness of the zinc is the same on both sides of the sheet, but sheets can be produced with the thickness on one side less than the other. Even one-side-only plating is possible.

The term *galvanneal* has been applied to hot-dip galvanized sheets that are subsequently annealed to allow the formation of Fe-Zn intermetallic compound.

Figure 21.13. Spangles on the surface of a steel sheet that was hot-dip galvanized. From *Making, Shaping and Treating of Steel, 9th Ed* (1971) United States Steel Corp.

Other types of plating are sometimes done. Tin plating is an example. However, "tin cans" today have little if any tin on them.

21.10 STAINLESS STEELS

Formable stainless steels fall into two classes: austenitic stainless steels and ferritic stainless steels. The austenitic stainless steels are fcc and contain from 17 to 25% Cr and 2 to 20% Ni with very low carbon. They form the 2xx and 3xx series. They are not magnetic. Austenitic grades are used for high temperature applications and where superior oxidation resistance is required. Austenitic grades work harden rapidly and consequentially have very high tensile elongations. The austenite in some grades is metastable and may transform to martensite during deformation. This partially accounts for the very high n-values. Like other fcc metals, the R-values are low.

Ferritic stainless steels are bcc and contain 12 to 18% Cr, less than 0.12% C and no nickel. They form the 4xx series. The ferritic grades are less expensive and are used widely for decorative trim. The mechanical properties of the ferritic grades are similar to those of low-carbon steel, except the yield and tensile strengths are somewhat higher. Ridging problems, noted in Section 21.2, can be controlled by high temperature annealing. Both grades may lose their corrosion resistance if welded. Figure 21.14 shows the tensile and yield strengths of typical austenitic and ferritic grades as a function of prior reduction.

As austenitic stainless steels like 301 and 304 are deformed at low temperatures, some of the austenite transforms to martensite. This accounts for their rapid work hardening. Figure 21.15 shows this transformation for 304 stainless. Note the amount of transformation decreases with temperature.

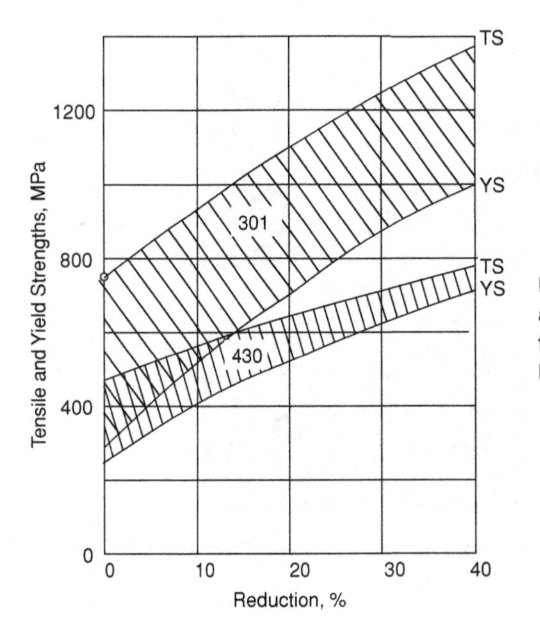

Figure 21.14. Tensile and yield strengths of an austenitic (301) and a ferritic (430) stainless steel as a function of cold-rolling reduction. Note the greater strain hardening of the austenitic grade.

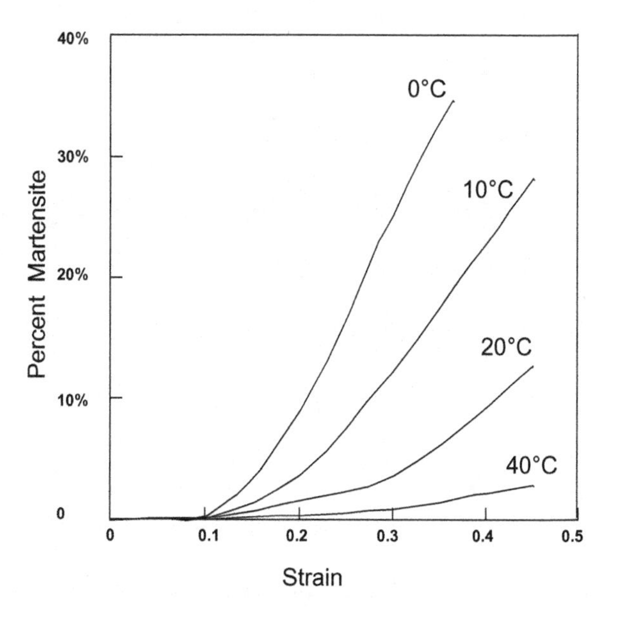

Figure 21.15. Formation of martensite in 304 stainless steel caused by strain. Data from J. Krauer and P. Hora in *Material Property Data for More Effective Numerical Analysis,* B. S. Levy, D. K. Matlock and C. J. Van Dyne eds. (2009).

21.11 ALUMINUM ALLOYS

There is a wide range of aluminum alloys available in sheet form. The formability varies with grade, but aluminum alloys generally are not as formable as low-carbon steel. Like other fcc metals the R-values are less than 1.0. The strain-hardening exponents of annealed grades tend to be between 0.2 and 0.3, but the strain-rate sensitivity is very low (negative in some cases). Springback is a severe problem in high-strength aluminum alloys as it is in high-strength steels because of the high strength to modulus ratio. If the same dies are used when aluminum is substituted for steel, there are often forming problems. However, they can usually be overcome with new dies.

The power-law expression for strain hardening does not apply well to aluminum alloys. The instantaneous strain-hardening exponent defined by

$$n = \mathrm{d}\ln\sigma/\mathrm{d}\ln\varepsilon \tag{21.3}$$

is not constant. It tends to decease at high strains as shown in Figure 21.16.

Wrought alloys are designated by four digits. The first indicates the primary alloying element. Copper-containing grades (2xxx), alloys with magnesium and more silicon than required to form Mg_2Si (6xxx) series and zinc alloys (7xxx) can be strengthened by heat treating. Commercially pure alloys (1xxx), those with manganese as the primary alloying element (3xxx) and those with magnesium as the primary alloy (5xxx) can be strengthened only by cold working.

The condition of the alloy is indicated by a temper designation (Table 21.3).

The 4xxx alloys are those with silicon as the principal alloying element, but none are produced in sheet form. The designation 8xxx is reserved for alloy with other elements as principal alloy (e.g., nickel, lithium).

The 3003 and 3004 alloys are very ductile and have strain-hardening exponents of about 0.25 in the annealed condition. Manganese provides solid solution strengthening. They find use as cooking utensils, roofing and siding. A special grade of 3004 with very

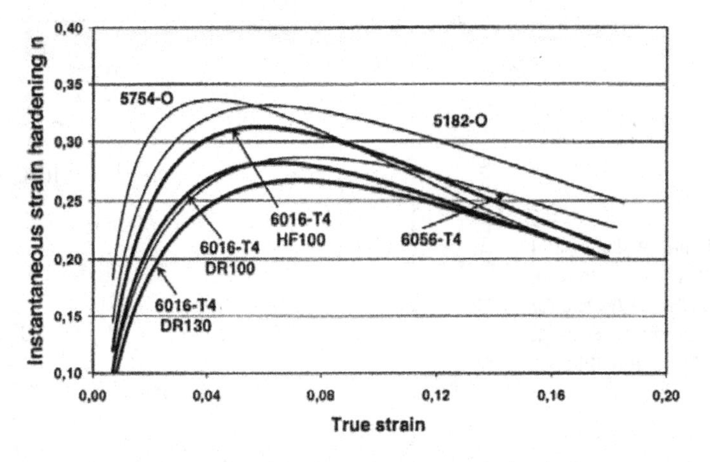

Figure 21.16. Decrease of the instantaneous strain-hardening exponent with strain. From D. Daniel, G. Guiglionda, P. Litalien and R. Shahani, *Materials Science Forum* (2006).

controlled limits on manganese and iron are used as stock for beverage cans. Strong can bottoms are assured by using starting stock that has been very heavily cold rolled (H-19 temper).

The addition of 2 to 5% magnesium (5xxx alloys) achieves greater solid-solution strengthening (see Figure 21.17). The formability and corrosion resistance are good, but these alloys are prone to develop stretcher strains. There are two types of stretcher strains: Type 1 are coarse like the Lüders strains in low-carbon steel that form at low strains; and Type 2, which are much finer like orange peel. They form at higher strains and are associated with dynamic strain aging that causes serrated stress-strain curves. Cold rolling of these alloys generates even higher strengths. Uses include automobile bodies (not the outer skin), trucks, and trailer parts. Fine grain 5083 is superplastic and is used in hard-to-form parts for autos and motorcycles (such as motorcycle gasoline tanks). Other applications for 5xxx alloys include canoes and boats. In highly stressed parts, the magnesium content is kept below 3.5% to avoid stress corrosion cracking.

Table 21.3. Temper designations

Designation	Meaning
F	As fabricated
O	Annealed
H	Strain hardened
H-1x	Strain hardened only
H-2x	Strain hardened and partially annealed
H-3x	Strain hardened and stabilized
The second digit (1 through 9) indicated the degree of strain hardening, 8 indicating a hardness achieved by a 75% reduction.	
W	Solution treated
T-3	Solution treated, strain hardened and naturally aged
T-4	Solution treated, and naturally aged to a stable condition
T-6	Solution treated, and artificially aged

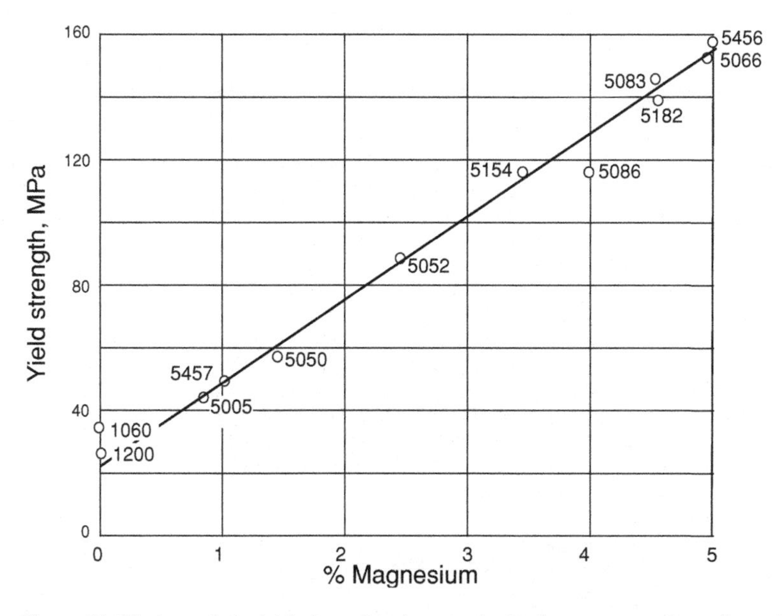

Figure 21.17. Annealed yield strength of several aluminum-magnesium alloys. Magnesium has a strong solid-solution strengthening effect in aluminum alloys. Data from *Aluminum and Aluminum Alloys, ASM Specialty Handbook,* ASM International, 1993.

Manganese is also a potent solid-solution strengthener as shown in Figure 21.18.

Aluminum alloys containing magnesium may have a negative strain-rate sensitivity. The result may be Lüders lines in formed parts. Figure 21.19 shows an example of Lüders lines on an aluminum-magnesium sheet.

The 2xxx alloys provide higher strengths and are used where the strength-to-weight ratio is important as in aircraft and trucks. However they have poor corrosion resistance. They may be clad with another aluminum alloy for galvanic protection.

The 6xxx alloys can also be hardened by heat treatment and have better corrosion resistance. They find major usage in automobile bodies. One canoe manufacturer uses

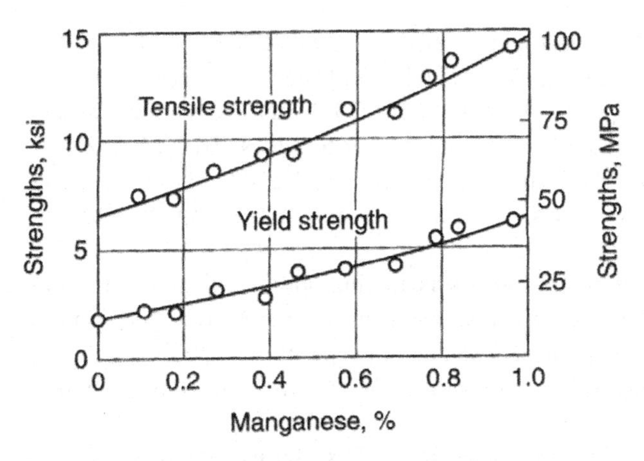

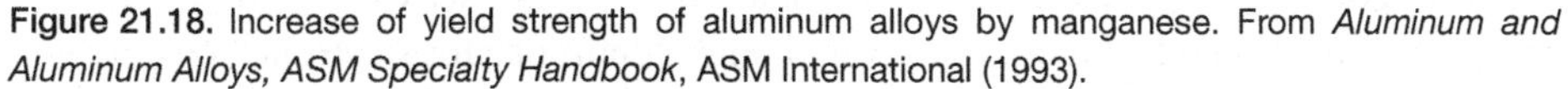
Figure 21.18. Increase of yield strength of aluminum alloys by manganese. From *Aluminum and Aluminum Alloys, ASM Specialty Handbook*, ASM International (1993).

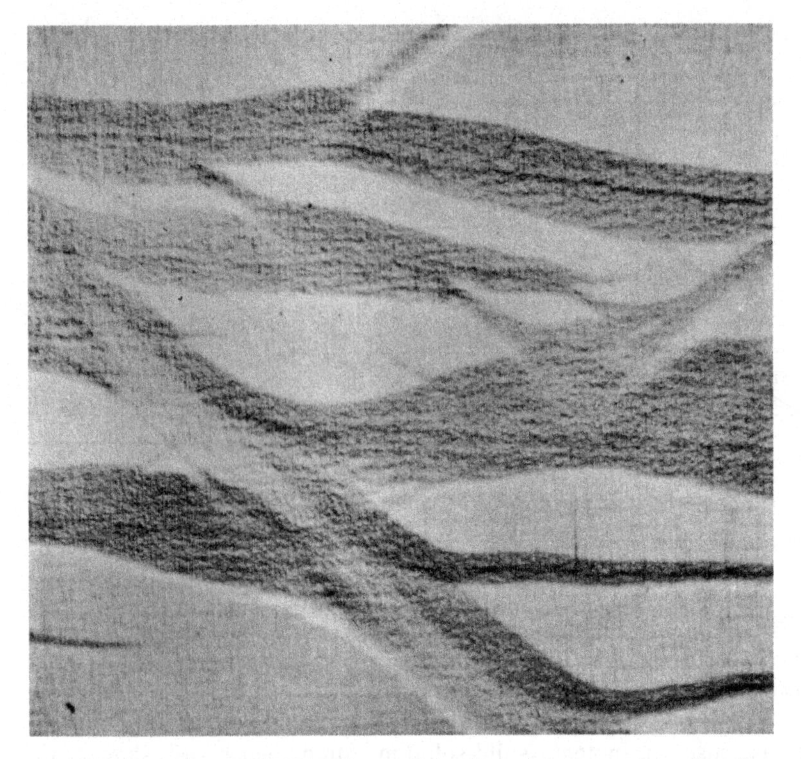

Figure 21.19. Lüders bands in a sheet of aluminum-rich magnesium alloy. *Properties and Physical Metallurgy*, ASM 1984, Courtesy of ALCOA.

the 6061 alloy and heat treats canoe halves before assembly. All automobile outer skins are made from 6xxx alloys. They are formed in the T-4 condition and further aged by the paint-bake cycle.

The highest strengths are achievable with 7xxx alloys containing up to 7% zinc and other elements. They find use in the aerospace industry. They have poor corrosion resistance and are susceptible to stress-corrosion cracking. These find applications as automobile bumpers.

In North America, the alloys most widely used for automobile bodies are Al-Mg alloys 5052 (O and various H tempers), 5754-O and 5152-O and in decreasing tonnage Al-Mg-Si alloys containing copper 6111-T4 and 6211-T4. In Europe the most widely used alloys are: Al-Mg 5754-O and 5182-O and Al-Mg-Si alloys 6016-T4, 6181-T4 and 6111-T4. Inner panels are made from 6xxx alloys and 5182. The 5xxx alloys are more formable than the 6xxx alloys. These all have strain-hardening exponents of between 0.18 and 0.25.

Superplastic forming of aluminum alloys is finding some application in auto-body panels. In the United States the interest is currently on fine grain 5083 and in Great Britain on 2004 alloy. The greatest advantage of superplastic forming is in cars produced in low quantities. Superplastic forming of Al 7475 is finding application in the aerospace industry.

Auto body parts are currently being made from a 5xxx series aluminum alloy by stamping at an elevated temperature. Even though conditions for superplasticity are not reached, the increased strain-rate sensitivity permits forming of more complex parts. A somewhat slower cycle time is compensated by lower forces, which permit the use of smaller and less expensive presses. However, lubrication is more difficult.

21.12 COPPER AND BRASS

Both annealed copper and brass work harden rapidly. The zinc addition in brass increases the yield and tensile strengths as shown in Figure 21.20. The strain-hardening exponent (0.35 to 0.5 for copper and 0.45 to 0.6 for cartridge brass) increases with zinc content, and larger grain sizes. However the larger grain size causes the orange peel effect. The strain-rate sensitivities for copper alloys are low (≤ 0.005) and the R-values are less than unity (typically 0.6 to 0.9). Annealing at a very high temperature after a very heavy cold reduction produced a cube texture in which <100> directions are aligned with the rolling, transverse and thickness directions. The resulting sheet has a low R_{av} and ΔR.

Brasses containing 15% or more zinc under stress are susceptible to stress-corrosion cracking in atmospheres containing ammonia. Tensile stresses across grain boundaries cause them to crack as shown in Figure 21.21. The susceptibility increases with zinc content. Unless the parts are stressed in service, the problem can be alleviated by stress-relief anneals. Brasses should be annealed between 200° and 300°C. Higher annealing temperatures are recommended for silicon bronze, aluminum bronze, and cupronickel.

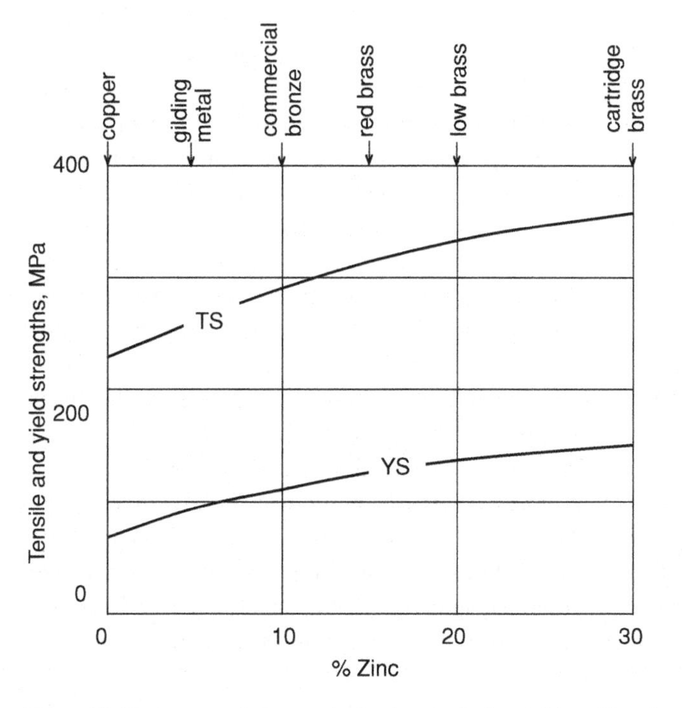

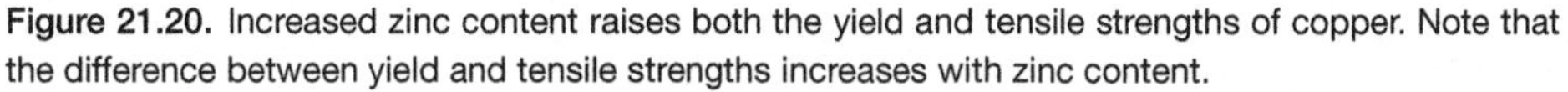

Figure 21.20. Increased zinc content raises both the yield and tensile strengths of copper. Note that the difference between yield and tensile strengths increases with zinc content.

Figure 21.21. Stress-corrosion cracking in brass. From *Metals Handbook*, 8th ed. v. 7 ASM 1972.

21.13 HEXAGONAL CLOSE-PACKED METALS

The crystallographic textures of hcp metal sheets tend to have the basal (0001) planes aligned with the plane of the sheet. With most, deformation is mainly by $<11\bar{2}0>$ slip, so high R-values are common.

In the past, there was very little forming of magnesium sheets because of limited ductility. Recently, however, there is an increased interest in magnesium because its low density can lead to weight saving. Magnesium sheets have strong crystallographic textures with the basal plane aligned with the plane of the sheet. As a result of this texture and easy $\{10\bar{1}2\} <\bar{1}011>$ twinning in tension this results in higher in-plane yield strengths in tension than in compression as indicated by Figure 21.22.

Basal slip and twinning are the primary deformation mechanisms. Above about 225°C, other slip modes can be activated. To have the necessary ductility, forming of magnesium alloy sheets is generally done at 200°C or higher. The most common sheet alloy, AZ31B contains about 3% Al and 1% Zn. Rolled sheets have shear bands inclined at about 60° to the sheet normal. These persist even after annealing at 350°C. Rolled sheets of an alloy AE21 (about 2% Al and 1% rare earths) have similar shear bands but these disappear after annealing at 350°C. Rolled ZEK100 (about 1% Zn, and less than 0.5% rare earths and zirconium) can be deep drawn at about 150°C.

It has been known for years that repeated roller leveling of AZ31B-O modifies its crystallographic texture and allows bends of 2 to 3 times sheet thickness instead of the usual 5.5 times sheet thickness. Recently it has been shown that a very fine grain size can be produced by severe deformation. This decreases the anisotropy and decreases the tendency for twinning so forming can be done at much lower temperatures.*

* Q. Yang and A. K. Ghosh, "Deformation Behavior of Ultrafine-Grain AZ31-B Magnesium Alloy at Room Temperature," *Acta Materialia*, v. 54, 2006, pp. 5159–70.

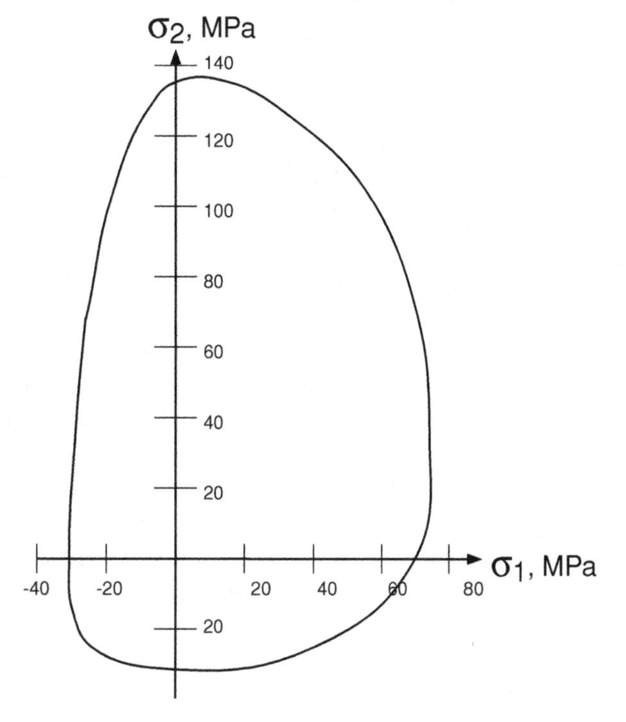

Figure 21.22. Yield locus of pure magnesium. Data from E. W. Kelly and W. F. Hosford, *Trans Met. Soc AIME* v. 242 (1968).

Zinc alloys have high m-values. The high m-value is to be expected because room temperature (293K) is 42% of the melting point (693K) of zinc. The strain-hardening exponent is very low.

Alpha-titanium alloys (hcp) typically have R-values of 3 to 5 and can be drawn into very deep cups. In contrast, β-titanium alloys (bcc) behave more like low-carbon steels. As discussed in Chapter 5, fine grain titanium alloys may be superplastically formed.

Beryllium is very brittle. Rolled sheets have very high R-values. Deformation is limited to $<11\bar{2}0>$ slip, so even bending is limited except for very narrow strips.

21.14 TOOLING

Dies for sheet forming are usually made from cast iron. Meehanite® with fine controlled graphite flakes is common. High wear locations may be flame hardened. For extreme conditions such as very high production, cast steel may be used. In all cases it is common to ion-nitride or chrome plate the die faces. Initial tryout dies may be made from a zinc alloy.

Flanging and trimming dies are made from tool steels.

21.15 PRODUCT UNIFORMITY

Variations of sheet thickness and properties are undesirable. Subtle changes can cause failures with tooling that has been adjusted to give good parts with a particular batch of steel. Problems may arise from differences from one coil of steel to another or differences between different mills. There may be nothing inherently bad about the steel that performs poorly. There are often differences between the edges and center of a coil. Such property differences may be caused by variations of grain size, texture, or composition.

During casting impurities segregate toward the center. When the cast slab is rolled, these may cause weakness at the centerline. Curved molds are frequently used in continuous casting. Because the last material to freeze in such molds is displaced somewhat from the centerline, failures may be caused by segregated inclusions not on the exact centerline.

21.16 SCRAP

There are two sources of scrap. *New* scrap is the offal of manufacturing processes resulting from trimming and occasional bad parts. Since the composition of new is well-known it can be re-melted to make more of the same alloy. Other scrap (*old* scrap) comes from cans, engine blocks, cylinder heads, buildings, and other sources. Its composition is less well-known.

A large fraction (42%) of steel is from recycling. Steel and iron can easily be separated magnetically from other scrap. The scrap is either re-melted in an electric arc furnace or added to pig iron in a basic oxygen furnace. All grades of steel can be recycled because most alloying elements are oxidized during processing. Tin and copper are the exceptions and there is concern in the steel industry about the gradual build up of these *tramp* elements in steel.

The distinction between new and old scrap is particularly important with aluminum alloys. Alloying elements cannot be removed from aluminum alloys during melting so old scrap must be used to produce alloys with less critical compositions. Aluminum scrap is shredded and any lacquer is removed from cans before re-melting. Re-melting aluminum requires only 5% of the energy to produce virgin aluminum from bauxite and emits only 5% of the CO2. The energy saving is 14 kwh/kg.

NOTES OF INTEREST

In 1886, both Charles M. Hall in the United States and Paul-Louis-Toussaint Héroult in France independently developed an economical process for producing aluminum. Before this, aluminum was more expensive than platinum, and for that reason was chosen to cap the Washington Monument. Both men were only 22. Hall was a recent chemistry graduate of Oberlin College, and Hérault had studied at the School of Mines in Paris. Within two years, aluminum production was in full swing in Europe and the United States. The American spelling (with "um" instead of the "ium" common to metals) comes from an advertisement of the Pittsburg Reduction Company. Whether the omission was intended or a mistake is unknown.

In the early 1850's, Henry Bessemer in England and William Kelley in the United States developed a scheme for making cheap steel. Both realized that the carbon content of pig iron could be reduced from about 4% to a low level by blowing air through it. Kelly's family and friends thought he was insane, but he perfected his process by 1851. Six years later, Bessemer received a U.S. patent for essentially the same process. Their steel replaced wrought iron as the strongest material. Although their process was later surplanted first by the open-hearth process and later by the basic oxygen process, the introduction of cheap steel made railroads and later automobiles possible.

W. Lüders first drew attention to the lines, which appeared on polished steel specimens as they yielded. *Dinglers Polytech., J.* Stuttgart, 1860.

REFERENCES

Aluminum: Properties and Physical Metallurgy, ASM, 1984.
W. F. Hosford. *Physical Metallurgy*, Taylor and Francis, 2005.
The Making, Shaping and Treating of Steel, U.S. Steel Corp. 1971.
Report on Advanced High Strength Steel Workshop. October 2006. Web site http://mse.osu.edu/~wagoner/AHSS/AHSSReportFINAL.pdf.

PROBLEMS

21.1. The table below from *The Making, Shaping and Treating of Steel* gives combinations of aging times and temperatures that result in equal amounts of strain aging in low-carbon steels.

a) From a plot of $\ln(t)$ versus $1/T$, where T is absolute temperature, determine the apparent activation energy for strain aging.

b) Explain the slope change between 0° and 21°C. (Consider how the data were obtained.)

Aging Times and Temperatures that Produce Equal Amounts of Aging

0°C	21°C	100°C	120°C	150°C
1 y	6 mo	4 h	1 h	10 min
6 mo	3 mo	2 h	30 min	5 min
3 mo	6 wk	1 h	15 min	2.5 min
1 wk	4 d	5 min		
3 d	36 h	2 min		

21.2. One engineer specified that a part be made from an extra-low-carbon grade of steel. Although it costs more than the usual grade, he thought that with the usual grade there might be an excessive scrap rate.

a) How could you determine whether the cheaper, usual grade could be used?

b) Would the substitution of a cheaper grade result in an inferior product?

21.3. With low-carbon steels, both n- and R-values can be raised by higher annealing temperatures. Why isn't this practice common?

21.4. The substitution of HSLA steels for aluminum-killed steels to achieve weight saving in automobiles is based on their higher yield strengths.

a) How does this affect the elastic stiffness of the parts?

b) In view of corrosion, how does this affect component life? (HSLA steels have about the same corrosion resistance as aluminum-killed steels.)

21.5. The table below lists properties of several sheet metals at the temperature they will be formed. Choose from these the appropriate material and state the reason for your choice.

a) Greatest LDR in cupping.
b) Most earing in cupping.
c) Greatest uniform elongation in tension.
d) Greatest total elongation in tension.
e) Excluding material E (because of its low yield strength) the material that could be drawn into the deepest cup by stretching over a hemispherical punch.

Properties							
Material	E (MPa)	$Y.S.$ (MPa)	R_0	R_{45}	R_{90}	n	m
A	210	200	1.9	1.2	2.0	0.25	.003
B	210	240	1.2	1.0	1.2	0.22	0.03
C	72	175	0.7	0.6	0.7	0.22	0.001
D	115	140	0.6	0.9	0.6	0.5	0.001
E	70	7	1.0	1.0	1.0	0.00	0.60

21.6. Explain why with deformation limited to $<11\bar{2}0>$ slip, wide sheets of beryllium can't be bent whereas narrow strips can.

Index

CPSIA information can be obtained
at www.ICGtesting.com
Printed in the USA
LVOW04s2011231217
560664LV00005B/91/P